Eric F. Lambin · Helmut Geist (Eds.)

Land-Use and Land-Cover Change

Local Processes and Global Impacts

With 44 Figures

 Springer

Editors

Eric F. Lambin

Department of Geography,
Catholic University of Louvain, Belgium

Helmut Geist

Department of Geography and Environment,
University of Aberdeen, United Kingdom

Library of Congress Control Number: 2006920338

ISSN   1619-2435
ISBN-10   3-540-32201-9   **Springer Berlin Heidelberg New York**
ISBN-13   978-3-540-32201-6   **Springer Berlin Heidelberg New York**

Springer is a part of Springer Science+Business Media
springer.com
© Springer-Verlag Berlin Heidelberg 2006
Printed in Germany

Cover design: Erich Kirchner, Heidelberg
Typesetting: Stasch · Bayreuth (stasch@stasch.com)
Production: Christine Jacobi
Printing: Stürtz AG, Würzburg
Binding: Stürtz AG, Würzburg

Printed on acid-free paper   30/2133/CJ – 5 4 3 2 1 0

Acknowledgments

This book is a synthesis of ten years of research conducted in the framework of the Land-Use and Land-Cover Change (LUCC) project of the International Geosphere-Biosphere Programme (IGBP) and International Human Dimensions Programme on Global Environmental Change (IHDP). The authors are grateful to a large number of people for contributions of various types. Past and current members of the LUCC Scientific Steering Committee and International Project Offices have contributed their expertise and enthusiasm to the project. As editors, we are most grateful for the support of the Federal Science Policy Office from the Services of the Prime Minister of Belgium. In August 2000, the Belgian Science Policy Office (www.belspo.be) made possible the establishment and operation of an International Project Office of the Land-Use/Cover Change (LUCC) project at the Department of Geography of the Université Catholique de Louvain at Louvain-la-Neuve, Belgium. This meant half a decade of productive land-use/cover change research and project management. The synthesis volume owes much to the benevolent financial support from the Belgian Science Policy Office and likewise to the efficient infrastructural services provided by the host university.

Obviously, the book has greatly benefited from ideas developed within the LUCC project that was the first joint project carried out under the auspices of the International Geosphere-Biosphere Programme (IGBP) and the International Human Dimensions Programme on Global Environmental Change (IHDP). Our special thanks go to the past and current chairs, executive directors and project liaison officers of IGBP and IHDP who – among many others at these sponsoring organizations – enabled the Scientific Steering Committee of LUCC to come together at its annual meetings and at dozens of regional and thematic workshops. Over one decade, the yearly LUCC Scientific Steering Committee meetings – besides fulfilling other functions – set the stage for the creation of concepts, databases and texts, finally maturing into the current synthesis volume.

In the order of individual chapters as they appear in the book, we want to convey acknowledgments that were brought to our awareness by individual book chapter lead authors. Given his long-standing involvement in the creation and development of LUCC as a project, the introductory chapter has benefited greatly from comments by B. L. Turner II, Clark University, USA. Numerous comments on earlier versions of Chap. 4 have helped sharpening the key messages on the manifold impacts of land-use/cover change; these have been comments by Katherine Homewood, University College London, UK, Kjeld Rasmussen, University of Copenhagen, Denmark, Jürgen Merz, International Centre for Integrated Mountain Development, Nepal, and Caroline Michellier and Sophie Vanwambeke, University of Louvain, Belgium. In Chap. 6, the Sect. 6.5 (main findings of scenarios) and 6.6 (towards better land scenarios) are based on discussions held at an IHDP/IGBP-sponsored workshop in Hofgeismar, Germany, titled "What have we learned from scenarios of land use and land cover?" in 2004. Each of the authors of Chap. 7 on linking science and policy brought years of experience working with colleagues and field teams on this subject, and wish to thank these fine teams for sharpening their thinking.

Finally, we owe thanks to Chantal Van Hemelryck from LUCC's International Project Office, for putting together the book's references, to Nicolas Dendoncker from the University of Louvain for helping out with illustrations, and, above all, to Armin Stasch, technical editor in Bayreuth, Germany, for bringing the book into the proper format.

A sterling staff and writing team surely made a difference, as did the comments by a large group of altogether 22 highly competent anonymous reviewers, with about three reviewers per chapter. In addition, the entire volume was reviewed by Darla Munroe and Karl Chang. These comments have led to some reorganizations of text sections, modifications and additions where appropriate, and undoubtedly to an overall improvement of the contents of the volume, thus sharpening the key messages. We thank all these scientists that took on their time to contribute to the book quality. Any remaining omission, inaccuracy or error fall under the sole responsibility of the book editors.

Eric F. Lambin
Helmut Geist

Contents

Authors

Chapter Lead Authors

Prof. Joseph Alcamo

University of Kassel
Center for Environmental Systems Research
Kurt-Wolters-Str. 3
34109 Kassel
Germany

Tel.:   +49-(0)561-804 3266
Fax: +49-(0)561-804 3176
E-mail: alcamo@usf.uni-kassel.de

Joseph Alcamo is Professor of Environmental Sciences and Director of the Center for Environmental Systems Research at the University of Kassel, Germany. He was member of the Scientific Steering Committee of the IGBP-IHDP Land-Use/Cover Change (LUCC) project, besides contributing to many other international scientific projects and initiatives. His research focuses on applied systems analysis and computer modeling. In 1998, he was awarded the Max-Planck research prize for pioneering the area of integrated global modeling of the environment and its application to international policy.

Gerald Busch

Büro for Applied Landscape Ecology
and Scenario Analysis
Am Weißen Steine 4
37085 Göttingen
Germany

Tel.:   +49-(0)1212-315 916 666
Fax: +49-(0)1212-624 124 124
E-mail: welcome.balsa@email.de

Gerald Busch is a senior researcher and works as a scientific consultant in international projects with a focus on land use change, carbon sequestration and scenario development. At the science-policy interface, he worked for the German Advisory Council on Global Change on topics of land use, biogeochemical cycles and environmental risks.

Dr. Abha Chhabra

Space Applications Centre
Indian Space Research Organisation
Ahmedabad – 380015
Gujarat, India

Tel.:   +91-79-2691 4371/4024
Fax: +91-79-2691 5823
E-mail: abha@sac.isro.gov.in

Abha Chhabra is Scientist at Remote Sensing Applications and Image Processing Area of the Space Applications Centre, Indian Space Research Organisation, Ahmedabad. She was member of the Scientific Steering Committee of the IGBP-IHDP Land-Use/Cover Change (LUCC) project. Her research focuses on assessment and understanding the impacts of landuse/ cover changes on terrestrial carbon cycle for India.

Prof. Helmut J. Geist

University of Aberdeen
Department of Geography and Environment
Elphinstone Road
Aberdeen AB24 3UF
United Kingdom

Tel.:   +44-1224-272 342
Fax: +44-1224-272 331
E-mail: h.geist@abdn.ac.uk

Helmut Geist is Professor of Human-Environment Interactions at the University of Aberdeen. He was Executive Director of the IGBP-IHDP Land-Use/Cover Change (LUCC) project at the International Project Office of the University of Louvain in Louvain-la-Neuve, Belgium, from 2000 until the project's completion in 2005. His research focuses on coupled economic-environmental changes.

Dr. Kasper Kok

Wageningen University and Research Center   Tel.:   +31-(0)317-485 208
Department of Environmental Sciences   Fax: +31-(0)317-482 419
P.O. Box 37   E-mail: Kasper.Kok@wur.nl
6700 AA Wageningen
The Netherlands

Kasper Kok is postdoc researcher at the Wageningen University, The Netherlands. He was science officer at the Focus 3 Office of the IGBP-IHDP Land-Use/Cover Change (LUCC) project on regional and global modeling. His research focuses on modeling and participatory scenario development.

Prof. Eric F. Lambin

Catholic University of Louvain   Tel.:   +32-10-474 477
Department of Geography   Fax: +32-10-472 877
3, Place Louis Pasteur   E-mail: lambin@geog.ucl.ac.be
1348 Louvain-la-Neuve
Belgium

Eric Lambin is Professor at the Department of Geography at the University of Louvain in Louvain-la-Neuve, Belgium. He was chair of the Scientific Steering Committee of the IGBP-IHDP Land-Use/Cover Change (LUCC) project, besides contributing to many other international scientific projects and initiatives. His research focuses on biomass burning, tropical deforestation, desertification and land use transitions.

Prof. Alexander S. Mather

University of Aberdeen   Tel.:   +44-1224-272 354
Department of Geography and Environment   Fax: +44-1224-272 331
Elphinstone Road   E-mail: a.mather@abdn.ac.uk
Aberdeen AB24 3UF
United Kingdom

Alexander Mather is Professor of Geography at the University of Aberdeen. He is chair of the International Geographical Union (IGU) Commission on Land-Use/Cover Change (LUCC). His research focuses on forest transition and rural development issues.

Dr. William McConnell

Michigan State University   Tel.:   +1-517-432 7108
Department of Fisheries and Wildlife   Fax: +1-517-432 1699
13 Natural Resources Building   E-mail: mcconn64@msu.edu
East Lansing, MI 48824
United States of America

William McConnell is Associate Director of the Center for Systems Integration and Sustainability, and Assistant Professor in the Department of Fisheries and Wildlife at Michigan State University. He was science officer at the Focus 1 Office of the IGBP-IHDP Land-Use/Cover Change (LUCC) project at the Anthropological Center for Training and Research on Global Environmental Change at Indiana University, Bloomington, United States. His research focuses on human-environment interactions in forest ecosystems.

Prof. Robert Gilmore Pontius Jr.

Clark University, School of Geography   Tel.:   +1-508-793 7761
Department of International Development,   Fax: +1-508-793 8881
Community & Environment   E-mail: rpontius@clarku.edu
950 Main Street
Worcester, MA 01610-1477
United States of America

Robert Gilmore Pontius Jr. is Associate Professor at Clark University in Worcester, Massachusetts, United States, where he coordinates the graduate program in Geographic Information Sciences for Development and Environment. His research foci include spatial statistics and modeling of land-use/cover change. His quantitative techniques are in the GIS software Idrisi.

Dr. Jörg A. Priess

University of Kassel   Tel.:   +49-(0)561-804 2496
Center for Environmental Systems Research   Fax: +49-(0)561-804 3176
Kurt-Wolters-Str. 3   E-mail: priess@usf.uni-kassel.de
34109 Kassel   URL: www.usf.uni-kassel.de
Germany

Jörg A. Priess is Assistant Professor at the Center for Environmental Systems Research at Kassel University, Germany. His is interested in modeling land use change and nutrient cycling in ecosystems. He has mainly published about his research in the tropics.

Dr. Navin Ramankutty

University of Wisconsin-Madison   Tel.:   +1-608-265 0604
Center for Sustainability and the Global Environment   Fax: +1-608-265 4113
Nelson Institute for Environmental Studies   E-mail: nramanku@wisc.edu
1710 University Avenue
Madison WI 53726
United States of America

From June 1, 2006: Department of Geography, McGill University
                      805 Sherbrooke St. W.
                      Montreal, QC H3A 2K6, Canada

Navin Ramankutty is Assistant Scientist at the Center for Sustainability and the Global Environment of the University of Wisconsin-Madison, United States. He was member of the Scientific Steering Committee of the IGBP-IHDP Land-Use/Cover Change (LUCC) project. He leads efforts on documenting global cropland patterns by merging satellite and socio-economic data. He will be Assistant Professor of Geography at McGill University starting June 2006.

Dr. Robin Reid

International Livestock Research Institute   Tel.:   +254-2-630 743
P.O. Box 30709   Fax: +254-2-631 499
Nairobi   E-mail: r.reid@cgiar.org
Kenya

Robin Reid holds a PhD in range science and is Senior Systems Ecologist and Project Leader at the International Livestock Research Institute in Nairobi, Kenya. She was member of the Scientific Steering Committee of the IGBP-IHDP Land-Use/Cover Change (LUCC) project. Her research focuses on the impacts of livestock production systems on the environment in the developing world.

Prof. Ronald R. Rindfuss

The University of North Carolina   Tel.:   +1-919-966 7779
Department of Sociology   Fax: +1-919-966 6638
CB 8120 University Square   E-mail: ron_rindfuss@unc.edu
Chapel Hill, NC 27516-2524
United States of America

Ronald R. Rindfuss is the Robert Paul Ziff Distinguished Professor of Sociology and Fellow of the Carolina Population Center, University of North Carolina at Chapel Hill. He was member of the Scientific Steering Committee of the IGBP-IHDP Land-Use/Cover Change (LUCC) project. His research focuses on household demography and environmental change.

Dr. Thomas P. Tomich

World Agroforestry Centre   Tel.:   +254-(20)-722 4139
Alternatives to Slash-and-Burn Programme   Fax: +254-(20)-722 4001
P.O. Box 30677   E-mail: t.tomich@cgiar.org
Nairobi
Kenya

Thomas P. Tomich is Principal Economist at the World Agroforestry Centre (ICRAF) in Nairobi, Kenya, where he is global coordinator of the Alternatives to Slash-and-Burn Program. His research focuses on integrated assessment of land use alternatives, economic and environmental impacts of land use change, land and tree tenure, underlying causes of fires, and national policies that affect upland resource management and tropical deforestation.

Prof. A. Veldkamp

Wageningen Agricultural University   Tel.:   +31-(0)317-484 410
Laboratory of Soil Science and Geology   Fax: +31-(0)317-482 419
P.O. Box 37   E-mail: tom.veldkamp@wur.nl
6700 AA Wageningen
The Netherlands

Antonie ("Tom") Veldkamp is Full Professor of Soil Inventory and Land Evaluation at Wageningen University, The Netherlands. He was member of the Scientific Steering Committee of the IGBP-IHDP Land-Use/Cover Change (LUCC) project, and leader of the LUCC Focus 3 Office on regional and global models. His research focuses on modeling of landscape processes and related land use dynamics.

Prof. Peter H. Verburg

Wageningen University & Research Center
Department of Environmental Sciences
P.O. Box 37
6700 AA Wageningen
The Netherlands

Tel.: +31-(0)317-485 208
Fax: +31-(0)317-482 419
E-mail: Peter.verburg@wur.nl

Peter Verburg is Assistant Professor at Wageningen University, The Netherlands. He was staff member of the Focus 3 Office of the IGBP-IHDP Land-Use/Cover Change (LUCC) project on regional and global models. His research focuses on the analysis and modeling of land-use/cover change.

Prof. Jianchu Xu

Water, hazard and environmental management
International Centre for Integrated Mountain Development
P.O. Box 3226
Kathmandu, Nepal

Tel.: +977-1-552 2839, 554 3227
E-mail: jxu@icimod.org

Jianchu Xu is Professor of Kunming Institute of Botany, the Chinese Academy of Sciences, currently Program Manager of Water, Hazards and Environmental Management of International Centre for Integrated Mountain Development. He was member of the Scientific Steering Committee of the IGBP-IHDP Land-Use/Cover Change (LUCC) project. His ethnoecological research focuses on cross-cultural comparisons in Southeast Asia and Himalayan Regions.

Co-Authors

Dr. Frédéric Achard

Joint Research Centre of the Eureopean Commission
Institute for Environment and Sustainability
21020 Ispra, Varese
Italy

Tel.: +39-0332-785 545, +39-0332-789 830
Fax: +39-0332-789 073 (9960)
E-mail: frederic.achard@jrc.it

Frédéric Achard is research Scientist at the Joint Research Centre (JRC) of the European Commission in Ispra, Italy. He has co-signed a number of scientific articles on the estimation of tropical and boreal forest cover from remote sensing. In 2002, he won the JRC best scientific publication award.

Prof. Samuel Babatunde Agbola

Department of Urban and Regional Planning
Faculty of the Social Sciences, University of Ibadan
Ibadan, Nigeria

Tel. (mobile): +234-80-3321 8243
E-mail: babatundeagbola@yahoo.com

Babatunde Agbola is Professor of Urban and Regional Planning in Nigeria's premier University where he teaches, land-use planning, environmental planning and management, housing and planning theory. He has been a member of the Scientific Steering Committee of the LUCC since 2000. He is a consultant ot the UN-Habitat, the WHO and many other national and international organizations.

Prof. Diógenes S. Alves

Instituto Nacional de Pesquinas Espaciais (INPE)
Av. dos Astronautas 1758
CEP 12201-010 São José dos Campos
Brazil

Tel.: +55-12-3945 6492, -6444
Fax: +55-12-3945 6468
E-mail: dalves@dpi.inpe.br

Diogenes Alves is Professor of Remote Sensing at the National Institute for Space Research of the Ministry of Science and Technology in São José dos Campos, São Paulo, Brazil, and member of the Scientific Steering Committee of the IGBP-IHDP Land-Use/Cover Change (LUCC) project. His research focuses on the study of land use dynamics in the Brazilian Amazon.

Dr. Ademola K. Braimoh

United Nations University
Institute of Advanced Studies
6F International Organizations Center
1-1-1 Minato Mirai, Nishi-ku
Yokohama 220-8502
Japan

Tel.: +81-45-221 2350
Fax: +81-45-221 2302
E-mail: braimoh@ias.unu.edu

Ademola Braimoh is a Postdoctoral Fellow at the United Nations University Institute of Advanced Studies in Japan. His research focuses on application of geospatial technology, and integration of biophysical and social datasets for modeling land use change.

Prof. Oliver T. Coomes

McGill University
Department of Geography
Burnside Hall, Rm 705
805 Sherbrooke Street West
Montréal, PQ H3A 2K6
Canada

Tel.: +1-514-398 4943
Fax: +1-514-398 7437
E-mail: coomes@felix.geog.mcgill.ca

Prof. Ruth DeFries

University of Maryland
Department of Geography
2181 Lefrak Hall
College Park, MD 20742
United States of America

Tel.: +1-301-405 4884
Fax: +1-301-314 9299
E-mail: rd63@umail.umd.edu

Bas Eickhout

Netherlands Environmental Assessment Agency (MNP)
Global Sustainability and Climate (KMD)
P.O. Box 303
3720 AH Bilthoven
The Netherlands

Tel.: +31-(0)30-274 2924
E-mail: bas.eickhout@mnp.nl

Bas Eickhout is scientist at the Netherlands Environmental Assessment Agency (MNP) of the National Institute of Public Health and Environment (RIVM). He works on integrated assessment and is responsible for the development and applicability of the IMAGE model.

Prof. Jon Foley

University of Wisconsin-Madison
Center for Sustainability and the Global Environment
1710 University Avenue
Madison WI 53726
United States of America

Tel.: +1-608-265 5144
Fax: +1-608-265 4113
E-mail: jfoley@wisc.edu

Jon Foley is the Director of the Center for Sustainability and the Global Environment (SAGE) at the University of Wisconsin, where he is also the Gaylord Nelson Distinguished Professor of Environmental Studies and Atmospheric & Oceanic Sciences. Foley's work focuses on the behavior of complex global environmental systems and their interactions with human societies. In particular, Foley's research group uses state-of-the-art computer models and satellite measurements to analyze changes in land use, ecosystems, climate and freshwater resources across local, regional and global scales. He and his students and colleagues have contributed to our understanding of large-scale ecosystem processes, global patterns of land use, the behavior of the planet's water and carbon cycles, and the interactions between ecosystems and the atmosphere.

Prof. Lisa Graumlich

Montana State University
Mountain Research Center
106 AJM Johnson Hall
Bozeman, MT 59717-3490
United States of America

Tel.: +1-406-994 5178
Fax: +1-406-994 5122
E-mail: lisa@montana.edu

Lisa Graumlich is Professor of Ecology and Director of the Mountain Research Center at the Montana State University, United States. She was member and vice-chair of the Scientific Steering Committee of the IGBP-IHDP Land-Use/Cover Change (LUCC) project. Her research focuses on mountain regions, climate variation and the boundary between forest and tundra.

Prof. Helmut Haberl

Klagenfurt University
Institute of Social Ecology
Schottenfeldgasse 29
1070 Vienna
Austria

Tel.: +43-1-5224 000 406
Fax: +43-1-5224 000 477
E-mail: helmut.haberl@uni-klu.ac.at

Helmut Haberl is Associate Professor of Human Ecology at the Institute of Social Ecology (Vienna) of Klagenfurt University in Austria. Originally trained as ecologist and mathematician, he has 15 years of experience in interdisciplinary co-operation. His research focused on energy to explore various aspects of society-nature interaction, in particular ecological energy flows resulting from land use. For more than a decade, he has been working on the significance of Human Appropriation of Net Primary Production (HANPP).

Maik Heistermann

University of Kassel   Tel.:   +49-(0)561-804 2341
Center for Environmental Systems Research   Fax: +49-(0)561-804 3176
Kurt-Wolters-Str. 3   E-mail: heistermann@usf.uni-kassel.de
34109 Kassel
Germany

Maik Heistermann is PhD candidate at the Centre for Environmental Systems Research of the University of Kassel in Germany. He is adjunct researcher at the International Max Planck Research School on Earth Systems Modeling in Hamburg. His scientific focus is on modeling crop production and agricultural change on the global scale.

Dr. Richard A. Houghton

The Woods Hole Research Center   Tel.:   +1-508-540 9900
149 Woods Hole Road   Fax: +1-508-540 9700
Falmouth, MA 02540   E-mail: rhoughton@whrc.org
United States of America

Dr. Kees Klein Goldewijk

Netherlands Environmental Assessment Agency (MNP)   Tel.:   +31-(0)30-274 5
Global Sustainability and Climate (KMD)   Fax: +31-(0)30-274 4464
P.O. Box 303, 3720 AH Bilthoven   E-mail: kees.klein.goldewijk@mnp.nl
The Netherlands

Kees Klein Goldewijk is a senior researcher at the Netherlands Environmental Assessment Office (MNP) in Bilthoven, The Netherlands. His research activities include integrated assessments such as the Global Environmental Outlook (GEO) of UNEP, contributions to the IMAGE model as input for the reports of the Intergovernmental Panel on Climate Change (IPCC), the Millennium Ecosystem Assessment, and several Dutch environmental Outlooks and Balances. He is the developer of the History Database of the Global Environment (HYDE). This database is a compilation of historical time series and geo-referenced data on several land use, population, and economic indicators for the last 300 years and may serve as input for integrated models of global change.

Jianguo Liu

Michigan State University   Tel.:   +1-517-355 1810
Department of Fisheries and Wildlife   Fax: +1-517-432 1699
13 Natural Resources Building   E-mail: jliu@panda.msu.edu
East Lansing, MI 48824
United States of America

Professor Andrew Millington

Texas A&M University, Department of Geography   Tel.:   +1-979-845 6324
3147 TAMU, College Station, TX 77843-3147   E-mail: millington@geog.tamu.edu
United States of America

Andrew Milligton is Professor of Geography at the Texas A&M University. He has researched extensively a wide range of issues combining natural resources and development in Africa, Asia and Latin America. His research into land cover change in the coca growing area of Bolivia was awarded the prize for the Best Scientific Paper at the 29th International Conference of Remote Sensing of the Environment in Buenos Aires in 2002.

Prof. Emilio F. Moran

Indiana University, Department of Anthropology   Tel.:   +1-812-856 5721
Bloomington IN 47405   Fax: +1-812-855 3000
United States of America   E-mail: moran@indiana.edu

Emilio Moran is Rudy Professor of Anthropology, Professor of Environmental Sciences, Director of the Anthropological Center for Training and Research on Global Environmental Change, Co-Director of the Center for the Study of Institutions, Population and Environmental Change of Indiana University, Bloomington, United States. He was member of the Scientific Steering Committee of the IGBP-IHDP Land-Use/Cover Change (LUCC) project, and LUCC Focus 1 Leader on land use dynamics.

Dr. Meine van Nordwijk

World Agroforestry Centre, South East Asia Regional Office   Tel.:   +62-251-6254 15, -17
P.O. Box 161   Fax: +62-251-6254 16
Bogor 16001   E-mail: m.van.nordwijk@cgiar.org,
Indonesia   icraf-indonesia@cgiar.org

Dr. Cheryl Palm

The Earth Institute at Columbia University
Tropical Agriculture Program, Lamont Campus
P.O. Box 1000
Palisades, New York, 10964
United States of America

Tel.:   +1-845-680 4462
Fax: +1-845-680 4866
E-mail: cpalm@iri.columbia.edu

Prof. Jonathan A. Patz

University of Wisconsin-Madison
Center for Sustainability and the Global Environment
1710 University Avenue
Madison WI 53726
United States of America

Tel.:   +1-608-262 4775
Fax: +1-608-265 4113
E-mail: patz@wisc.edu

Prof. Kjeld Rasmussen

University of Copenhague, Geographical Institute
Østervoldgade 10
1350 Copenhague
Denmark

Tel.:   +45-35-322 563
Fax: +45-35-322 501
E-mail: kr@geogr.ku.dk

Prof. Dale S. Rothman

The Macaulay Institute
Socio-Economic Research Programme (SERP)
Craigiebuckler, Aberdeen AB15 8QH
UK

Tel.:   +44-(0)1224-49 8200 (ext. 2336)
Fax: +44-(0)1224-49 8205
E-mail: d.rothman@macaulay.ac.uk
Web: http://www.macaulay.ac.uk

Prof. Mark Rounsevell

University of Louvain, Department of Geography
1348 Louvain-la-Neuve
Belgium

E-mail: rounsevell@geog.ucl.ac.be

Mark Rounsevell is Professor of Geography at the Université catholique de Louvain, Belgium. He has undertaken research on the analysis and modeling of land use change within Europe within several projects for the European Commission. Since 1994 until the present he has been a lead author to the Intergovernmental Panel on Climate Change (IPCC) second, third and fourth assessment reports on climate change impacts and adaptation.

Prof. Thomas Rudel

The State University Rutgers, Department of Human Ecology
55 Dudley Road
New Brunswick, NJ 08901-8520
United States of America

Tel.:   +1-732-932 9169
Tel.:   +1-732-932 6667
E-mail: rudel@aesop.rutgers.edu

Dr. David E. Thomas

World Agroforestry Centre
c/o Chiang Mai University, Faculty of Agriculture
Chiang Mai 50200
Thailand

Tel.:   +66-(0)53-35 7906, -7907
Fax: +66-(0)53-35 7908
E-mail: d.thomas@cgiar.org

Prof. B. L. Turner II

Clark University, George Perkins Marsh Institute
950 Main Street
Worcester, Massachusetts 01610-1477
United States of America

Tel.:   +1-508-793 7325
Fax: +1-508-751 4600
E-mail: BTurner@clarku.edu

B. L. Turner II is the Higgins Professor of environment and society at Clark University in Worcester, Massachusetts, United States. His research focuses on human-environment relationships from ancient to contemporary times, and he is involved in the development of land-change science. He is a member of the U.S. National Academy of Sciences and the American Academy of Arts and Sciences.

Prof. Paul L. G. Vlek

University of Bonn, Center for Development Research
Walter-Flex-Str. 3
53113 Bonn
Germany

Tel.:   +49-(0)228-73 1866
Fax: +49-(0)228-73 1889
E-mail: p.vlek@uni-bonn.de

Paul Vlek is professor and Director General of the Center for Development Research of the University of Bonn. He also serves on the Scientific Committee of the IHDP and of the German Committee of Global Change Research. He leads a series of transdisciplinary projects in Central Asia and Africa dealing with sustainable use and management land and water in the context of global environmental change.

Box Authors

Prof. Arild Angelsen

Norwegian University of Life Sciences
Department of Economics and Resource Management
P.O. Box 5033
1432 Ås
Norway

E-mail: arild.angelsen@umb.no

Dr. François Bousquet

CIRAD – Tere Green Cirad-Tera-Green
Campus Baillarguet, TA 60/15
73, rue Jean François Breton
34398 Montpellier Cedex 5
France

E-mail: francois.bousquet@cirad.fr

Prof. Daniel G. Brown

University of Michigan
School of Natural Resources and Environment
Ann Arbor, MI 48109-1041
United States of America

E-mail: danbrown@umich.edu

Dr. Keith Clarke

University of California, Geography Department
Santa Barbara, CA 93106-4060
United States of America

E-mail: kclarke@geog.ucsb.edu

Dr. Tom Kram

Netherlands Environmental Assessment Agency (MNP)
Global Sustainability and Climate (KMD)
P.O. Box 303
3720 BA Bilthoven
The Netherlands
Tel: +31 (0)30 274 274 5

E-mail: tom.kram@mnp.nl

Koen P. Overmars

Wageningen University and Research Center
Department of Environmental Sciences
P.O. Box 37
6700 AA Wageningen
The Netherlands

Tel.:   +31-(0)317-485 611
Fax: +31-(0)317-482 419
E-mail: Koen.Overmars@wur.nl

Prof. Dawn Cassandra Parker

George Mason University
Department of Environmental Science and Policy
4400 University Drive
Fairfax, Virginia 22030-4444
United States of America

E-mail: dparker3@gmu.edu

Dr. Suzanne Serneels

Catholic University of Louvain, Department of Geography
3, Place Louis Pasteur
B-1348 Louvain-la-Neuve
Belgium

E-mail: serneels@geog.ucl.ac.be

Prof. Stephen J. Walsh

University of North Carolina, Department of Geography
Chapel Hill, NC 27599-3220
United States of America

Tel.:   +1-919-962 3867
Fax: +1-919-962 1537
E-mail: walsh@geog.unc.edu

Chapter 1

Introduction: Local Processes with Global Impacts

Eric F. Lambin · Helmut Geist · Ronald R. Rindfuss

1.1 A Research Agenda for a Primary Driver of Global Change

1.1.1 Introduction

Concerns about land-use/cover change emerged in the research agenda on global environmental change several decades ago with the realization that land-surface processes influence climate. In the mid-1970s, it was recognized that land-cover change modifies surface albedo and thus surface-atmosphere energy exchanges, which have an impact on regional climate (Otterman 1974; Charney and Stone 1975; Sagan et al. 1979). In the early 1980s, terrestrial ecosystems as sources and sinks of carbon were highlighted; this underscored the impact of land-use/cover change on global climate via carbon cycle (Woodwell et al. 1983; Houghton et al. 1985). Decreasing the uncertainty of these terrestrial sources and sinks of carbon remains a serious challenge today. Subsequently, the important contribution of local evapotranspiration to the water cycle – that is, precipitation recycling – as a function of land cover highlighted yet another considerable impact of land-use/cover change on climate, at a local to regional scale in this case (Eltahir and Bras 1996).

A much broader range of impacts of land-use/cover change on ecosystem goods and services were further identified. Of primary concern are impacts on biotic diversity worldwide (Sala et al. 2000), soil degradation (Trimble and Crosson 2000), and the ability of biological systems to support human needs (Vitousek et al. 1997; Millennium Ecosystem Assessment 2003). Land-use/cover changes also determine, in part, the vulnerability of places and people to climatic, economic, or sociopolitical perturbations (Kasperson et al. 1995; Turner et al. 2003a; Kasperson et al. 2005). When aggregated globally, land-use/cover changes significantly affect central aspects of Earth System functioning (DeFries et al. 2004c; Cassman et al. 2005). All impacts are not negative though as many forms of land-use/cover changes are associated with continuing increases in food and fiber production, in resource-use efficiency, and in wealth and well-being.

Understanding and predicting the impact of surface processes on climate required long-term historical reconstructions – up to the last 300 years – and projec-

tions into the future of land-cover changes at regional to global scales (Ramankutty and Foley 1999; Taylor et al. 2002b). Quantifying the contribution of terrestrial ecosystems to global carbon pools and flux required accurate mapping of land cover and measurements of land-cover conversions worldwide (Dixon et al. 1994; Houghton et al. 1999; McGuire et al. 2001). Fine resolution, spatially explicit data on landscape fragmentation were required to understand the impact of land-use/cover changes on biodiversity (Margules and Pressey 2000; Liu et al. 2001). Predicting how land-use changes affect land degradation, the feedback on livelihood strategies from land degradation, and the vulnerability of places and people in the face of land-use/cover changes requires a good understanding of the dynamic human-environment interactions associated with land-use change (Kasperson et al. 1995; Turner et al. 2003a; Kasperson et al. 2005). Sustainable land use refers to the use of land resources to produce goods and services in such a way that, over the long term, the natural resource base is not damaged, and that future human needs can be met. The time horizon of the concept covers several generations.

Over the last few decades, numerous researchers have improved measurement of land-cover change, the understanding of the causes of land-use change, and predictive models of land-use/cover change, in part under the auspices of the Land-Use and Land-Cover Change (LUCC) project of the International Geosphere-Biosphere Programme (IGBP) and International Human Dimensions Programme on Global Environmental Change (IHDP) (Turner et al. 1995; Lambin et al. 1999). This work, part of an international effort that has helped to propagate the emergence of "land-change science," has taken on the task of demonstrating the role of land change in its own right within the Earth System. An "integrated land science" has emerged, uniting environmental, human, and remote sensing/GIS sciences to solve various questions about land-use and land-cover changes and the impacts of these changes on humankind and the environment (Turner 2002). This science has demonstrated both the pivotal role of land change in the Earth System and its complexities that transcend such simplifications as unidirectional and permanent land-cover change caused by immediate population or consumption changes, replacing them by a representation of a much more complex process of land-use/cover change (Lambin et al. 2001; Turner 2002; Lambin et al. 2003; Steffen et al. 2004; Gutman et al. 2004; Moran and Ostrom 2005). The new Global Land Project (Ojima 2005) is developing further land-change science based on the foundations generated by LUCC and by other projects on terrestrial ecosystems. The objective of this introductory chapter is to set the stage for the following book chapters through a brief review of the main accomplishments of the LUCC project, and a discussion of the need for an overarching theory of land-use change. A more detailed history of the LUCC project is provided in Moran et al. (2004).

1.1.2 Development of the Land-Use/Cover Change (LUCC) Project

Following preparatory work in the early 1990s (Turner et al. 1990, 1993b; Turner and Meyer 1994; Turner et al. 1994), the Land-Use/Cover Change (LUCC) project as a worldwide, interdisciplinary joint core project of IGBP and IHDP was formalized in 1995 through the publication of its science/research plan (Turner et al. 1995) and reached the stage of implementation four years later (Lambin et al. 1999). The present book presents some of the main scientific results produced by the project during a decade of operation (Lambin and Geist 2001, 2005; LUCC Scientific Steering Committee 2005).

The three missions of the LUCC project have been to build a compendium of information about local to global land-use and land-cover dynamics, to identify a small number of robust principles that can better knit together local insights into a predictive science, and to foster the development of common models which may then become widely available to scientists and stakeholders. In order to implement the project's science plan, six broad research questions were formulated (Lambin et al. 1999). These overarching science questions basically relate to the past 300 years as well as to the next 100 years (Steffen et al. 2004; Turner et al. 1990) – see Box 1.1.

Box 1.1. Science questions of the Land-Use/Cover Change (LUCC) project

- How has land cover been changed by human use over the last 300 years?
- What are the major human causes of land-cover change in different geographical and historical contexts?
- How will changes in land use affect land cover in the next 50–100 years?
- How do human and biophysical dynamics affect the coupled human-environment system?
- How might changes in climate (variability) and biogeochemistry affect both land use and land cover, and *vice versa*?
- How do land uses and land covers affect the vulnerability of the coupled human-environment system?

In order to implement the project's science plan, a series of tasks and activities was set up to meet the broad objectives. Most importantly, three interlocking strategies were encapsulated in three research foci – see Table 1.1.

Among these interlocking strategies have been the development of case studies to analyze and model the processes of land-use change and land management in a range of geographic situations (Focus 1: land-use dynamics). Two complementary strategies were applied – see Fig. 1.1. First, a global network of new case studies was established focused on land-use and -cover change, seeking to understand its dynamics as they operate in different regions of the world. Some of these efforts followed a study-based protocol in an effort to elicit lessons or principles about land change. A prominent example is the study on the institutional, demographic, and biophysical dimensions of forest-ecosystem change in the Western Hemisphere (Moran and Ostrom 2005). Second, meta-analytical frameworks have increasingly been applied, involving comparisons of disparate case studies to detect problems resulting from the use of interdisciplinary, team-based analytical frameworks, to improve land-change analysis, and to provide general insights about land-change dynamics at the meso- and macroscales while preserving the descriptive richness of local studies (Rindfuss et al. 2004b). Prominent examples are the exploration of synergistic causal combinations in tropical deforestation as well as desertification (Geist and Lambin 2002, 2004), the study of agricultural change through intensification in core cropland areas of the world (McConnell and Keys 2005), and an investigation into the conditions for a transition towards sustainable land use (Lambin 2005). These strategies helped to improve the understanding of the causes of land-use change and to move from simplistic representations of two or three driving forces to deeper understanding involving situation-specific interactions among a large number of factors at different spatial and temporal scales. Although, the richness of explanations has greatly increased, this often occurred at the expense of generality of the explanations.

The second major theme or strategy was the development of empirical, diagnostic models of land-cover change through direct observations and measurements of the explanatory factors (Focus 2: land-cover dynamics). In addition to the effort to understand the multiple causes of land-use/cover change, there has been a concomitant rapid expansion in the availability of data and information on land-cover dynamics. In particular, remote sensing data and analysis made important contributions in documenting the actual changes in land cover at regional and global spatial scales from the mid-1970s onwards. A standardized land-cover classification system has been developed for application at multiple scales (Di Gregorio and Jansen 2000a). Several products are now available that depict global land

Table 1.1. Series of activities related to the various focus areas of LUCC

Focus 1: Land-use dynamics *Comparative analysis*	Focus 2: Land-cover changes *Direct observations and diagnostic models*	Focus 3: Regional and global models *Integrative assessments*
Activities	Activities	Activities
1. Understanding land-use decisions	1. Land-cover change hot-spots and critical regions	1. Comparison of past and current regional land-use/cover change models
2. From process to pattern: linking local land-use decisions to regional and global processes	2. Linking people to pixels	2. Methodological issues in regional land-use/cover change models
3. Sustainability and vulnerability scenarios	3. From patterns to processes of land-cover change	3. Land-use/cover change and the dynamics of interrelated systems
		4. Scenario development

cover, and the same is true for snapshots of many important regions with substantial land-cover change. The TREES project, for example, using satellite data observed changes in humid tropical forest worldwide (Achard et al. 2002, 2004; Achard 2006). A complementary approach is based on continuous fields of biophysical attributes at the global scale (DeFries et al. 2002b). The international BIOME 300 project reconstructed historical changes in cropland at a global scale during the last 300 years, compiling various contemporary and historical statistical inventories on agricultural land (such as census data, tax records, and land surveys) and applying different spatial analysis techniques to reconstruct historical maps of agricultural areas (Ramankutty and Foley 1999; Klein Goldewijk 2004). Attention also turned to the status and trends in terrestrial land cover, to produce the most reliable synthesis of documented rapid land-cover change worldwide over the period 1981 to 2001 (Lepers et al. 2005) – in collaboration with the Global Observation of Forest and Land Cover Dynamics (GOFC-GOLD) project for the Millennium Ecosystem

Assessment (MA). Together, these studies are leading the community to a consensus about the rates and locations of some of the main land-change classes underway globally, including tropical deforestation, cropland expansion and contraction, dryland degradation, and urbanization.

The third major theme was the development of integrated models as well as of prognostic regional and global models (Focus 3: Integrated modeling). Modeling, especially if done in a spatially-explicit, integrated and multi-scale manner, is an important technique for the projection of alternative pathways into the future, for conducting experiments that test our understanding of key processes, and for describing the latter in quantitative terms. Land-use change models offer the possibility to test the sensitivity of land-use patterns to changes in selected variables. They also allow testing of the stability of linked social and ecological systems, through scenario building (Veldkamp and Lambin 2001). To be able to generate reliable projections into the future or, in backward mode, to the past, a model must link dynamically

Fig. 1.1.
Case study comparison. Precursors of this approach are Brookfield (1962) and Turner et al. (1977)

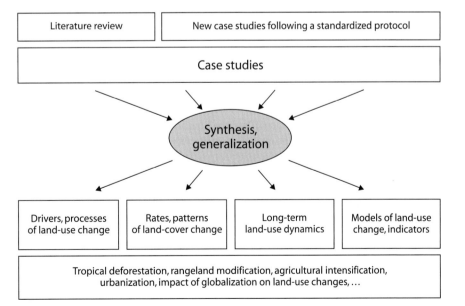

the processes of land-use change to biophysical processes, to represent biophysical feedbacks to land-use changes and land-use adaptations to biophysical changes. Today, only a very few models of land-use change can generate long-term, realistic projections of future land-use/cover changes at regional to global scales. The last decade, however, has witnessed innovative methodological developments in the modeling of land-use change at local to regional scales at a decadal scale (Liu 2001; Veldkamp and Lambin 2001; Parker et al. 2003). While some models are focussed on predicting the rates (or quantities) of change, others put more emphasis on spatial patterns. A fundamental difference in modeling tradition between different disciplines concerns the use of process-based (or structural) models *versus* statistical (or reduced form) models. Before the emergence of new modeling approaches, land-use change modeling was either dominated by economic theory (Fischer 2001) or was data-driven (Veldkamp and Fresco 1996). Also, a number of scenarios of future land use, at the global, national or local scales, have been formulated. Recent experiences involve policy-makers and stakeholders to define and negotiate relevant scenarios, in participatory approaches. There has been a shift from more physical and data-driven approaches to more human decision oriented approaches such as agent-based modeling (Veldkamp and Verburg 2004). Finally, a solid framework for a systematic validation of projections generated from land-use change models has been an essential component of this research field.

The three major themes or foci of the LUCC project – land-use change, land-cover dynamics, and integrated modeling – were meant to be interlocking strategies to understand land-use/cover changes (Lambin et al. 1999). These perspectives have been combined in various ways in integrated, place-based research on causes and impacts of land-use change. Prominent examples are studies such as the southern Yucatán peninsular region (SYPR) project (Turner et al. 2004), the Serengeti-Mara ecosystem project (Homewood et al. 2001), and the Nang Rong District project in northeastern Thailand (Entwisle et al. 1998; Walsh et al. 1999). Other integrated land-change studies over a particular geographical region include the Ecuadorian Amazon (Walsh et al. 2002), western Honduras (Nagendra et al. 2003), the Yaqui Valley in Mexico (Riley et al. 2001; Turner et al. 2003c), and the Southeast Asia Land-Use/Cover Change project (Rice et al. 2004). More recently, elaborate studies have been performed for the European Union (ATEAM, EUruralis and PRELUDE projects are examples of such large integrated projects). In addition, there are several other integrated land-change studies not carried out under the umbrella of the LUCC project such as those on the African Sahel (Raynaut 1997; Mortimore and Adams 1999).

To meet its mission goals, the LUCC project reached out to a large scientific community to generate a wealth of results on its fundamental science questions. Two particular concerns had been the facilitation of interdisciplinary research work between the social and natural sciences, and to globalize research on land-change processes by contrasting results obtained from a variety of regions and geographic situations.

1.2   The Distinction between Land Use and Land Cover

Well before the implementation of the Land-Use/Cover Change (LUCC) project, it was recognized that land-use and land-cover change observed at any spatio-temporal scale involves complex synergy with changes observed at other analytical scales (e.g., Marsh 1864; Thomas 1956; Turner et al. 1990, 1994; Turner and Mayer 1994; Meyer and Turner 2002). The LUCC science plan and implementation strategy sought multiple ways to deal with this reality, the first being to distinguish land "use" from land "cover" (Turner et al. 1995; Lambin et al. 1999).

The terrestrial surface, or land covers of the Earth and changes therein, is central to a large number of the biophysical processes of global environmental change, qualifying "land change as a forcing function in global environmental change" (Turner 2002, 2006). Land cover has been defined by the attributes of the Earth's land surface and immediate subsurface, including biota, soil, topography, surface and groundwater, and human (mainly built-up) structures. Land-cover conversions constitute the replacement of one cover type by another and are measured by a shift from one land-cover category to another, as is the case of agricultural expansion, deforestation, or change in urban extent. Land-cover modifications, in contrast, are more subtle changes that affect the character of the land cover without changing its overall classification (Turner et al. 1995; Lambin et al. 2003). Whatever the type of changes in land cover, they encompass changes in biotic diversity, actual and potential primary productivity, soil quality, runoff and sedimentation rates, and other such attributes of the terrestrial surface of the Earth (Steffen et al. 2004; DeFries et al. 2004c). Land covers and changes in them are sources and sinks for most of the material and energy flows that sustain the biosphere and geosphere, including trace gas emissions and the hydrological cycle (Vitousek et al. 1997; Meyer et al. 1998; Haberl et al. 2004b; Kabat et al. 2004; Crossland et al. 2005b; Canadell et al. 2006).

Contemporary land-cover change is generated principally by human activity, activity directed at manipulating the Earth's surface for some individual or societal need or want, such as agriculture (Turner et al. 1990; Ojima et al. 1994; Walker et al. 1999; Cassman et al. 2005). Land use has been defined as the purposes for which humans exploit the land cover. It involves both the manner in which biophysical attributes of the land are manipulated and the intent underlying that manipulation, i.e., the purpose for which the land is used. Exemplary classes denoting intent or purpose are forestry, parks, livestock herding, suburbia and farmlands. Land management, biophysical manipulation or the techno-managerial aspect of a land-use system, by contrast, refer to the specific ways in which humans treat vegetation, soil, and water for the purpose in question. Examples are the use of fertilizers and pesticides, irrigation for mechanized cultivation in drylands, or the use of an introduced grass species for pasture, and the sequence of moving livestock in a ranching system (Turner et al. 1995; Lambin et al. 2003).

In methodological terms, land cover and changes are visible in remotely-sensed data or by generating evidence from secondary statistics, such as (agricultural) census data. Such data require interpretation and ground truthing. Land-use as well as land-management information, in contrast, is mainly gained through detailed ground-based analysis, although land use can be inferred in remotely-sensed data under certain circumstances. Regardless, land cover and land use are so intimately linked that understanding of either has required approaches for linking household and community surveys, demographic and agricultural censuses, and market data, among others, to remote sensing and geographical information systems (Fox et al. 2003). Dating back to the mid-1990s at least, a distinguished feature of the LUCC project has been a unique strain of research aiming at linking people to pixels – "socializing the pixel" and "pixelizing the social" (Liverman et al. 1998; Geoghegan 2006a,b). In a further conceptual advance, the intimate linkage between land use and land cover has called for a coupled human-environment or social-biophysical system analysis or models in a much broader Earth System perspective as adopted by the new Global Land Project (Moran et al. 2004; Walsh and Crews-Meyer 2002).

1.3 Theoretical Foundations for Land-Change Science: Multiple Theories but Not "Atheoretical"

The complexity of causes, processes and impacts of land change has so far impeded the development of an integrated theory of land-use change. The need to distinguish between land use and land cover to account for interactions between socio-economic and biophysical processes is one source of complexity. Moreover, land-use change processes are dominated by multiple agents, multiple uses of land, multiple responses to social, climatic and ecological changes, multiple spatial and temporal scales in the causes of and responses to change, multiple connections in social and geographical space, and multiple ties between people and land. Causes and consequences of land-use change depend on the social, geographic and historical context. Theory building has thus been – and still is – a difficult task

From the beginning of LUCC, and recognized by the joint sponsorship by IHDP and IGBP, researchers from

multiple disciplines spanning the physical, spatial, and social sciences were involved. These researchers brought with them the methods and theories of the disciplines in which they were trained. Much of the early work on integration across studies involved either data and methodological issues (e.g., Liverman et al. 1998; Fox et al. 2003; Rindfuss et al. 2004b) or substantive empirical results (e.g., Geist and Lambin 2002, 2004; Moran and Ostrom 2005). While a core of LUCC researchers focused on understanding land change in its own right, an equal number sought to use land change to understand core disciplinary issues, be they demographic or ecological in kind. As LUCC progressed, however, more attention was given to understanding and explaining land change *per se*. While no unified theory of this change has emerged, much progress has been made in understanding under what conditions different theoretical orientations prove useful, and the community increasingly recognizes the need to address land change as a coupled human-environment system or societal-ecological system.

The empirical work, especially the case studies, was guided by multiple theories (van Wey et al. 2005), with the specific mix primarily determined by the disciplinary origins of the investigators on the team. Below we illustrate a few of them to provide a sense of their diversity. The empirical work was also guided by a variety of box and arrow type diagrams that provided theoretical guidance but were not theories, *per se* (Green et al. 2005). Rather, they tended to have a more *ad hoc* quality which recognized the underlying complexity of the determinants and consequences of land-cover/use change. We begin with a discussion of the box and arrow frameworks, then illustrate two of the specific theories that have been used, concluding with a discussion of the potential emergence of an overarching land-change theory.

1.3.1 Box and Arrow Frameworks

There are several factors that led to the prominence of box and arrow type frameworks to guide land-change science research. First, and perhaps most important, even the most cursory and casual thought would lead to the recognition that there are both natural/physical elements and social elements affecting land cover and use. Both need to be accounted for in any explanation of land change, and hence diagrammatic frameworks tend to start with at least three boxes (natural, social, and land change) with arrows from natural and social to the land-change box, and quite typically feedback arrows from land change to the natural and social boxes. There are complex systems within the natural and social boxes, and these boxes are often partitioned to reveal that complexity until chart clutter reaches the point where the designers of the figure have to stop.

A second reason for the inevitable emergence of these box and arrow frameworks is the multi-disciplinary nature of most land-change research teams and various scientific committees, including LUCC's Scientific Steering Committee. The box and arrow frameworks help multi-disciplinary teams fix terms, crystallize differences, and establish the types of data that need to be gathered and analyzed, especially in the absence of a commonly accepted, overarching land-change theory.

Finally, the box and arrow diagrams provide a convenient mechanism for communicating to readers. They summarize the main points, show the hypothesized direction of effects, and quickly illustrate whether the framework lends itself to conventional statistical analyses or whether modeling techniques, such as agent based models or cellular automata, are needed.

Figure 1.2 shows an example of a box and arrow model. This figure is clearly theoretical and capable of providing guidance to empirical researchers. It makes clear that researchers need to consider both social and biophysical factors. It shows the central role of land managers and highlights the importance of feedbacks. But from a theoretical and empirical perspective, Fig. 1.2 is also problematic. For example, the boxes themselves are sufficiently broad ("social systems," "ecological systems," and "land-use system") that, while providing guidance, they are also vague. Or to take another example, almost all the arrows are double-headed or involve feedbacks. There is a tendency for these diagrams to show that everything

is connected to, influences, and is influenced by everything else. While this is in keeping with the complexity that exists in the real world, it makes empirical estimation exceedingly difficult unless the analyst makes strong, and some would say "heroic," assumptions. Further, it is difficult to link the figure to theories that arise from such disciplines as demography, ecology, geography, or economics. Thus such figures are simultaneously helpful and problematic, have an *ad hoc* flavor, and should be viewed as a heuristic device. Box and arrow frameworks reflect the infancy of theoretical studies, and were designed to facilitate the quest for general principles and integrated theories.

1.3.2 Disciplinary Theories

The richness of the disciplinary backgrounds of those who have been examining land change is both an important asset to emerging land-change science and has acted as a constraint. The disciplinary make-up of research teams has a profound influence on the theories that guide data collection and analysis (Entwisle and Stern 2005). Land-change scientists run the gauntlet from anthropologists to zoologists, and within most of these disciplines there are multiple theoretical orientations and interests guiding research. A complete review of all these theories is clearly beyond the scope of this introductory chapter. Instead, two quite different yet compatible theories (multiphasic response theory and complexity theory) are illustrated, indicating their implications for land-change research, both separately and jointly.

Originally based in demography, multi-phasic response theory argues that when faced with sustained high rates of natural increase, households tend to use all possible demographic means such as delayed marriage, contraception, abortion, and migration in order to protect their relative status in society and to maximize new opportunities (Davis 1963). In the context of agricultural land use, various conceptualizations employ population change registered through land pressures as a mechanism to explain cropping strategies. Popularized by Boserup (1965, 1981), drawing on concepts developed by Chayanov (1966), induced intensification theories have emerged to explain peasant or smallholder household land change (e.g., Turner and Ali 1995).

Applied more broadly to include induced intensification, multi-phasic response theory can help us understand how household actions/decisions impact land-use and land-cover changes (Bilsborrow 1987). Households respond not only to the increase in number of surviving children but also to changes in other demographic and socioeconomic characteristics (such as number of adult males in the household, education level of the household head, and household economic status). Responses to changes in these household characteristics may include

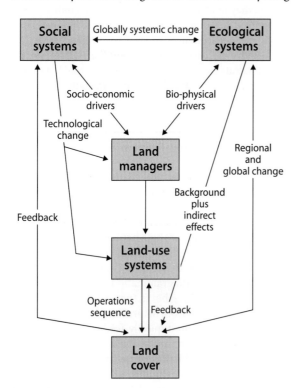

Fig. 1.2. Framework for understanding land-use/cover situations. *Source:* Turner et al. (1995), p. 35

land-use activities (such as more intensified use of land, decisions to plant an orchard or grow a grain crop, and collecting biomass for fuelwood) that may alter land cover as well as land use. The nature of such responses depends upon household resource endowments, availability of appropriate socioeconomic institutions, and biogeophysical conditions such as soil fertility, rainfall, temperature, slope, and elevation. This, in turn, has led to such themes as household life cycling as it affects rates of deforestation, forest succession, and cropping intensity in tropical frontiers (Walker et al. 2000).

Complexity theory, which is a relatively broad systems theory, states that complex systems are systems that contain more possibilities than can be actualized (Cilliers 1998; Luhman 1985). The goal of complexity theory is to understand how simple, fundamental processes can combine to produce complex holistic systems (Gell-Mann 1994). Non-equilibrium systems with feedbacks can lead to non-linearity (Bak 1998). Some systems contain multiple variables with apparent complexity. But a system based on multiple variables does not, in itself, guarantee complexity. Complex systems generally exhibit processes and relationships that are non-linear as well as embody hierarchical linkages that operate at different spatial and temporal scales. Research framed within complexity theory can address the rates and patterns of land-cover dynamics as well as possible non-linear feedbacks between the processes of change and existing patterns of land use/cover. Changes may depend partly on the existing patterns of land use, which may involve critical points where a small amount of land-use change significantly alters feedback processes and leads to a new pattern or equilibrium. Research on the resilience of socio-ecological systems draws heavily on this theoretical framework.

Note that both multi-phasic response theory and complexity theory have some common implications for data collection and analysis. Both are concerned with change over time, requiring longitudinal data and analysis techniques that address change. Both would be amenable to either statistical analytical procedures or model based approaches such as cellular automata or agent based models.

1.3.3 Potential for an Overarching Theory

The time is getting ripe for one or more overarching theories of land change to emerge, theories that incorporate insights from multiple social and natural sciences, and theories that explain change in the behavior of people as well as land-cover/use change. Why is this claim made? First, there is evidence that a land-change science has emerged, building on the foundations generated by LUCC, and transitioning to the Global Land Project (Ojima 2005). We are seeing a steady increase in conferences devoted to land change, journal articles reporting land-change science results, journals focused on land change (e.g., *Land Use Policy; Journal of Land Use Science*), and funding available to pursue land-change science. Emerging sciences need their own theories. Second, empirical research, guided by box and arrow frameworks and disciplinary theories, have started to produce a number of stylized facts that can serve as grist for more general theorizing. Third, the practical issues to which land-change science has been responding (e.g., climate change and biodiversity loss) demand more comprehensive theories so that we can better understand the past and predict the future. Any new endeavor such as land-change science requires an inductive stage first. As the body of understanding grows, one expects synthesis to take place, theoretically or otherwise. However, because of the high complexity of land change, simplifications other than at the abstract systems level will likely be difficult to achieve.

While we are not ready to propose an overarching theory of land change, we are in a position to understand some aspects and issues that such a theory must address. First, an overarching land-change theory needs to engage both the behavior of people and society (agency and structure) and the uses to which land units are put, as well as feedbacks from one to the other. To facilitate discussion, the smallest land units are referred to as "pixels," abstracting from the complexity that pixel size varies across data sets and sensors. Just as individuals and societies have characteristics (respectively, education, age, sex, and wealth, and institutions and policies) that affect land-use decisions, pixels also have characteristics (elevation, slope, aspect, and expected climate) that affect the range of cover types that will grow on them and the likelihood of being converted from one use to another. Unlike people, pixels as land units, do not move, and hence location is a critical aspect of pixels. Determinants and consequences of land change, of necessity, involve the characteristics of both people and pixels.

Second, an overarching theory of land change needs to be multi-level with respect to both people and pixels, recognizing that they can combine in ways that affects their behaviors, as a single unit or collectively. Individuals combine into households, and households are more likely to be land-use decision making units than individuals. Households are typically combined into villages or towns, which are further aggregated into larger geopolitical units. Policies, customs and markets operating at a higher level of organization can affect, additively and interactively, the land-use decisions made by units at lower levels of organization. People can also be members of organizations (corporations, religions, or voluntary organizations) that make or influence land-use decisions. Similarly, pixels can be combined into various biophysical (watersheds, valleys, or large unfragmented forests) and geopolitical (districts, provinces and countries) units that affect, additively and interactively, the use to which a pixel is put.

Third, an overarching theory of land change would need to incorporate the extent to which people and pixels are connected to the broader world in which they exist, both currently and in the past. Quite frequently, rapid land-cover/use change is associated with a change in the connection of those pixels to the broader geopolitical and economic world, and the direction can be towards either more or less integration with that broader world. The fall of the Maya in Southern Yucatán (Turner et al. 2004) and the Angkor Empire in northwest Cambodia (Rindfuss et al. 2004b) led the conversion of many managed agricultural pixels back to "unmanaged" forest. Deforestation in those areas today, as with most of the hot spots of rapid land-cover/use change, are the result of integration going in the opposite direction, that is, pixels and their managers becoming more integrated into the local, regional, national and world geopolitical organization and economies. As pixels become more integrated into the broader world, their locational and biophysical aspects likely become more important predictors of how they are used.

Finally, an overarching theory of land use will need to incorporate time, both past time (history) and the future. The history of a system, people and pixels have a profound influence on the likelihood of any given pixel changing into a different land-use type. The influence of past land uses on options for future uses is referred to as path dependency. Perhaps this is most obvious in the case of large urban areas. Once land has been converted into urban use, it is difficult for that land to be converted to agricultural or relatively unmanaged uses – but the Mayan and Angkor examples cited above remind us that it does happen.

1.4   Objectives and Structure of the Book

This volume describes how our understanding of land-use/cover change has moved from simplicity to greater realism and complexity over the last decade, with the overall goal to extract from this complexity a general framework for a more realistic understanding of land-use/cover change. This is achieved by first presenting latest findings on rates, causes and impacts of land-use/cover change, discussing how this new understanding is captured in models and scenarios of land-use/cover change, identifying relevant links between land-change science and policy, and then highlighting important topics at the research frontier.

In Chap. 2 (on recent progress and remaining challenges in the detection of global land-cover change), historical changes in land cover as well as the most recent estimates of the magnitude of land-use/cover change are summarized for some of the most important land classes such as forests, agricultural areas, pastoral areas, urban zones and drylands. The complex nature of land-cover change is discussed to emphasize the need to integrate all scales and processes of change. The power and limitations of remote sensing as a promising tool are raised in a concluding comment.

In Chap. 3 (on the causes and trajectories of land-use/cover change), a synthesis of recent case study evidence on the causes of land-use change is presented, with emphasis on the mode of interaction between diverse causes and dominant pathways of change.

In Chap. 4 (on the multiple impacts of land-use/cover change), a trade-off approach has been adopted to come closer to a balanced view of the "positive"impacts such as continuing increases in food and fiber production, resource-use efficiency, wealth, livelihood amelioration, and human well-being *versus* the "negative" impacts such as those on climate and the provision of ecosystem goods and services.

In Chap. 5 (on modeling land-use/cover change), the role of models in land change and related sciences and the diversity of modeling approaches are discussed, including the spatial and temporal dimensions of modeling as well as techniques of calibration and validation.

In Chap. 6 (on the search for the future of land through global to local land scenarios), results from major, worldwide available scenarios of future land change are presented, compared, and synthesized under short-term *versus* long-term perspectives.

In Chap. 7 (on the current lessons and future integration of linking policy and land-use/cover change science), key public policy lessons are presented, successes and failures of the impact of land-change science are discussed, and a road map is sketched of how land-change science can be more useful in the policy process.

Finally, in Chap. 8 (on research frontiers), concluding remarks highlight the dynamic nature of coupled human-environment systems and the conditions for a transition towards sustainability in relation to land-use/cover change, further outlining the frontiers of research with a particular perspective upon the Global Land Project (GLP) under the auspices of the Earth System Science Partnership (E-SSP).

Chapter 2

Global Land-Cover Change: Recent Progress, Remaining Challenges

Navin Ramankutty · Lisa Graumlich · Frédéric Achard · Diogenes Alves · Abha Chhabra · Ruth S. DeFries
Jonathan A. Foley · Helmut Geist · Richard A. Houghton · Kees Klein Goldewijk · Eric F. Lambin
Andrew Millington · Kjeld Rasmussen · Robin S. Reid · Billie L. Turner II

2.1 Introduction

Since time immemorial, humankind has changed landscapes in attempts to improve the amount, quality, and security of natural resources critical to its well being, such as food, freshwater, fiber, and medicinal products. Through the increased use of innovation, human populations have, slowly at first, and at increasingly rapid pace later on, increased its ability to derive resources from the environment, and expand its territory. Several authors have identified three different phases – the control of fire, domestication of biota, and fossil-fuel use – as being pivotal in enabling increased appropriation of natural resources (Goudsblom and De Vries 2004; Turner II and McCandless 2004).

The first stage in human history, the Paleolithic age, was characterized by the use of stone tools and the control of fire. The control of fire and its use by paleolithic hunters changed habitats and was partly responsible for the extinction of megafauna at the beginning of the Holocene (Barnosky et al. 2004). This innovation also enabled humans to expand their territory by migrating from their origins in East Africa to Eurasia, Australia and the Americas. Observations of large-scale landscape burning by humans can be traced to antiquity, as in a Carthaginian reference to western Africa some 500 years before the birth of Christ:

> "By day we saw nothing but woods, but by night we saw many fires burning ... we saw by the night the land full of flame and in the midst of lofty fire ... that seemed to touch the stars." (in Stewart 1956, p 119).

The next stage of human history began with the domestication of plants and animals, termed the Neolithic Revolution, which began roughly 10 000 years ago in several places around the world – in Mesopotamia, in China, eastern U.S., New Guinea, and the Sahel, and later on in Mesoamerica and the Andes. The advent of sedentary agriculture matched and exceeded the land changes wrought by fire. Both Plato and Aristotle commented on the soil erosion and deterioration of the hills and mountains of Greece. Plato, wrote in Critias 2 400 years ago, that

> "... what now remains ... is like the skeleton of a sick man, all the fat and soft earth having wasted away, and only the bare framework of the land being left".

Agricultural lands today occupy roughly a third of the planet's land surface (Ramankutty and Foley 1998; Klein Goldewijk 2001).

The third stage of human history was marked by the human appropriation of energy stored in fossil fuels. This stage began roughly 300 years ago, and was characterized by the rise of globalization, the dominance of capitalism, and the advent of Industrial Revolution technologies. During this period, the world's human population expanded exponentially. While many economically developed nations achieved the demographic transition of low birth rates and low death rates during the late 20th century, this is only just happening in most of the developing world.

The extent and pace of human activities on the land surface also accelerated during the last 300 years. More land was converted for human use than before, and already converted land was managed more intensively to increase the yields of agricultural and forest products. As early as 1864, George Perkins Marsh, in his book *Man and Nature*, documented his observations of landscape changes resulting from human activities. Richards (1990) estimated that more forests were cleared between 1950 and 1980 than in the early 18th and 19th centuries combined. Such accelerated changes have been accompanied by local and global environmental problems, and various writers such as John Muir, Henry David Thoreau, Rachel Carson, Aldo Leopold, and Paul Ehrlich have contributed to the rise of consciousness about environmental issues. In 1956, the book *Man's Role in Changing the Face of the Earth* (Thomas Jr. 1956), the outcome of an international conference, documented major changes of the planet's landscapes. More recently, books such as *The Earth as Transformed by Human Action* (Turner II et al. 1990), *Mappae Mundi* (Goudsblom and De Vries 2004), and various reports of the World Resources Institute (e.g., *People and Ecosystems: The Fraying Web of Life*, http://pubs.wri.org/pubs_pdf.cfm?PubID=3027), the United Nations Environment Program (e.g., *One Planet, Many People: Atlas of our Changing Environment*, http://www.na.unep.net/OnePlanetManyPeople/index.php; *Global Environmental Outlook 3*, http://www.unep.org/geo/geo3/), the World Watch Institute (e.g., State of the World 1996; Brown et al. 1996) and the Millennium Ecosystem Assessment (*Ecosystems and Human Well-being*, Millennium Ecosystem Assessment 2005) have lent further credence to the notion that one of the most obvious global changes in the last three centuries has been the direct human modification and conversion of land cover.

A recent study by the World Conservation Service estimated that the "human footprint" covers 83% of the global land surface (Sanderson et al. 2002). However, the presence of humans does not necessarily imply that landscapes are degraded, and that the ecosystem services they offer are diminished. Indeed, another study more optimistically estimated that roughly half of the world's land surface is still covered by "wilderness" areas (Mittermeier et al. 2003). Therefore, it is becoming increasingly obvious that we need to move beyond subjective terms such as "wilderness" and "human footprint", and evaluate the trade-offs between ecosystem goods and services extracted by humans through their land-use practices, and any resulting ecological degradation – see Box 2.1 (Millennium Ecosystem Assessment 2003; DeFries et al. 2004b).

Several estimates have been made of the natural resources consumed by humans through their land-use practices. Several authors (Vitousek et al. 1986; Haberl et al. 2004b; Imhoff et al. 2004) have estimated that roughly 20–40% of global net primary productivity is being co-opted (not available to other species) by humans. However, Rojstaczer et al. (2001) estimated a wider uncertainty range of 10% to 55%. Postel et al. (1996) estimated that about half of the global renewable freshwater supply was being co-opted. It should be noted that such human use of natural resources is not uniform globally, but varies spatially depending on the levels of socio-economic development, lifestyles, and cultures of different nations of the world. To measure such differences in consumption patterns, Wackernagel et al. (1997) developed the concept of "ecological footprint" to denote the amount of land area needed to produce the resources consumed and absorb the waste generated by human societies. A recent estimate by Venetoulis et al. (2004) indicates that while an average Bangladeshi has a footprint of 0.5 ha/person, an Italian has a footprint of 3.3 ha/person, and an American of 9.6 ha/person.

In this chapter, we will review the major landscape changes resulting from human land-use activities. We will initially review, albeit briefly, the millennial timescale changes covering the first and second phases of human history. However, our major focus in this chapter will be on reviewing the land-use changes that occurred in the last 300 years; this was one of the major tasks identified by the LUCC implementation plan – see Chap. 1. Over the years, environmental historians and historical ecologists have reconstructed fairly accurate depictions of landscape change around the world; however, these local studies did not comprehensively cover the entire globe and could not be pieced together to get a global synoptic view. With the advent of remote sensing, it became possible to obtain a consistent, global picture of the world's landscapes. However, global remotely-sensed data are only available for three decades into the past. One of the major tasks facing LUCC was to bridge both the global synoptic spatial scale, as well as the centennial timescale perspective. The BIOME 300 project, a joint LUCC-PAGES initiative oversaw the creation of such a global historical land-cover database – see Fig. 2.1. Furthermore, the Millennium Ecosystem Assessment (MA) supported a LUCC study to map regions of the world undergoing

---

Box 2.1. The myth of the "natural"

Until recently, many discussions of land-use and land-cover change have treated it as a dichotomy between the "natural" and "human dominated" portions of the biosphere. However, this is an artificial distinction. There are differing degrees of human activity across the landscapes of the globe, ranging from the extreme transformation of urban environments, to the intensive management of agricultural areas, or the careful protection of recreational areas and parks. And even the most remote and isolated landscapes are still affected by human actions; changes in atmospheric chemistry (including changes in CO_2 concentrations, with their effects on plant physiological processes) now mean that there are no "natural" landscapes left on the planet (McKibben 1989).

Instead of labeling landscapes as "natural" or "human dominated", it might be useful to think of land-use and land-cover change as a continuum (Theobald 2004). Naturally, we immediately recognize land-use and land-cover change in the most extreme cases, such as urban areas and intensively managed croplands. But less intensive land-use practices can be subtle or confused as "natural"; even the explicit decision to restore "natural" areas is a land-use practice in itself.

While the nature of land-use practices vary greatly across the world, the ultimate purpose and result of these practices is generally the same: ecosystem goods and services are used in order to meet immediate human needs, often at the expense of degrading ecosystem conditions in the long term. Land-use practices thus present us with a trade-off (Millennium Ecosystem Assessment 2003; DeFries et al. 2004b; Foley et al. 2005). On one hand, land-use practices are essential for the ongoing success of our civilization, as they provide a steady stream of critical natural resources, such as food, freshwater and fiber. On the other hand, many forms of land use are disrupting environmental systems and simultaneously diminishing the capacity of ecosystems to sustain the flow of services – such as food production, maintaining freshwater and forest resources, regulating climate and air quality, and mediating infectious diseases (DeFries et al. 2004b; Foley et al. 2005).

A major challenge to the research community will be to develop analytical frameworks for assessing and managing the trade-offs between meeting immediate human needs and maintaining the capacity of ecosystems to provide goods and services in the long term (DeFries et al. 2004b). Such assessments of land-use trade-offs must draw both from social sciences and natural science disciplines, and recognize that land use provides crucial social and economic benefits even while leading to long-term negative consequences for ecosystem functioning and human welfare (Foley et al. 2005) – see Box 2.3 and Chap. 4.

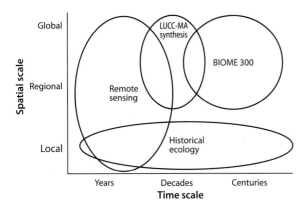

Fig. 2.1. Spatial and temporal scales addressed by different projects or fields of study of land-cover change. The BIOME 300 project uniquely addressed both large spatial scales and long temporal scales in describing changes in agricultural land. The LUCC-MA study conducted a detailed global synthesis of areas of major change in forest cover, agricultural land, degraded land, and urban areas over the recent decades

rapid land-cover change today. This chapter will review the historical and contemporary land-cover changes on the basis of these studies, and the lessons learned from them. It will then discuss remaining challenges in quantifying and understanding land-cover change.

2.2 Historical Changes in Global Land Cover

The notion that tropical rainforests are being cleared at a rapid pace, with enormous loss of biodiversity is common lore today. Not so well appreciated, however, are two countervailing observations: (a) large portions of some tropical forests were deforested in ancient times, most of which have reverted back to forest cover; and (b) the current magnitude and pace of tropical deforestation is a relatively recent phenomenon and apparently unprecedented in scope. Therefore, to understand the significance of present-day tropical deforestation, it will be useful to develop a "baseline" of historical land-cover changes. Furthermore, the study of historical land-cover changes can generate general principles to understand the conditions under which past land-use systems collapsed or sustained. It is in this context that LUCC, in collaboration with PAGES, launched the BIOME 300 project (see Box 2.2) to reconstruct historical land cover over the last 300 years. The 300-yr timeframe captures the period of greatest and most rapid transformation of global land cover with measurable impacts on today's landscapes. This section describes historical land-cover changes as described by the BIOME 300 land-cover data set. However, in many parts of the world, human impact on the land dates back beyond the last 300 years. Therefore, we first briefly consider land-cover changes that occurred in ancient times.

2.2.1 Global Land-Cover Changes over the Last Millennium

Humans have actively managed and transformed the world's landscapes for millennia. Students of prehistory and the paleosciences have discovered and described numerous examples of rapid or extensive modifications of the environment by ancient cultures (Redman 1999). In this section, we provide a broad overview of the evidence of early human modification of land cover, focusing on evidence for intensive modification as well as extensive, globally pervasive changes.

In areas where local records are rich with multiple lines of evidence, we catch tantalizing glimpses of human-nature interactions that provide early examples of human dependence on ecosystem services. A classic example of human alteration of landscapes comes from the Near East – the cradle of civilization. A case study of Ain Ghazal in southern Jordan suggests that this village and its neighbors were abandoned around 6000 B.C. due to deterioration in the natural vegetation (Redman 1999). Rollefson and Kohler-Rollefson (1992) attribute the land-cover change to a prolonged drought coupled with a change toward a home construction technique – lime plaster – that required larger quantities of fuelwood than was used previously. In the decades before site abandonment, the evidence for forest-cover change is clear, and is reflected in the size and quantify of timbers used in buildings at the site. A decline in the variety of wild animals eaten provides further evidence of ecological disruption associated with forest changes. At about 6000 B.C. a slight deterioration in climate appears to have been the proximate factor for the abandonment of the village. This case study portends our current understanding of how climate change interacts with other factors in making a landscape more or less vulnerable to change (see Chap. 3).

Another particularly rich account of land-use change and its social precursors and long-term consequences comes from the Yucatán Peninsula of Mexico. Sediment records from lakes in the region provide strong evidence that most of the Yucatán forests were altered by human activities as early as 3 000 to 4 000 years ago (Redman 1999). The lowland Maya cleared substantial portions of the tropical forests of their greater Yucatán Peninsula homelands between 1 000 to 2 000 years ago. Deforestation and cultivation in the early phases of the growth in the central and southern heartland of the civilization led to considerable erosion and sedimentation of coastal wetlands. Improved land management in their Classic phase sustained extremely large populations for hundreds of years, but at the cost of major forest losses and the ecosystem services they rendered. The famous collapse of the Classic Period Maya at the end of the 10th century apparently involved synergistic links between severe land stress, prolonged climatic desiccation (perhaps amplified by regional forest losses), and socioeconomic disruption. Interestingly, the long-term abandonment of the central and southern heartlands permitted the return of the forest, although its species abundance was apparently altered by past Mayan activities (Turner II et al. 2003b).

Perhaps the largest early transformation of the landscape occurred in the Indus Valley, where an ancient civilization flourished from before 6000 B.C. to at least 1500 B.C. in what today is Pakistan and northwest India. The Indus Valley or Harappan Civilization was the largest and oldest urban civilization in the world. Archaeological explorations have revealed impressive ruins in parts of the Indus River and the extinct Saraswati/Ghaggar River valleys in the Indian sub-continent (Thakker 2001). The slow desertification of the region may have led to the decline of the Harappan Civilization after 3500 B.C., although recent evidence suggests that the decline may have been related to the waning of monsoon rainfall (Tripathi et al. 2004).

Numerous other studies document environmental deterioration, and associated social disruption, with intensification of land use (e.g., Tainter 1990). Examples include the salinization of farmlands due to irrigation in southern Mesopotamia 4000 years ago (Jacobsen and Adams 1958; Redman 1999). Another classic example is the deforestation of the cedar forests of Lebanon around 2600 B.C. Cedar wood was highly valued in Mesopotamia, and the intensity of trade reduced the cedar woodlands, which once covered 500 000 ha, to the four small groves found in Lebanon today (Oedekoven 1963). A particularly intriguing account comes from the now barren Easter Island. Pollen records from sediment cores document severe deforestation several hundred years after colonization of the island around A.D. 400. Archaeological accounts reveal that forest cover deteriorated further as a consequence of excessive use of forest resources by the islanders who used them as rollers for transporting heavy statues that they built as part of their ceremonial activities. When the ~7 000 people were unable to support themselves, the society collapsed quickly around A.D. 1700 (Diamond 1995).

Despite the richness of local and regional accounts (Redman 1999), a truly global view of early rates of land transformation has eluded us until recently. Evidence for a global signal of human impacts on ecosystems has been derived from estimates of atmospheric trace gases trapped in ice cores extracted from Antarctica (Ruddiman 2003). Ruddiman's argument for human impacts is based on analyzing changes in CO_2 and CH_4 concentrations over the last four glacial-interglacial cycles, as estimated from measurements in ice-core records. He finds that from roughly 5 000 years ago onwards, Holocene CH_4 atmospheric concentrations diverge from what would be expected based on the orbital insolation cycle. Similarly, from roughly 8 000 years ago, Holocene CO_2 atmospheric

concentrations diverge from the patterns observed in the previous three interglacials. In both cases, the divergence is to values higher than those expected. He therefore suggests that deforestation 8 000 years ago, and the emergence of paddy rice cultivation 5 000 years ago may have been responsible for the marked divergence of the CO_2 and CH_4 records from the expected trends. Not all authorities accept Ruddiman's interpretation of these Holocene trace gas records (see e.g., Joos et al. 2004). Although other modeling studies also suggest that past land-use/cover changes have impacted global climate (e.g., Bauer et al. 2003) there is, as yet, no clear consensus on the extent to which they have been responsible for the trends noted by Ruddiman.

2.2.2 Global Land-Cover Changes over the Last 300 Years

Agriculture has been the greatest force of land transformation on this planet. Nearly a third of the Earth's land surface is currently being used for growing crops or grazing cattle (FAO 2004a). Much of this agricultural land has been created at the expense of natural forests, grasslands, and wetlands that provide valuable habitats for species and valuable services for humankind (Millennium Ecosystem Assessment 2003). It is estimated that roughly half of the original forests (ca. 8 000 years ago) have been lost (Billington et al. 1996).

The pace of agricultural land transformation has been particularly rapid in the last 300 years. The BIOME 300 project (see Box 2.2) focussed on describing these historical changes in agricultural land. Results from the project indicate that global cropland area increased from ~3–4 million km^2 in 1700 to ~15–18 million km^2 in 1990 (Fig. 2.2a,b, Ramankutty and Foley 1999; Klein Goldewijk 2001). The area of grazing land, around which there is greater uncertainty, increased from ~500 million km^2 in 1700 to 3100 million km^2 today. Much of the expansion of croplands came at the expense of forests, while much of today's grazing land was formerly grasslands; although there are notable exceptions to these trends – for e.g., the North American Prairies were lost to croplands, and many Latin American forests are being cleared for ranching. Subsequently, the forest area globally has decreased from ~53 million km^2 in 1700 to ~43–44 million km^2 today, while the area of savannas and grasslands has decreased from 30–32 million km^2 to 12–23 million km^2.

The expansion of agriculture has shifted spatially over time, following the general development of human settlements and the global economic order (Richards 1990). Much of the large-scale cultivation in 1700 was concentrated in the Old World, specifically in Europe, the Indo-Gangetic Plains, eastern China, and Africa. Roughly 2–3% of the global land surface was cultivated at that time. Since then, the rate of cropland expansion increased with

European colonization and increasing globalization of world markets. New settlement frontiers were established in North America, Latin America, South Africa, and the Former Soviet Union (FSU). North America and the Former Soviet Union experienced their most rapid expansions of cultivated lands starting around 1850. Latin America, Africa, and South and Southeast Asia experienced slow cropland expansion until the 20th century, but have seen exponential increases in the last 50 years. China had a steady expansion of croplands throughout most of the last three centuries.

In the last 50 years, several regions of the world have seen cropland areas stabilize, and in some there has even been a decrease. In the United States of America, as cultivation shifted from the east to the Midwest, croplands were abandoned along the eastern seaboard around the turn of the 20th century, and this has been followed by a regeneration of the eastern forests during the 20th century. Similarly, cropland areas have decreased in southern China and Western Europe. Another new trend is the loss of prime farmland areas to urban expansion. Because cities were often founded near the prime farmland areas, expansion of cities due to population growth leads to an encroachment of built-up areas on to some of the world's best agricultural soils. While this is not significant yet on a global scale, regional-scale trends are alarming. It is estimated that the United States of America paved over roughly 2.9 million ha of agricultural land between 1982 and 1997, and that ~30% of the increase in developed land during 1982–1997 occurred on prime farmland (NRCS 2001). China lost nearly 1 million ha of its cultivated land to expansion of infrastructure (both urban and rural) between 1988 and 1995 (Heilig 1999; Seto et al. 2000). Some rough estimates indicate that 1 to 3 million ha of cropland may be taken out of production every year in developing countries to meet the land demand for housing, industry, infrastructure, and recreation (Döös and Shaw 1999).

Asia

Extensive research on land-use changes in tropical Asia is available for the period 1880–1980 (Flint and Richards 1991). This involves an area of 8 million km^2 and 13 countries (India, Sri Lanka, Bangladesh, Myanmar, Thailand, Laos, Cambodia, Vietnam, Malaysia, Brunei, Singapore, Indonesia and the Philippines). In this area as a whole, forest/woodland and wetlands declined over the hundred year period by 131 million ha (47%). At the same time, cultivated area increased by 106 million ha, nearly double that of 1880. Thus, 81% of the forest and wetland vegetation appears to have been converted during the expansion of agricultural land. Intensified timber extraction for domestic and export markets and the exploitation of firewood, fodder and forest products all contributed to deforestation in this part of the world. Some examples from several Asian countries are presented below with a view to elucidate the different processes of land-use changes in the past (also see Chap. 3).

China. Settled agriculture may have begun in China as early as 10 000 years ago, contemporaneous with its origin in Mesopotamia. The earliest cultivation began in the middle and lower reaches of the Yellow River and radiated outwards. Since those early origins the area under cultivation in China has generally increased, although not uniformly. Cropland areas have contracted as well as advanced in different regions and time periods in response to invasions, wars, environmental catastrophes, and political programs. The Loess Plateau in the north of China was largely stripped of natural vegetation by the Western Han dynasty (206 B.C. to A.D. 8) (Fang and Xie 1994). Much of the vegetation in the region returned between the third and sixth centuries when nomads from Mongolia drove out the farmers and replaced farmlands with grazing lands; but croplands expanded again when farmers moved back after the sixth century. In southeastern China as well, the cultivated area expanded and contracted, with peaks in the area under cultivation around A.D. 1200 and A.D. 1600, and large decreases as a result of wars and foreign conquests in the 14th and 17th centuries (Mongol and Manchu invasions, respectively). By 1800,

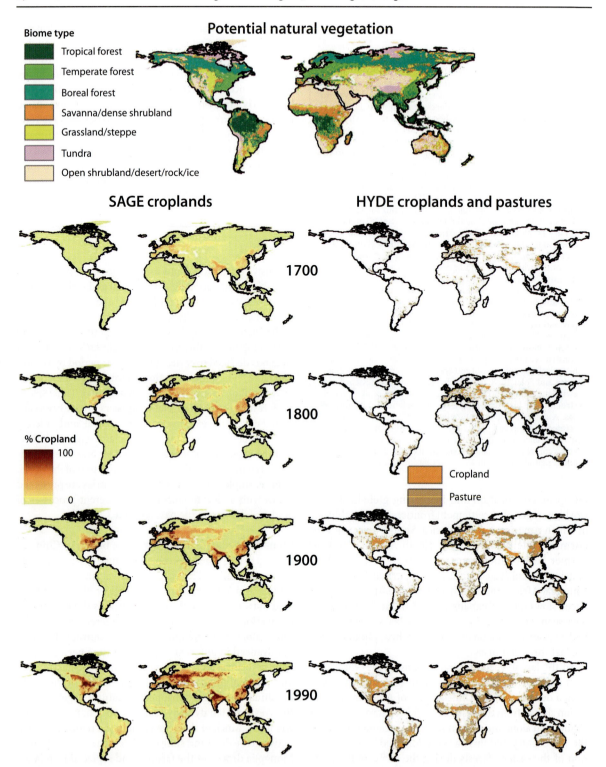

Fig. 2.2a. The BIOME 300 project presented two independent, global, geographically-explicit reconstructions of historical changes in agricultural land from 1700 to 1990. It also provided a global data set of potential natural vegetation

however, the area under cultivation in the region was twice as large as at any time previously, and by 1853, all of the cultivable land in Guangdong province had been cleared (Marks 1998).

The overall trend in cultivated land area in China has been one of expansion. By 1400 croplands covered an estimated 25 million ha. This number had increased to 50 million ha around 1700, to 100 million ha by 1935 (Per-

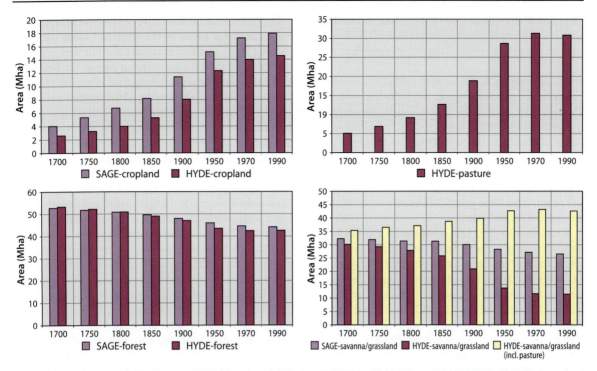

Fig. 2.2b. Intercomparison of the SAGE and HYDE data sets of agricultural land, forest land, and savanna/grassland areas. The forest land area and savanna/grassland area were inferred by subtracting agricultural land area from the potential natural vegetation data set

kins 1969), and to 130–140 million ha in the mid-1990s (Yang and Li 2000). The current estimate is 30–40% higher than the officially published value (Heilig 1999). In recent decades, the total area of cropland has declined by about 4.7 million ha (4.45%), with large losses in southeastern China being partially offset with increases in the north and southwest (Yang and Li 2000). In the context of land-use change, it is worth pointing out that the current area under crops in China accounts for only half of the long-term loss of forests (Houghton and Hackler 2003), suggesting that unsustainable logging practices, burning or other activities have led to a long-term increase in lands that are not well defined or quantified.

Malaysia. Southeast Asia has a long human history. Wet rice cultivation was certainly prevalent by the 14th century in Malaysia. Early colonial influence was limited, but the need for control of mineral production (tin) and land for plantations led to spread of colonial administration (Brookfield et al. 1990). In the 19th century, under British rule, rubber and oil palm were introduced to Peninsular Malaysia for commercial purposes. Over the course of time, Malaysia became the world's largest producer of rubber and oil palm. As a result, perennial crops constituted around 80% of the agricultural land in Malaysia by 1961 (FAO 2004a). Agricultural land expanded from covering 21% of the Malay Peninsula in 1966 to 39% in 1982. Consequently, forest cover in the peninsula was reduced from around 73% in the early 1950s to ~51% in 1982 (Brookfield et al. 1990). Loggers were well estab-

lished in Peninsular Malaysia by 1920 and have now cut over virtually all of the forest/woodland vegetation outside the limits of protected reserves. Logging has also been the main source of massive deforestation in East Malaysia, in Sarawak and Sabah. However, with the increasing pressure to protect the remaining forests, oil palm has emerged as a major commercial crop.

Indonesia. Evidence for agriculture in Sumatra goes back to 7 000 years before present. Land-use history in Indonesia was influenced by the Dutch colonial influence (Frederick and Worden 1992). In the late 19th century, central and east Java formed the center of Dutch sugar cultivation, supported by labor from the dense population. Java was the center of paddy rice cultivation (sawah), and while sugar was grown on the most fertile lands, sufficient rice was also grown. Not surprisingly, Java had been largely deforested before 1880. In contrast, the islands of Sumatra, Kalimantan (Borneo) and Sulawesi supported relatively sparse populations practicing slash-and-burn agriculture, and were scarcely exploited before 1920 (Richards and Flint 1994). Commercial logging started in Kalimantan during the 1960s, and was boosted in 1967 through huge financial investments by Japanese companies. By 1980, one-fifth of Kalimantan's forests had been logged. Additional forest losses occurred during the droughts of the 1982–1983 El Niño, when almost 3.5 million ha were burnt (Malingreau et al. 1985), and again during the 1997–1998 El Niño when almost 5 million ha of forests and 4 million ha of agricultural lands were damaged (Barber and Schweithelm 2000). Studies have concluded that much of the fires were attributable to side effects of logging (Siegert et al. 2001).

India. Agriculture in the form of settled cultivation began around 7 000 years ago in the Indo-Gangetic Plain. The rich fertile soil of the Indo-Gangetic alluvium contributed significantly to the growth of one of the world's oldest civilizations. Records from the Mughal Empire (1526–1857) indicate that much of the land in the Indo-Gangetic Plains was already under use (Abrol et al. 2002). With the establishment of British colonial rule in early 19th century, land-use practices became much more extensive and intensified.

Richards and Flint (1994) estimated a forest loss of 40% over India during 1880 to 1980. Important driving forces were an increase in cultivated area by more than 42 million ha (40%), a staggering tripling of the population (increase of half a billion people), and livestock increase of 193 million head (105%). Increasing food demand was often met through an expansion of cultivated area, often at the expense of forests. However, crop production increased through intensification of production after the Green Revolution in 1965, with the introduction of high-yield seed varieties after 1965, the use of irrigation and fertilization, and introduction of double cropping practices (see Box 3.1). Indeed, between 1950 and 1997, irrigated areas tripled, fertilizer consumption increased from 66 000 t to 16 million t, and multiple cropping index (ratio of gross area sown to net area sown) increased from 111% to 134% (DES 2004; FAO 2004a).

North America

Numerous studies have described historical changes in North American agriculture (e.g., Bailey 1909; Helfman 1962; Menzies 1973; Schlebecker 1973, 1975; Yates 1981; Richards 1990; Riebsame 1990; Cronon 1991; Sisk 1998; U.S. Department of Agriculture 1998).

In the United States of America, the Homestead Act of 1862 (wherein 160 acres of government land were given free to those settling and cultivating it for at least 5 years) led to a rapid settlement of public lands in the following decades. This was further stimulated by the end of the civil war and the disbanding of the armies. In particular, the Great Plains looked promising to people who had lost everything in the war. The increasing flow of immigration added further to the movement of people into the Midwest. Furthermore, the building of canals in the early 1800s, and subsequent expansion of railroads facilitated the rapid transport of goods from the Midwest.

The corn and wheat belts began to develop in the 1850s. Wheat cultivation was constantly forced westward by the rising price of land and by corn encroaching from the east. By 1870, the Corn Belt had moved westward and stabilized

in its present location. Extensive agricultural settlement in the Great Plains began in the 1870s and 1880s. Dryland farming in the semiarid regions of the Midwest began in the 1880s. In 1902, the government passed the Reclamation Act of 1902 to provide irrigation resources to small farmers, which further encouraged the agricultural development of the Midwest. Since 1900, cropland area mostly increased in the Great Plains region at the expensive of grasslands. The period from 1898 to 1914 is sometimes known as the "Golden Age of American Agriculture". By 1920, grain production had reached the most arid regions of the Great Plains, and cotton had moved into western Texas and Oklahoma. In the 1930s, prolonged drought combined with poor agricultural soil management led to the "Dust Bowl" in the southern Great Plains. For eight years, dust storms blew away topsoil in this region until the drought ended. Between the 1930s and the 1950s, the federal government sponsored large irrigation projects in the west, which led to subsequent agricultural development of California and other western states. Around the 1940s, crop acreage in the United States of America began to stabilize. In the 1960s, soybean acreage expanded in the Great Plains, as an alternative to other crops. The early 20th century also saw the abandonment of croplands and regrowth of forests in parts of eastern United States of America, starting in New England, followed by the Mid-Atlantic States, and more recently in the Southeast. The abandonment of croplands in the eastern United States of America was partly due to competition from more fertile regions of the Midwest, and also due to competing demands on land within the east from rapid urbanization.

In Canada, railways reached Winnipeg by 1885, providing easy access to the prairies. Roughly a decade later, immigration into the Canadian West reached huge proportions. Innis (1935) observed that roughly one million settlers came to the Canadian West from the United States of America, and several thousand others from Europe. The Prairie Provinces were mostly cultivated by 1930. The expansion of this frontier finally ended with settlement in the Peace River Valley in Alberta in the 1930s.

Europe and the Former Soviet Union

Land-cover change in Europe and the Former Soviet Union has a long history. The agricultural revolution of the Middle East spread into Europe between 7000 B.C. and 5500 B.C. Extensive forest clearing in the Mediterranean Basin has been ascribed to economic and political activities during the Greek and Roman Empires. Deforestation continued through to the medieval period in temperate western and central Europe, driven by the increase in population from 18 million in ca. A.D. 600, to 39 million in A.D. 1000, to 76 million in the early 13th century. Forests and swamps decreased from roughly 80% of the land area to about half between A.D. 500 and

A.D. 1300 (Williams 2003). In Russia, there is some evidence for clearing accompanying the eastward movement of Slavs into the mixed-oak forest zone; but the clearing rates were slow and less extensive in comparison to central and Western Europe. The continued deforestation of Europe was interrupted significantly by the spread of the bubonic plague between 1347 and 1353. The total population decreased from 76 million in 1340 to around 50 million in 1450. The effects of the population decline caused by the "Black Death" on land cover was that between one-fifth and one-fourth of all settlements were abandoned across the continent, and forests regenerated (Williams 2003). Forests continued to reestablish themselves during the following century, as wars ravaged much of the continent.

The Renaissance of the 16th century was accompanied by the expansion of commerce. Initially, the sea-based economies of the Western European nations intensified land-use practices – mostly urban – at their cores. These influences spread to the peripheries in support of the core economies. European overseas expansion, especially the establishment of trading posts along the coastline of Africa and European "discovery" of the Americas in 1492, resulted in the continued exploitation of natural resources and corresponding changes in land cover, both in Europe and abroad. In Europe, the forests that had regenerated since the "Black Death" were cleared once again, partly to build ships but also to provide land for agriculture (Williams 2003). Much of this land was mixed crop cultivation and animal rearing.

By 1700, croplands and grasslands for grazing were widespread in Europe and parts of the FSU. The advent of the Industrial Revolution in the middle of the

---

Box 2.3. Land-use change history near La Roche-en-Ardenne, Belgium

Antoine Stevens · Université catholique de Louvain, Louvain-la-Neuve, Belgium

The maps in Fig. 2.3 have been digitized from 1 : 10 000 topographic sheets from a region near La Roche-en-Ardenne in the Belgian Ardennes. During the 19th century, heathlands were converted to deciduous forests and arable land. Afterwards, deciduous forests were converted to rapidly growing coniferous plantations. Concurrently, as a result of the increase in agricultural yields and the specialization in breeding of dairy and beef cattle, the largest part of arable land was converted to grassland. Peatlands as well as wet meadows have experienced a slow and constant decline. They are often severely degraded or threatened, and are nowadays subject to conservation and restoration measures. This illustrates that, even in this rural area considered by many people as "natural", landscape has been strongly modified by human activity (see Box 2.1). Conversion from natural to managed land is likely to have a large impact on ecosystems attributes (e.g. carbon content, biodiversity) (see Chap. 4).

Fig. 2.3. Land-use change history of the Belgian Ardennes (1868 to 1973). Source: IGN (Institut Géographique National)

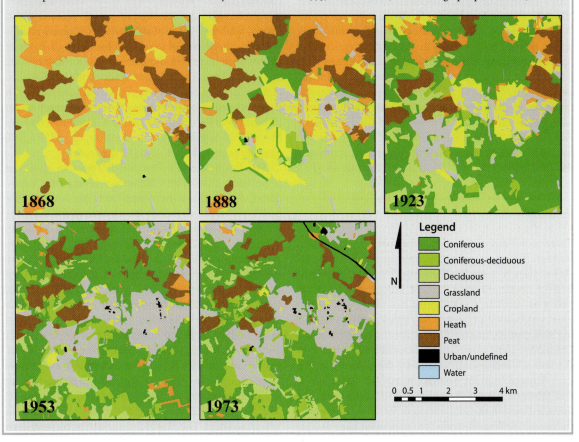

Legend
- Coniferous
- Coniferous-deciduous
- Deciduous
- Grassland
- Cropland
- Heath
- Peat
- Urban/undefined
- Water

0 0.5 1   2   3   4 km

18th century stimulated further land-cover changes. Agriculture continued to expand during the late 18th and 19th centuries as the rapidly growing industrial cities created new demands on food supply. A further, significant, land-use change was that some rural areas, for example, the west midlands of England – the cradle of the Industrial Revolution – became predominantly urban and industrial. In Russia agricultural expansion occurred at the expense of steppe grasslands and new frontiers of settlement were also established elsewhere in Russia and Siberia.

The agricultural revolution that began in Europe in the 1700s, continued through the 18th and 19th centuries with the widespread introduction of irrigation, the development of chemical fertilizers and selective crop and animal breeding. As a consequence agricultural production intensified. Cropland expansion had significantly slowed down in Europe by the second decade of the 20th century. However, there were some counter-trends during the 20th century. For example, in Britain much grassland was brought into cultivation to try and alleviate food shortages during the Second World War; and during the 1950s, the steppes of Russia and Kazakhstan were extensively cleared for cultivation as a result of the *Virgin and Idle Lands* Program of 1954–1960, initiated by Khrushchev. This plan was designed to increase grain production; but by the early 1960s much of the land, which had only been grazed before, had become another "Dust Bowl" and was abandoned. By the 1960s, crop production increases in the temperate zones of the world came almost solely from increases in yields. New agricultural settlement and expansion had stopped in the former temperate forest and steppe zones, which were for all intents and purposes "deforested" in many Western and central European countries. Since the 1980s, agricultural areas have started to contract in Europe and the Former Soviet Union; in Western Europe much of this contraction has been driven by European Union (EU) agricultural and environmental policies (McNeill 2000; Williams 2003).

Australia

There is widespread evidence for forest modification through the aboriginal use of fires on the savanna-forest interface in Australia (Pyne 1991). Landscape modification continued with the first European settlers arriving in Australia in 1788. The initial phase of development was based on commercial cattle ranching, exporting beef, mutton, and wool to Europe and North America (Richards 1990). Large-scale wheat cultivation began in the late 19th century, stimulated by the growing urban food demand from the eastern United States of America and Western Europe (Richards 1990). Cultivation was initially based in the settlements of Victoria, South Australia and Western Australia, but expanded later to New

South Wales with the construction of railways. The expansion of cultivation resulted in the modification and loss of the dense eucalyptus forests of southern and eastern Australia. It is estimated that 69% of the vegetation in Victoria and 50% of the vegetation of New South Wales were modified since 1780 (Wells et al. 1984).

Africa

Before the 19th century, land in Sub-Saharan Africa was used largely for hunting, gathering, herding, and shifting cultivation (Kimble 1962). Some settled agriculture existed in Africa long before the imposition of colonial rule in the late nineteenth century, but in the pre-colonial period, demographic and economic needs allowed for land cleared for cultivation to be left fallow for long periods or abandoned as cultivators moved on and cleared new land. Estimates of cropland areas before ~1900 are variable, in part because of the lack of data and in part because "croplands" were part of a shifting cultivation rotation, where the distinctions between cropped areas and fallows are unclear (Kimble 1962). Shifting cultivation included annual clearing of 0.5–3.0 ha of forest per family creating a mosaic of cropped fields intermingled with fields 2–3 years old, fallows, and stands of secondary and mature forest. Clearing of previously cultivated areas (old fallows of 10–50 years) was generally preferred over clearing old-growth forests. Before ~1900, land use had probably been in a "quasi-equilibrium" for thousands of years (Kimble 1962). Changes included both increases and decreases as a result of wars, epidemics, famines, and slave trade (both intra-African and trans-Atlantic). In fact, populations are thought to have declined somewhat during the 19th century. Between 1850 and 1900 European colonization introduced changes; but with a few exceptions, the most rapid and dramatic changes occurred after 1930 (Kimble 1962).

Two factors led to the cropland expansion after 1930: population growth and European demand for export crops. Populations increased from improved public health provision, as well as the absence of the wars, epidemics, and famines that had characterized the late 19th century (Kimble 1962). The area under export crops expanded significantly because colonial governments needed the revenues they provided to recover from worldwide depression of the 1930s. In addition, by the 1930s, the railroads and most of the other major transport routes were in place in colonial Africa, and it became feasible to begin development of areas that had hitherto been inaccessible. This combination of demographic pressure and economic incentive has continued to the present (see Chap. 3). Cropland area in Sub-Saharan Africa is estimated to have been 119 million ha in 1961 and 163 million ha in 2000 (FAO 2004a), an increase of 37% in 40 years. The rate of forest clearing for long-term shifting cultivation has been even greater than the rate of clearing for permanent croplands in recent decades (Houghton and Hackler 2006).

Biggs and Scholes (2002) reconstructed land-cover change for South Africa for the period 1911–1993. The area under cultivation more than tripled during the last century, while the plantation area increased more than tenfold. They found that expansion of the cultivated area was highly correlated to total domestic population growth until the 1960s; since then increasing food demand has been met through increasing yields from fertilizer use and irrigation, and not through continued expansion of the cultivated area.

South America

Forest exploitation in Brazil started with Brazil-wood extraction along the Atlantic Coast nearly 500 years ago soon after the first Portuguese arrived. The discovery of gold in Minas Gerais in 1690 stimulated mining here and in other scattered pockets elsewhere, which led to large-scale destruction of forests (Williams 1990). This was followed by large, export-oriented sugar cane plantations, strengthening Portugal's supremacy in the world sugar trade. Large-scale forest conversion started with coffee plantations in the 19th century that resulted in the nearly complete loss of the Brazilian Atlantic Forest. The Atlan-

tic forest once stretched all the way from Rio Grande do Norte to Rio Grande do Sul; three million ha had been converted into coffee plantations during the nineteenth century, and more than 90% had been cleared by the end of the last century.

The first large-scale migration into Amazonia was motivated by the rubber boom; the population of Brazilian Amazon increased four-fold between 1870 and 1910 (Salati et al. 1990). Livestock production accompanied the rubber exports, but was not very successful. From 1914 until World War II, little development occurred in the Amazon Basin. During the last decades of the 20th century, massive conversion of land for large-scale cattle ranching occurred. In addition, large-scale infrastructure projects like the Trans-Amazonian Highway opened-up pristine tropical forest areas, often followed by the influx of settlers. Between 1850 and 1985, 370 million ha of forest in Latin America (~28%) was converted into other land uses (Houghton et al. 1991). Most of this deforestation was due to the expansion of cattle ranching (Lambin and Geist 2003b). However, in the recent decades, deforestation is being increasingly dominated by soybean expansion (Laurance et al. 2004). Soybean areas in Brazil increased from 240 000 ha in 1961 to 8.8 million ha in 1980, to 21.5 million ha in 2004 (FAO 2004a), and Brazil is on the verge of becoming the largest exporter of soybeans in the world today. Logging is another major cause of deforestation in Brazil, the extent of which has been un-

Box 2.4. Review of data sets of historical land cover

Data on land-use and land-cover change can be gleaned from various sources such as tax records, land surveys, periodic censuses, forest inventories, paleo records, reconstructions by historical geographers, and remote sensing. Using such sources, numerous local-to-continental scale data sets of land-cover change have been developed (Table 2.1). Here we briefly review some of the major sources of such data.

Ground-based data

Data on land-use and land-cover change were collected systematically over the last century through censuses. The first World Census of Agriculture was conducted in 1930, and since then, the FAO has promoted a worldwide census every 10 years. FAO has also compiled national-level data on agricultural land use annually since 1961 (the FAOSTAT database), and has also performed periodic global inventories of forests (Forest Resources Assessment). These data are reported to the FAO by the member nations. The quality of data is only as good as the quality of monitoring and reporting by the various countries to FAO. For example, nations with poor infrastructure, or that are ravaged by civil wars, are incapable of performing the systematic observations, and therefore have unreliable data. Data on land use are also available at the subnational level from various national census organizations at roughly five-to-ten-year intervals (e.g., U.S. Department of Agriculture; Fundação Instituto Brasileiro de Geografiae e Estatística (IBGE), Brazil; Directorate of Economics and Statistics, India, etc.). A first global synthesis of such subnational agricultural census data is currently being accomplished through the AgroMAPS project (http://www.fao.org/landandwater/agll/agromaps/).

Unfortunately, the quality of census data prior to World War II is very poor, and other sources of data necessarily need to be used.

The British Colonies kept extensive tax records and conducted frequent cadastral and forest surveys. Furthermore, historical maps, aerial photographs, pollen records, and land-use models have been used to reconstruct historical land-use information. The effort by Flint and Richards (1991) to reconstruct land-cover changes in Southeast Asia from 1880 to 1980 is a good example of this type of work (other examples can be found in Table 2.1). Recently, the International Geographical Union Commission on Land Use and Land Cover Change (IGU-LUCC) published a four-volume Atlas titled "Land use/cover changes in selected regions in the world", which presents regional maps of land cover through the last century (Himiyama et al. 2001, 2002, 2005). Another simple proxy often used to reconstruct historical land-cover change is the total human population numbers. Prior to the 19th century, technology played a minor role in resource extraction, and therefore the extent of human activities was likely well correlated to human population growth, the number of livestock, etc., for which better data are available. Two examples of such use of data are Houghton et al. (1983) in their effort to reconstruct a 300-year global history of land-cover change, and the more recent effort by Stéphenne and Lambin (2001) to reconstruct land-use changes in Sudano-Sahelian Africa.

Remotely-sensed data

In the last three decades, the advent of the remote sensing satellites has led to the development of instruments to systematically monitor land cover from space. While satellite data present a useful baseline for historical reconstructions, they are, by themselves, not useful to study land-cover change before the 1970s. Therefore, we will not discuss them further in this section. In Sect. 2.5, we review the development of remotely sensed data on land-use and land-cover change.

derestimated until recently as shown by a study by Asner et al. (2005).

In the mid 19th century, much of Argentina was devoted
to the grazing of sheep and cattle, with very little arable
cultivation (Grigg 1974). Argentina emerged to become a
major agricultural nation in the late 19th century, with the
introduction of agricultural technology and integration
into the world economy. Investment, mostly by the British,
as well as migrant workers from Spain and Italy, helped
this development. Italian sharecroppers started growing
wheat, which became the major crop in the humid pampa.
Cropland areas increased from 0.3 million ha in 1870, to
6 million ha in 1900, 19 million ha in 1910, and 24 million ha
in 1930 (Grigg 1987). By 1900, Argentina became one of the
leading exporters of wheat in the world. In addition, to
wheat, maize also began to be grown in the 1890s, and vegetables, dairying, and other intensive agricultural practices
also took hold after World War I. The 1930s was the peak
of Argentinean agriculture. Since 1930, agriculture became
stagnant with the great depression and cropland areas stabilized. However, in the recent decades, as in Brazil, soybean production has exploded in Argentina, increasing
from 26 000 ha in 1970 to 14.3 million ha in 2004 (FAO
2004a). Soy is expanding not only at the expense of other
crops, but is also causing deforestation in the foot of the
Andes, and in Chaco (see Box 4.9).

2.2.3   What Makes the 20th Century Unique?

Land-use change increased markedly in the 20th century,
both in terms of extent and intensity (see Chap. 3). As noted
earlier, more forests were cleared between 1950 and 1980
than in the early 18th and 19th centuries combined. However, the late 20th century also saw a shift in agriculture away
from expansion toward intensification. Increasingly, global food production is coming from the intensification of
production on existing croplands, rather than expansion
of croplands. In *State of the World 1996*, Lester Brown referred to the "Acceleration of History", where he observed
that world energy use has accelerated dramatically through
much of the world in the last 50 years. Indeed, between
1961 and 2002, while cropland areas increased by only 15%,
irrigated areas doubled, world fertilizer consumption increased 4.5 times, and the number of tractors used in agriculture increased 2.4 times (FAO 2004a).

In addition to the increased intensity of land use, the
20th century is unique in terms of the scale of land use.
Land-use changes, often thought of as a local problem, have
now accumulated to become a global problem, on par with
other global problems such as climate change and stratospheric ozone depletion (see Chap. 4). Indeed, it is very
likely that our next major global pollution problem may

Table 2.1. Examples of geographically explicit studies of historical land use

Authors	Spatial characteristics	Temporal characteristics
Local/national level		
Bicik (1995)	Czech Republic	1845, 1948, 1990
Bork et al. (1998)	Germany	7th century–present
Cousins (2001)	Southeast Sweden	17th and 18th century, 1946, 1980
Crumley (2000)	Burgundy (France)	Iron Age–present
Himiyama (1992)	Japan	1850, 1900, 1980
Larsson and Frisk (2000)	Sweden	ca. 1700–present
Manies and Mladenoff (2000)	Sylvania Wilderness Area, USA	Pre-settlement
Odgaard and Rasmussen (2000)	Denmark	Past 2 millennia
Petit and Lambin (2001)	Belgium Ardennes	1700–present
Serneels and Lambin (2001)	Mara region, Kenya	1975–present
White and Mladenoff (1994)	Northern Wisconsin, USA	1860s, 1931, 1989
Continental level		
Darby (1956)	Central Europe	900, 1900
Williams (2000)	Western Europe	11th–13th century
Geoscience Australia (2004a,b)	Australia, scale 1 : 20 000 000	Pre-settlement (1788), 1988
Maizel et al. (1988)	Conterminous United States	1850–1990
Waisanen and Bliss (2002)	Conterminous United States	1850–1997
Richards and Flint (1994)	Southeast Asia	1880, 1920, 1950, 1970, 1980
Global level		
Houghton and Hackler (2001)	9 regions of the world	1850–2000
Richards (1990)	10 regions of the world	1700, 1850, 1920, 1950, 1980

be related to nitrogen pollution of the world's waterways through excess fertilizer application. Therefore, land-use changes in the 20th century are, unlike any earlier time in history, accelerating in intensity throughout the world.

2.3 Most Rapid Land-Cover Changes of the Last Decades: Rapid and Extensive

With the recognition that land use is an important driver of global environment change, numerous studies in the last two decades have estimated the rates of tropical deforestation and other kinds of land-cover change around the world. Remote sensing has played a critical role in documenting these changes (Mollicone et al. 2003), and there are multiple examples of studies and resultant databases of rapid land-cover change and ecosystem disturbances in important regions of the world: deforestation in the pan-tropical forest belt; snapshots of land cover in European Russia, continental U.S. and Canada; fire frequency globally and regionally in South America, Southern Africa, and parts of Russia; and the influence of urbanization in selected cities around the world. While most studies were at the local-to-regional scales, global land-cover data sets were developed, using different methodologies, for the early 1990s using AVHRR satellite data (Loveland et al. 2000) and 2000–2001 using MODIS satellite data (Friedl et al. 2002) and SPOT VGT data. There has also been a profusion of information and studies based on data sources other than remote sensing.

Despite the plethora of land-cover change studies and global remote sensing observations, a systematic, global synthesis and review of the major trends in land-cover change was not conducted until recently. The Millennium Ecosystem Assessment, in collaboration with LUCC, recently undertook a synthesis of the regions undergoing rapid land-cover change around the world (see Box 2.5, Lepers et al. 2005). Here, on the basis of this synthesis, and other publications, we review the recent changes in land cover around the world, particularly focusing on the following types of land-cover change: deforestation and forest degradation, changes in croplands and grazing lands, urbanization, and changes in drylands.

2.3.1 Recent Forest-Cover Changes

Deforestation, one of the most commonly recognized forms of land-cover change, is nevertheless plagued by inconsistencies in definitions (Williams 2003). The Food and Agriculture Organization (FAO) of the United Nations defines deforestation as occurring when tree canopy cover falls below 10% in natural forests (or when a forest is transformed to other land uses even if tree canopy cover remains higher than 10% – e.g., shifting cultivation). On the basis of this definition, and using country forest inventories, expert es-

timates, forest-plantation data, and an independent remote sensing survey, the Global Forest Resources Assessment 2000 (FAO 2001a) [FRA 2000 hereafter] estimated a net decrease in forest area of 9.4 million ha yr^{-1} from 1990 to 2000. This change was a result of a 12.5 million ha yr^{-1} net

---

Box 2.5. The LUCC-MA rapid land-cover change assessment

Recently, the Millennium Ecosystem Assessment (MA), an international program designed to assess the status and trends in the global ecosystem change, also recognized the importance of land-use and land-cover change (Millennium Ecosystem Assessment 2005). The LUCC project was commissioned by the MA to assess places in the world undergoing the most rapid land-cover changes. The tremendous advance in scientific analysis of land-cover change over the last decade made such a synthesis both possible and timely. The assessment used both remote sensing information available in widely scattered literature as well as sub-national, national, and regional inventory data on land-cover change.

The rapid land-cover change assessment identified forty-nine data sets at the national and global scale showing either rates of land-cover change, or "hot spots" of land-cover change over the last two decades. The types of change (or proxy variables for change) included in the analysis were: *(a)* deforestation and forest degradation; *(b)* degraded lands in the drylands and hyper-arid zones of the world (referred to here as desertification, even though most definitions of desertification do not include hyper-arid zones); *(c)* cropland expansion and abandonment; and *(d)* urban settlements. Some important land-cover changes were not included because of absence of reliable data. For instance, no spatially-explicit data sets of reliable quality on afforestation and reforestation or on changes in pastoral lands are available at a regional-to-global scale. Data limitations also precluded the analysis of questions such as where future land-cover changes are likely to occur, or where ecosystem impacts are large even though the extent of land-cover change may be small.

The synthesis had to overcome several challenges:

- Some of the data sets identified "hot spots" of land-cover change directly while others provided estimates of rates of change. For the latter, rapid land-cover change areas were identified as those with rates of change above a certain threshold percentile value. Threshold values were chosen separately for each of these data sets, and type of land-cover change.
- Different data sources are not based on standard definitions, even though some definitions are more commonly accepted. For this synthesis, areas with the highest rates of land-cover change were determined given the definition adopted for a particular data set, rather than attempting to harmonize the definitions among all data sets.
- The different data sets had different spatial resolutions – the finest one being the remote sensing-based data (in the order of one km^2) and the coarsest one being the (sub)-national statistics (in the order of hundreds to thousands km^2). Depending on the scale of the source data sets, this led to commission and omission errors.
- Not all data sets covered the 1980–2000 time period chosen for the synthesis. Therefore, the final maps provide no detailed information on the time period during which a particular area experienced rapid land-cover change, nor on the frequency of disturbances.
- Some parts of the world were covered by several data sets whereas, for others, only national statistics were available. Consequently, some areas appear to be more affected by rapid land-cover change simply because they have been studied more intensively. To indicate this bias, the map legend provides additional information on the number of data sets covering an area.

decrease in natural forests (comprising deforestation of 14.6 million ha yr^{-1}, conversion to forest plantation of 1.5 million ha yr^{-1}, and regeneration of 3.6 million ha yr^{-1}), and 3.1 million ha yr^{-1} net increase in forest plantations (1.5 million ha yr^{-1} converted from natural forests, and 1.6 million ha yr^{-1} of afforestation). Most of the deforestation occurred in the tropics, while most of the natural forest regrowth occurred in Western Europe and eastern North America; the total net forest change was positive for the temperate regions and negative for the tropics (Fig. 2.4a).

The overall estimates of forest-cover change from FRA 2000, cited above, were a combination of national data adjusted using information from the FRA remote sensing survey, and forest-plantation data (see Chap. 1 of FAO 2001a, pages 8–10). One of the major reasons for this adjustment was because FAO recognized the unreliability of country data, especially in tropical Africa. Indeed, a comparison between the FAO country data and the independent remote sensing survey showed reasonably good agreement in Latin America and tropical Asia, but poor agreement in tropical Africa – see Table 2.2. Two recent studies using remote sensing (Landsat derived estimates from the Tropical Ecosystem Environment Observations by Satellite (TREES) project of Achard et al. (2002), and Advanced Very

High Resolution Radiometer (AVHRR)-based estimates (DeFries et al. 2002b) also lend credence to the idea that the FRA 2000 country estimates of deforestation are too high, especially in tropical Africa (see Table 2.2, DeFries and Achard 2002). Achard et al. (2002) estimated deforestation rates for the humid tropics that were 23% lower than FRA while DeFries et al. (2002b) estimated deforestation rates for the entire tropics that were 53–62% lower. A simple comparative analysis of these two remote sensing estimates to the FAO country data undertaken by Houghton and Goodale (2004) suggests that FAO overestimated deforestation rates by roughly 30% if dry tropical Africa is ignored. In tropical Africa, decreases in net forest area from the FAO country studies are a lot higher than any of the remote sensing estimates, and much of the difference seems to arise in dry tropical forests (Table 2.2). Indeed, even the FRA 2000 study suggests that the country reports seem too high for certain countries (e.g., Sudan and Zambia). Another significant difference between the AVHRR-based estimates and the FAO estimates is that the former suggests that deforestation rates were higher in the 1990s compared to the 1980s (see Table 2.2), while FAO found no statistically significant trends, except for a decreasing rate of deforestation in tropical moist deciduous forests (FAO 2001a).

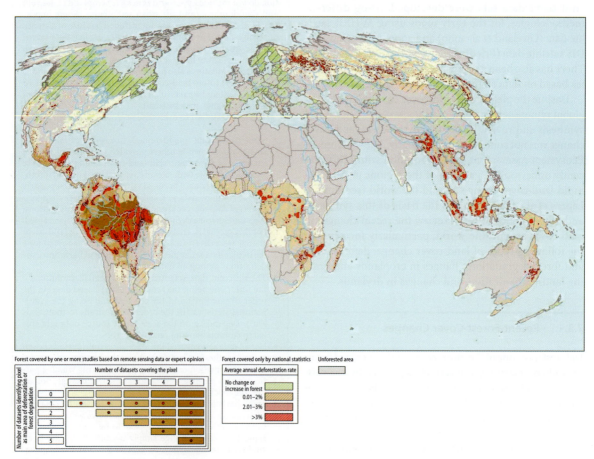

Fig. 2.4a. Results from the LUCC-MA Rapid Land-Cover Change Assessment showing major areas of forest-cover change in the world between 1980 and 2000

Table 2.2. Comparison of tropical deforestation rates (million ha yr^{-1})

Net forest change	All tropics			Humid tropics	
	FAO country survey	FAO remote sensing	AVHRR	FAO country survey	TREES
1990s					
Tropical Asia	−2.4	−2.0	−2.0	−2.5	−2.0
Tropical Africa	−5.2	−2.2	−0.4	−1.2	−0.7
Tropical Latin America	−4.4	−4.1	−3.2	−2.7	−2.2
Pantropics	−12.0	−8.3	−5.6	−6.4	−4.9
1980s					
Tropical Asia	−2.4		−1.2		
Tropical Africa	−3.9		−0.3		
Tropical Latin America	−7.1		−3.6		
Pantropics	−13.4		−5.1		

While gauging these estimates, a caveat to note is that changes in dry tropical forests are difficult to estimate from coarse-resolution remote sensing imagery, and while country estimates may have overestimated deforestation, coarse-resolution remote sensing estimates may underestimate deforestation (FAO 2001a). Indeed for Africa, the AVHRR-based estimate of net forest change in the 1990s for both humid and dry tropics is less than for the Landsat-based estimate of net forest change in the humid tropics alone, indicating that the AVHRR data likely underestimate deforestation in Africa where size of clearings are relatively small. The estimation of deforestation in such situations may require optimally chosen Landsat images supplemented with very-high resolution data such as from the IKONOS satellite, as well as *in-situ* surveys.

The TREES data is considered more reliable because of its high resolution as well as being restricted to more easily observed humid tropical forest changes; however, it only covers a statistically-selected subset of the total area (6.5%) based on a stratification determined from expert opinion. According to the TREES data, 5.8 ±1.4 million ha of humid tropical forest were lost each year between 1990 and 1997 (Achard et al. 2002). Over the same period, forest regrowth was estimated to be 1.0 ±0.32 million ha yr^{-1}, leading to a net change of 4.9 ±1.3 million ha yr^{-1} (0.43% per year). An additional 2.3 ±0.7 million ha of forest were visibly degraded (but this does not include forests affected by selective logging). Southeast Asia experienced the highest rate of net forest-cover change (0.71% per year), with the highest rates of deforestation being estimated for central Sumatra (3.2–5.9% per year). Africa and Latin America had lower estimates of net forest change (0.36% and 0.33%, respectively). However, in terms of extent, Latin America lost about the same area of forest as Southeast Asia between 1990 and 1997. Forest degradation was most extensive in Southeast Asia (0.42% per year), lowest in Latin America (0.13% per year), and in-

termediate in Africa (0.21% per year). Forest regrowth was more extensive, both in absolute and relative terms, in Southeast Asia than in the other humid tropical regions (0.19% for Southeast Asia, 0.04% for Latin America, and 0.07% for Africa).

Deforestation is not widespread throughout these regions, but rather is largely confined to a few areas undergoing rapid change, with annual rates of deforestation ranging from 2% to 5% (Achard et al. 1998; Lepers et al. 2005) – see Fig. 2.4a). The largest deforestation front is the well-known arc of deforestation in the Brazilian Amazon (see Box 2.6). Recently, deforestation has extended outside Brazil, to the eastern foothills of the Andes where illegal coca cultivation has promoted deforestation (Steininger 2001; Millington et al. 2003), and along the road from Manaus to Venezuela (Sierra 2000). More scattered areas of forest loss are detected in the Chaco and Atlantic forest areas in South America. Central America has significant deforestation fronts in the Yucatán Peninsula and along the Nicaraguan border with Honduras and Costa Rica. In Africa, forest-cover change is very rapid in small scattered hot spots in Madagascar, Côte d'Ivoire, and the Congo Basin. In Southeast Asia, several deforestation fronts are found around Sumatra, Borneo, Vietnam, Cambodia, and Myanmar. Relative to the tropics, there are fewer data sets on deforestation covering the temperate or boreal forests (Wagner et al. 2003; Hansen and DeFries 2004). Less is therefore known about forest-cover changes in Canada or Siberia. However, forest degradation in Eurasia, resulting from unsustainable logging activities and an increase in fire frequencies, has been growing over the recent years. Fire frequency has increased dramatically in Siberia in particular; over 7.5 million ha yr^{-1} of Russian forests were burnt over a 6-year period in the late 1990s (Sukhinin et al. 2004). Although deforestation is one of the best studied processes of land-cover change, it is clear that regional gaps in spatially explicit data persist.

Box 2.6. Deforestation in the Brazilian Amazon

The Brazilian Amazon has experienced some of the highest deforestation rates in the world. Total deforested area in the region increased from approximately 10 million ha in the 1970s to more than 60 million ha at the turn of the century, after development policies strongly based on road building and Government-directed colonization were put into place (see Chap. 3). Government-led projects lead to the concentration of forest clearing in the vicinity of roads and colonization areas (see Box 4.10). In fact, colonization and forest clearing were sustained mostly in the southern and eastern flanks of the Amazon, in regions closely linked to markets in other parts of Brazil (Alves 2002a,b); 90% of all deforestation in the 1991–1997 period was found within 100 km of main roads (see Fig.7.3); at the same time, 87% of this deforestation was observed within 25 km of areas cleared during the 1970s (Alves 2002a,b).

The effects of roads have been the focus of many research efforts. Opening a road through an unexplored region attracts new settlers that initiate deforestation. On a second phase of colonization, concentration of farms and forest clearing lead to increasing road density and intensification of land use, while deforestation expands beyond the limits allowed by the Brazilian Forest Code (Chomitz and Thomas 2001; Alves et al. 2003; Pacheco 2006c).

Expansion of deforestation into new areas is frequently linked to the illegal appropriation of public land in more remote areas where roads are often opened by loggers, farmers and unlawful tenants. Sayago and Machado (2004), based on data from a recent Federal Government Census, reported that half of all farm land in the Brazilian Amazon had been illegally appropriated in Brazilian Amazon, showing the importance of this issue to understand the driving forces behind Amazonian deforestation.

Despite much progress in mapping deforestation in the Amazon in the closed forests, rates of forest regeneration following abandonment, land-cover modification by selective logging and land-cover conversion in the Cerrado areas still lack systematic efforts to enable a more complete understanding of land-cover/use changes in the region (Schimel et al. 1995; Alves 2001a). Also, research focused on the role and functioning of institutions, as well as driving forces and actors behind the deforestation process is generally recognized as lacking (Alves 2001a; Mahar 2002; Walker 2004).

2.3.2 Recent Changes in Agricultural Areas

Historically, humans have increased agricultural output mainly by bringing more land into production. The greatest concentration of farmland is found in Eastern Europe, with more than half of its land area under crops (Ramankutty et al. 2002). In the United Kingdom, about 70% of its area is classified as agricultural land (cropland, grassland/rough grazing), with agriculture and areas set aside for conservation or recreation intimately intertwined (Hails 2002). Despite claims to the contrary, the amount of suitable land remaining for crops is very limited in most developing countries where most of the growing food demand originates (Döös 2002). Where there is a large surplus of cultivable land, land is often under rain forest, permanent pastures, or in ecologically marginal areas (Young 1999; Döös 2002).

Southeast Asia witnessed the greatest expansion of croplands in the past few decades (Fig. 2.4b). The other main areas of recent cropland expansion have been in Bangladesh, along the Indus Valley, in parts of the Middle East and Central Asia, in the Great Lakes region of east Africa, along the southern border of the Amazon Basin, and in the Great Plains region of the United States (although much of what is defined as croplands in the Great Plains is in a soil conservation program, and is not sown). Extensive abandonment of croplands occurred in North America (lowlands of south eastern United States), eastern China, and parts of Brazil and Argentina.

Since 1960 we have witnessed a decoupling of the increase in food production from cropland expansion. The 1.97-times increase in world food production from 1961 to 1996 was associated with only a 10% increase of land under cultivation but also with a 1.68-times increase in the amount of irrigated cropland and a 6.87- and 3.48-times increase in the global annual rate of nitrogen and phosphorus fertilizer use (Tilman 1999). In 2000, 271 million ha were irrigated (FAO 2004a). Globally, the cropland area per capita decreased by more than half in the twentieth century, from around 0.75 ha per person in 1900 to only 0.35 ha per person in 1990 (Ramankutty et al. 2002). Note, however, that national statistics in developing countries often substantially underreport agricultural land area (Young 1999; Ramankutty et al. 2002), e.g., by as much as 50% in parts of China (Seto et al. 2000).

The mix of cropland expansion and agricultural intensification has varied geographically (Ramankutty et al. 2002). Tropical Asia increased its food production mainly by increasing fertilizer use and irrigation. Most of Africa and Latin America increased their food production through both agricultural intensification and extensification. Western Africa is the only part of the world where, overall, cropland expansion was accompanied by a decrease in fertilizer use (–1.83% per year) and just a slight increase in irrigation (0.31% per year compared to a world average of 1.22% per year). In 1995, the global irrigated areas were distributed as follows: 68% in Asia, 16% in the Americas, 10% in Europe, 5% in Africa, and 1% in Australia (Döll and Siebert 2000). In Western Europe and the northeastern United States, cropland decreased during the last decades after abandonment of agriculture or, in a few cases, following land degradation mostly on marginal land. Globally, this change has freed 222 million ha from agricultural use since 1900 (Ramankutty et al. 2002).

2.3.3 Recent Changes in Pastoral Areas

Natural vegetation covers have given way not only to cropland but also to pasture, defined as land used perma-

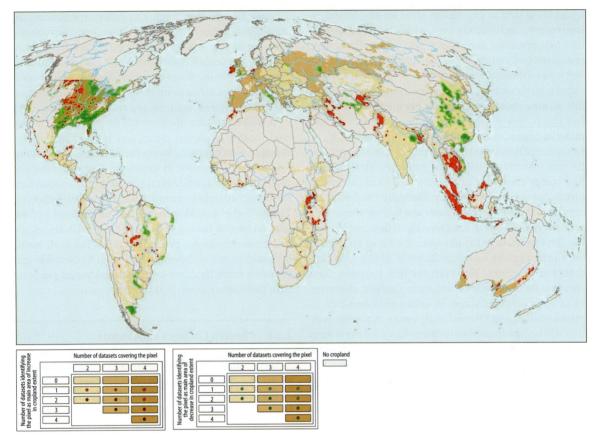

Fig. 2.4b. Results from the LUCC-MA Rapid Land-Cover Change Assessment showing areas of the world that underwent most rapid cropland change between 1980 and 1990

nently for herbaceous forage crops, either cultivated or growing wild (FAO 2004a). The distinction between pasture and natural savannas or steppes is not always clear. In many parts of the world, such landscapes are "multifunctional", making it difficult to classify them for inventories. Therefore, the LUCC-MA assessment – see Box 2.5 – did not deal with grazing land changes. Nevertheless, broad patterns can probably be derived from the FAO statistics, which show that most pastures are located in Africa (26% of the global total of ~35 million ha) and Asia (25%), and only a small portion is located in North America (8%) and Europe (2%) (FAO 2004a). Latin America and the Caribbean have 18% of the world's pastures, while the FSU nations have 10%, and Oceania has 12%. During the last decade, pastures increased considerably in Asia and the FSU (6.8% and 10%, respectively), whereas the largest decreases were seen in Europe and Oceania. Data suggest that pasture land has apparently decreased in eastern Africa; however, as eastern Africa recorded a large increase in head of cattle over this period (872 000 additional head of cattle per year between 1992 and 1999, according to FAO (2004a)), it is likely that many pastoral areas in this part of Africa are classified as natural vegetation.

2.3.4 Recent Changes in Urbanization

In 2000, towns and cities housed more than 2.9 billion people, nearly half of the world population (United Nations Population Division 2002). Urban populations have been growing more rapidly than rural populations worldwide over the last two decades, particularly in developing countries. According to the UN Population Division (United Nations Population Division 2002), the number of megacities, defined here as cities with more than 10 million inhabitants, has increased from one in 1950 (New York) to 17 in 2000, the majority of which are in developing countries. Urban form and function have also changed rapidly. Built-up or impervious areas are roughly estimated to occupy between 2% to 3% of the Earth's land surface (Grübler 1994; Young 1999). This relatively small area reflects high urban population densities: for example, in 1997, the 7 million inhabitants of Hong Kong lived on as little as 120 km^2 of built-up land (Warren-Rhodes and Koenig 2001). However, urbanization affects land in rural areas through the ecological footprint of cities (see Chap. 3 and 4). This footprint includes, but is not restricted to, the consumption of prime agricultural land in peri-urban areas for residential, infrastructure, and amenity uses, which blurs the distinction between cities and countryside, especially in western developed countries. Urban inhabitants within the Baltic Sea drainage, for example, depend on forest, agriculture, wetland, lake, and marine systems that constitute an area about 1 000 times larger than that of the urban area proper (Folke et al. 1997). In 1997, total non-food material resources consumed in Hong Kong (i.e., its urban material metabolism) were nearly 25 times larger than the total material turnover of the natural ecosystem. Fossil fuel energy consumed in this city (i.e., its urban energy metabolism) exceeded photosynthetically fixed solar energy by 17 times (Warren-Rhodes and Koenig 2001). Time series of global maps of nighttime lights detected by satellite (Elvidge et al. 2001) illustrate the rapid changes in both urban extent and electrification of the cities and their surroundings. However, the link between these coarse scale observations and more detailed characteristics of structural changes in urban environments remains challenging (Herold et al. 2003). Another question still being debated is whether urban land use is more efficient than rural land use and, therefore, whether urbanization saves land for nature (see Chap. 7).

The most densely populated clusters of cities are mainly located along the coasts and major waterways – in India, East Asia, on the east coast of the U.S., and in Western Europe (Fig. 2.4c). The cities experiencing the most rapid change in urban population between 1990 and 2000 are mostly located in developing countries (Deichmann et al. 2001). It is estimated that 1 to 2 million ha of cropland are being taken out of production every year in developing countries to meet the land demand for housing, industry, infrastructure, and recreation (Döös 2002). This is likely to take place mostly on prime agricultural land located in coastal plains and in river valleys. For example, a recent study in the Pearl River Delta in China found a 364% increase in urban area between 1988 and 1996 (Seto et al. 2002). About 70% of this new urban land was converted from farmland. Another study of the Beijing-Tianjin-Hebei corridor found that urban land had expanded by 71% between 1990 and 2000, with about 74% being converted from prime farmland (Tan et al. 2006). It should be noted, however, that rural households may consume more land per capita for residential purposes than their urban counterparts (Döös 2002).

2.3.5 Recent Changes in Drylands

Desertification is a difficult process to evaluate because of its varying definitions and perceptions (see Sect. 2.4.1 on desertification). The United Nation's Convention to Combat Desertification (UNCCD) defines desertification as

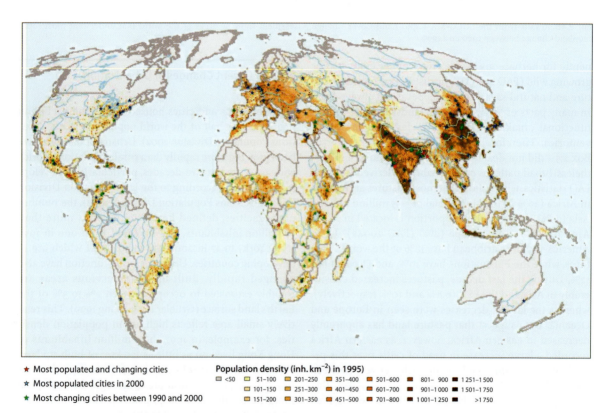

★ Most populated and changing cities
★ Most populated cities in 2000
★ Most changing cities between 1990 and 2000

Population density (inh. km^{-2}) in 1995)
<50   51–100   201–250   351–400   501–600   801– 900   1 251–1 500
101–150   251–300   401–450   601–700   901–1 000   1 501–1 750
151–200   301–350   451–500   701–800   1 001–1 250   >1 750

Fig. 2.4c. Results from the LUCC-MA Rapid Land-Cover Change Assessment showing population density in 1995 and most populated and changing cities over 750 000 inhabitants between 1980 and 2000

"land degradation in arid, semi-arid and dry sub-humid areas resulting from various factors, including climatic variations and human activities". Land degradation is defined as the decrease or destruction of the biological productivity of the land, including vegetation degradation, water and wind erosion, and chemical and physical deterioration, or a combination of these processes (Geist 2005).

The LUCC-MA synthesis of the main areas of degraded drylands was constrained by lack of reliable data. Most available data were heterogeneous in terms of the monitoring methods or the indicators used. The study found that the main areas of degraded dryland lie in Asia (Fig. 2.4d). The synthesis did not support the claim that the African Sahel is a desertification "hot spot" at the present time. However, it found major gaps in desertification studies, including around the Mediterranean Basin, in eastern Africa, in parts of South America (in northern Argentina, Paraguay, Bolivia, Peru and Ecuador) and in the United States of America. If dryland degradation data were available in compatible format for all the continents, the global distribution of the most degraded drylands could be different, but the patterns observed in Asia would most likely remain the same.

While there continue to be major gaps in our understanding of the rates of desertification, there have been various attempts in the past to assess the magnitude of the problem and provide a baseline for monitoring (e.g.,

Lamprey 1975; Dregne 1977, 1983; Mabbutt 1984). The *World Atlas of Desertification*, published by the United Nations Environment Program (UNEP), estimated that global drylands cover about 5 160 million ha, and that 70% of all susceptible drylands suffer from some form of land degradation in varying degrees (Middleton and Thomas 1992). The second edition of the *World Atlas of Desertification* (Middleton and Thomas 1997) incorporated vegetation changes (in addition to the soil degradation information in the first edition). UNCCD estimated that 20–25% of the Earth's land surface is affected by desertification.

The Global Assessment of Human-induced Soil Degradation (GLASOD), an effort led by the International Soil and Reference Information Center (ISRIC), was the last global survey conducted under the sponsorship of UNEP (Oldeman et al. 1991). It was a qualitative assessment based on the opinions of about 250 regional soil degradation experts, showing degradation type, extent, degree and human causes (see Sect. 4.6). Although it may be "the best representation of world soil degradation" (Dregne 2002), it has been criticized widely (Biswas et al. 1987; Reynolds 2001; Reynolds and Stafford-Smith 2002; Prince 2004). Another example of exaggerated claims of advancing deserts stems from the alleged advance of the Sahara southwards in the last 17 years (Lamprey 1975, 1988; Desert Encroachment Control and Rehabilitation Programme 1976;

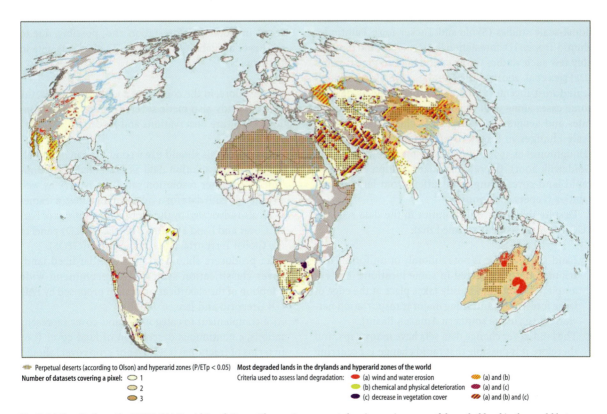

Perpetual deserts (according to Olson) and hyperarid zones (P/ETp < 0.05)   **Most degraded lands in the drylands and hyperarid zones of the world**

Number of datasets covering a pixel: ◯ 1   Criteria used to assess land degradation: ● (a) wind and water erosion   〰 (a) and (b)
◯ 2   ● (b) chemical and physical deterioration   ● (a) and (c)
◯ 3   ● (c) decrease in vegetation cover   〰 (a) and (b) and (c)

Fig. 2.4d. Results from the LUCC-MA Rapid Land-Cover Change Assessment showing major areas of degraded land in the world between 1980 and 2000

Smith 1986; Suliman 1988). This claim has also been discredited by the studies of Tucker et al. (1991) and Prince et al. (1998), although there are some areas showing signs of degradation (Ringrose and Matheson 1992). Following an analysis of all available global information sources, Dregne (2002) concluded that there is still a pressing need for more reliable data. Another recent data-driven effort is the work of Eswaran et al. (2003) who mapped global vulnerability to desertification using the FAO/UNESCO Soil Map of the World (FAO 1992). Using a vulnerability assessment methodology (Eswaran and Reich 1998), the map shows vulnerability to, but not actual desertification. Therefore, it is another type of expert interpretation that does not allow for change detection and, thus, monitoring.

2.3.6   Summary and Conclusions

The emergence of remote sensing has proved to be a valuable tool to monitor large-scale changes in land cover. Coarse and fine spatial resolution satellite sensors have been used to measure tropical deforestation (FAO 2001a; Achard et al. 2002; DeFries et al. 2002b), and change in nighttime city lights, which is a proxy for changes in urban extent and electrification (Elvidge et al. 1997, 2001). While numerous local-scale studies have mapped and quantified land-cover change using fine resolution remote sensing data, there are a few subnational- to national-scale studies (Skole and Tucker 1993; Pathfinder Humid Tropical Deforestation Project 1998), and remarkably few such studies at the continental to global scales (DeFries et al. 2002b). A few studies have used national-scale forest inventory and agricultural census data, in some cases along with remote sensing data, to estimate rates and geographic patterns of continental-to-global scale changes in forest cover and agricultural lands (Houghton et al. 1999; Ramankutty and Foley 1999; Klein Goldewijk 2001). Overall, the quantification of areas of rapid land-cover change still suffers from large uncertainties (Lepers et al. 2005).

Despite the major uncertainties in the data sets, the LUCC-MA assessment revealed that:

- Land-cover change is not randomly or uniformly distributed but is clustered in some locations. For example, deforestation mostly takes place at the edge of large forest areas and along major transportation networks (e.g., in the southern Amazon Basin).
- The land-cover change data sets have many gaps, and therefore it is possible that rapid change is occurring in many parts of the world but is not identified on the synthesis maps. Moreover, it is also possible that many locations not identified as undergoing rapid land-cover change are experiencing other ecological impacts.

- Different parts of the world are experiencing different phases of land-cover transition (e.g., a decrease in cropland in temperate zones and an increase in the tropics).
- Dryland changes are the most poorly understood, because of difficulties in conceptualization, difficulties in satellite interpretation in these regions, and an inability to distinguish human-induced trends from climate-driven interannual variability in vegetation cover.
- The Amazon Basin is a major hot spot of tropical deforestation. Rapid cropland increase, often associated with large-scale deforestation, is most prominent in Southeast Asia. Forest degradation in boreal Eurasia is increasing rapidly, mostly as a result of logging and increased fire frequency. The southeastern U.S. and eastern China have experienced a rapid decline in cropland area. Asia currently has the greatest concentration of areas undergoing rapid land-cover changes, in particular dryland degradation. Existing data do not support the claim that the African Sahel is a desertification hot spot. Many of the most populated and rapidly changing cities are found in the tropics.
- Much of our information on tropical land-cover change comes from remotely-sensed land-cover data, while information on change in the extra-tropical regions comes predominantly from census data. Systematic analyses to identify land-cover change has been predominantly done in the tropics, because of the interest in tropical deforestation and, possibly, due to the lower availability and reliability of census data in the tropics.

Furthermore, in the context of a global synthesis, the LUCC-MA study also made several general recommendations for future observations and research:

- Future synthesis should use a combination of global-scale coarse-resolution data as presented here, combined with finer resolution satellite imagery as well as ground-truth data for a subset of locations, a framework also advocated by the Global Observations of Forest Cover and Land Dynamics (GOFC-GOLD) panel of the Global Terrestrial Observing System (GTOS).
- Data producers should adopt a standardized land-cover classification system. We recommend wide adoption of the classification system proposed by FAO (Di Gregorio and Jansen 2000b).
- As a complement to categorical land-cover representations, a continuous description of land cover (i.e., in terms of fractional tree cover or crop cover) should be more widely adopted whenever possible as it offers greater ease for comparison of different databases, and the ability to identify land-cover modifications, in addition to conversions (DeFries et al. 1995a; Ramankutty and Foley 1998).

- Operational monitoring of land cover should be extended to regions that are not considered "hot spots" today, but where rapid changes may still take place and catch the scientific community by surprise.
- Systematic, consistent measurements of soil properties should be undertaken at a global scale, at a relatively fine resolution, since soil attributes are an important component of land cover (see Sect. 4.6).
- New empirical work is required based on conceptual advances in dealing with definitions of desertification (Reynolds and Stafford-Smith 2002) and urbanization.
- There is an urgent need for systematic observations on the still poorly measured processes of land-cover change (see Sect. 2.4.1).

2.4   The Complexity of Land-Cover Change

2.4.1   The Poorly Documented Changes

The BIOME 300 project and the LUCC-MA Rapid Land-Cover Change Assessment provided initial estimates of the rates of change of several important land-cover types at the global scale, over the last 300 years and over the last two decades, respectively. However, local-to-national scale studies highlight the significance and ecological importance of other forms of rapid land-cover change that are thought to be widespread but still poorly documented globally. Both the extent and the precise location of these changes, often modifications rather than conversions, are barely understood. Prominent among these are changes in the (sub)-tropical dry forests; forest-cover changes resulting from selective logging, fires, and insect damage; drainage or other forms of alteration of wetlands; soil erosion and degradation in croplands; changes in the extent and productive capacity of pastoral lands; and dryland degradation, also referred to as desertification (Lambin et al. 2003). Other poorly documented changes at the global scale relate to (sub)urban and other infrastructure extension and the expansion of non-food crops such as sugar, tea and tobacco, primarily driven by shifts in lifestyles (Heilig 1994).

Changes in (Sub-)Tropical Dry Forests

Globally, dry forest or woodland ecosystems cover a greater area than humid forests but changes are not well documented. More so than in rain forest zones, the dry forest life zone is greatly affected by human activities including conversion for agricultural uses and overexploitation through fuelwood and polewood collection, even in protected areas (Janzen 1988; Solbrig 1993). While dry forests and woodlands in South and Southeast Asia have been mostly converted into other covers (including bare, eroded soils), large portions of the (sub)tropical dry forest life zone still exist in Africa and Latin America. Local- to subna-

tional-scale studies highlight the significance and ecological importance of rapid land-cover change there.

The Miombo woodlands in the Central African Plateau form the world's largest contiguous area in dry forest (Mayaux et al. 2004). Local evidence suggests exceptionally high rates of change, especially in the late 20th century. For example, Lusitu in southern Zambia experienced an annual rate of land-cover change of 4.0% between 1986 and 1997 (Petit et al. 2001), while in the Lake Malawi National Park of southern Malawi, massive wood extraction for rural industries caused a rapid loss of closed canopy cover between 1982 and 1990, replaced by sparse woodland cover which increased by almost 300% (Abbot and Homewood 1999).

Much of the rapid deforestation in the 1990s was in subtropical South America, particularly in the Brazilian Cerrado (Jepson 2005) (see Box 4.9). Rapid deforestation in the Santa Cruz department of Bolivia, in the vicinity of Mennonite settlements and more recently in industrial soybean farms, as well as in the Chaco region of Argentina have been documented (Steininger 2001; Zak and Cabido 2002). Estimates of land-cover loss for the Brazilian Cerrado vary between 40 and 80% of the original cover. In addition to soybeans, cattle ranching and mechanized commercial agriculture including cotton, rice and maize, have expanded rapidly during the past three decades. Exceptionally high conversion and modification was seen since 1970 in Rio Grande do Sul and Paraná, and since 2003 in Mato Grosso, and further stimulated new agricultural frontiers in the states of Minas Gerais, Goiás, Bahia, Tocantins and Maranhão (Jepson 2005). Exceptionally high rates of deforestation have been demonstrated for North-Central Yucatán, a region dominated by tropical dry forest vegetation, experiencing exponential population growth and land-use development (Sohn et al. 1999). Similarly, significant deforestation has taken place in southern Yucatán, leading to a landscape increasingly dominated by secondary forests and opened lands (Turner II et al. 2004).

In summary, land-cover changes in (sub)tropical dry forest and woodland ecosystems are thought to be widespread, but continental-to-global scale estimates are not available at this time. Remote sensing techniques, even 30-m resolution Landsat TM data, have difficulty in distinguishing different land-cover types in the dry season, while most rainy season images are contaminated by cloud cover (Asner 2001). Some potential solutions are the inclusion of cloud-free rainy season images and sampling cloudy areas with very high-resolution images from the IKONOS™ satellite (Sánchez-Azofeifa et al. 2003), or the use of Synthetic Aperture Radar (SAR) data (e.g., Grover et al. 1999; Saatchi et al. 2000). Landsat TM thermal band 6 data, using continuous data rather than discrete classes, measure the emission of energy from the land surface and allow for the differentiation between successional stages of forest growth in dry forest ecosystems (Southworth 2004).

Logging, Fire and Insect Damage Triggering Forest Change

Forest degradation following fires and (selective) logging has been termed "cryptic deforestation" (Nepstad et al. 1999). In the Brazilian Amazon, Nepstad et al. (1999) have estimated that every year forest impoverishment caused by selective logging and fires affects an area at least as large as the area affected by forest-cover conversion. Similarly, biomass collapse around the edges of forest fragments has also been estimated to be a significant contributor to land-cover change in Amazonia (Laurance et al. 1997). All of these subtle changes are not well documented at the global scale, although new estimates of logging from the Amazon indicate that they are very important (Asner et al. 2005). This is also true for losses of forest associated with forest-management practices (e.g., Canada and Russia), insect damage, and large fires (e.g., Indonesia in 1997, Siegert et al. 2001; Page et al. 2002), and in Russia in the late 1990s (Sukhinin et al. 2004).

Fires play a significant, yet complex, role in their relationship to land-use and land-cover change (see Box 2.7). While the dynamics of fire and how they interact with humans, climate and vegetation is poorly understood, remote sensing has made rapid progress in documenting fires at a global scale (Dwyer et al. 2000), both for the mostly anthropogenic fires in tropical regions (Pereira et al. 1999) and the mostly natural fires in boreal regions (Kasischke et al. 2002). At least three major efforts have been undertaken to document fires at the global scale. The Global Burnt Area (GBA 2000) data set derived from SPOT Vegetation satellite data provides the first independent estimate of the area of vegetation burnt at the global scale in the year 2000 (Tansey et al. 2004). Another global inventory is the *ATSR World Fire Atlas*, a database of monthly global fire maps from 1995 to the present, produced using the Along Track Scanning Radiometers (ATSR) on board of successive European Space Agency ERS and ENVISAT satellites. The product has already been used to compute the emissions of greenhouse gases and aerosols from biomass burning and explore the impacts on tropical ozone levels (Schultz 2002; Duncan et al. 2003). Finally, GLOBSCAR is the European Space Agency's initiative to map the global distribution of burnt area using the ATSR-2 instrument on-board of the ERS-2 satellite (Simon et al. 2004). The product has been developed at a spatial resolution of one km at monthly time intervals. GLOBSCAR and GBA 2000 are complementary data sets.

Wetland Alterations

Changes in wetlands, including extensive drainage, are not well documented at the global scale. Wetlands provide a valuable ecosystem service in terms of maintaining and improving water quality (Steffen et al. 2004). In the devel-

oped world, wetlands have been lost at a rapid rate historically, and although measures are actively being sought today to stem wetland losses and restore wetland acreage, the rate of wetland conversion can still be dramatic (Dahl 1990). Maintenance and restoration of wetlands in developed countries has undoubtedly contributed to the improvement of water quality there, while the dominant trend in developing countries is still towards the conversion of wetlands into other land uses (Steffen et al. 2004) (see Sect. 4.7).

An attempt at attaining a global wetlands inventory has been initiated. GLOBWETLANDS is an ongoing project of the European Space Agency to develop remote sensing methods for a range of different wetland types focusing on five continents (North and South America, Africa, Asia, Europe). Satellite imagery, inventory maps and digital elevation models of approximately 50 wetlands and surrounding catchments are being analyzed. The project, which is supporting the 1971 RAMSAR Convention on Wetlands, will provide information about difficult and inaccessible terrain, describe the local topography, map the types of wetland vegetation and monitor land-use/cover changes there.

Soil Erosion and Degradation in Cultivated Lands

Soil erosion and degradation in cultivated lands is poorly documented at the global scale, and remains a controversial issue (see Sect. 4.6). The Global Assessment of Human-induced Soil Degradation (GLASOD) is the only global survey of soil degradation for both arid and humid regions, including the type, degree, extent, rate and even cause of soil degradation. The GLASOD database reflects the informed opinion of hundreds of scientists (Oldeman et al. 1991), and is still by far the best representation of world soil degradation (Geist 2005; Dregne 2002). However, in the context of drylands, the GLASOD database has been heavily criticized (e.g., Reynolds 2001). Furthermore, any documentation of soil erosion and degradation suffers from methodological inadequacies. For example, the severity of soil erosion and its impacts in the United States has been debated because of discrepancies between estimates based on models and observed sediment budgets (Trimble and Crosson 2000).

The situation is further complicated by the fact that natural environmental change and variability interacts with human causes to trigger erosion and degradation. Highly variable environmental conditions amplify the pressures arising from high demands on land resources. For example, in the Iberian Peninsula during the 16th and 17th centuries, the peak of the Little Ice Age coincided with large-scale clearing for cultivation following the consolidation of Christian rule over the region. This cultivation triggered changes in surface hydrology and significant soil erosion (Puigdefábregas 1998). Moreover, cultivation does not always result in deterioration of soil conditions, as evidenced by recent study evidence that highlights situations across several African countries where population growth and agricultural intensification have been accompanied by improved soil (and water) resources (Tiffen et al. 1994a; Mortimore and Adams 2001). Such issues have made it difficult to document soil degradation resulting from cultivation at a global scale.

Changes in Extent and Productivity of Grazing Lands

Worldwide, only few grazing lands are edaphically or climatically determined natural entities that, in the absence of human impact, would persist unchanging within climate epochs. In contrast, most grazing lands are maintained in their current state by the interaction of human and biophysical drivers (Solbrig 1993; Sneath 1998). In the temperate and tropical zones, rangelands are both highly dynamic and also resilient, moving through multiple vegetation states, either as successional sequences or by shifting chaotically in response to the random interplay of human and biophysical drivers. Grazing lands are increasingly seen as non-equilibrium ecosystems (Walker 1993). Furthermore, grazing lands are often also multi-functional landscapes, with grazing occurring on croplands after harvesting, grazing in wooded lands, grazing on natural pastures *versus* planted pastures, etc. These features and the wide spectre of definition render estimates of extent and productivity difficult to make. Few, complex pathways of grazing land modification can be identified qualitatively, but it remains difficult to attach quantitative figures to them (Lambin et al. 2001). Therefore, data provided by the Food and Agriculture Organization of the United Nations, for example, are to be treated with caution.

Dryland Degradation (Desertification)

As with soil degradation on croplands, desertification is still unmeasured or poorly documented at the global scale (Lambin et al. 2003; Geist 2005). The notion of "desertification" resonates with public perception of land change in drylands, especially in the Sahel-Sudan zone of Sub-Saharan Africa. Desertification has become the dominant theme of an environmental convention – the United Nations Convention to Combat Desertification (UNCCD) – that emerged from the Rio-summit of 1992.

However, the definition of desertification, its causes, and extent remain widely disputed (Reynolds and Stafford-Smith 2002). The concept covers a wide variety of environmental change processes, taking place at a range of spatial and temporal scales. This makes it difficult to measure desertification and even more difficult to provide a general accounts of its causes – see Chap. 3. Here we briefly highlight several of the major sources of uncertainty and disputes surrounding desertification.

- Desertification has most often been reported to occur because of estimated "undesirable" changes in soil and vegetation properties. However, "undesirability" is a matter of perception. Cultivators are likely to have a different perspective than those herding or utilizing forest products. Soil scientists and geomorphologists are likely to have different perspectives from botanists, ecologists, foresters and agronomists, not to mention economists.

- Since the Sahel drought in the late 1960s, there has been discussion on whether environmental change caused by climate change/variability should be included. The UNCCD definition (see Sect. 2.3.5) now explicitly acknowledges both "natural" (climatic) and "human" induced factors. However, there is still the problem of measuring desertification against a "baseline". Multiple studies point out that dryland ecosystems may seldom be perceived as being in an equilibrium state, rather they fluctuate widely in response to climate change and other external and internal controls (Behnke and Scoones 1993). For example, the Sahel drought of the seventies, which was thought to have led to permanent desertification, is now believed to have been a temporary multi-decadal climate anomaly, and the vegetation is now rebounding (Tucker et al. 1991).

- Another issue is whether evidence for desertification should be derived on the basis of long-term responses such as soil degradation, and not of short-term ecosystem dynamics such as vegetation changes. For example, the GLASOD assessment (see Sect. 2.3.5) estimated that 19.5% of drylands worldwide were suffering from desertification, while another survey carried out by the International Center for Arid and Semiarid Land Studies (ICASALS) estimated a much higher figure of 69.5% due to their inclusion of vegetation changes in addition to those areas affected by soil erosion (European Commission 2000).

- Another controversy surrounding desertification relates to the notion of "irreversibility". Should the concept be reserved for environmental changes that are irreversible or only very slowly reversible, relative to a "human" time scale? The currently used UNCCD definition does not include this requirement. Some scientists claim that most, if not all, desertification, especially in an advanced state, "is often essentially irreversible" (Phillips 1993). This runs contrary to evidence that "for the vast majority of the drylands the wasteland end point never occurs" (Dregne 2002). It has been estimated that the very severe or irreversible desertification class includes only about 1.5% of the global drylands (Dregne 1983; Dregne and Chou 1992). It is clear that changes in soil physical properties, such as massive loss of fine-grained material from the topsoil due to wind erosion, may be only very slowly reversible, while vegetation may recover rapidly.

In summary, many indicators of desertification have been suggested, including changes in both vegetation and soil properties, responding to both climate and human induced changes. However, these processes do not necessarily occur in parallel – vegetation changes occur over annual-to-decadal timescales, while soil property changes occur over decadal-to-centennial timescales; similarly, the timescales of the climatic and human drivers differ. Therefore, the use of different indicators, different timescales, and different perspectives often leads to different interpretations. For instance, it is possible that one indicator (e.g., greening vegetation) suggests decreasing desertification, while another indicator (e.g., soil erosion) suggests increasing desertification in the same location. These two observations are not contradictory, yet the example indicates the inherent problem with operationalizing the concept of desertification.

Extension of Industrial and Service Economy Infrastructure

Apart from agricultural land uses, a broad range of other land uses have gained importance since the middle of the 18th century. This has mainly been due to the transition from an agricultural to an industrial society, and later in some countries, to a service economy. Land uses related to non-agrarian modes of land management include human habitation, manufacturing and industrial facilities, water and energy supply infrastructure, (mass) tourist facilities, waste deposition and sanitation facilities, transport infrastructures, military establishments, bureaucratic and communication facilities, and many more (Heilig 1994).

Industrialization has affected practically every region of the world, especially after World War II. The degree of industrialization, however, differs widely across countries, and is still an ongoing process in many parts of the world (see Sect. 3.3.2). In the form of built-up or paved-over areas, infrastructure is estimated to occupy only 2–3% of the Earth's surface (Grübler 1994; Young 1999), but land uses related to industrialization have clearly outpaced conventional agricultural land uses in terms of the speed at which they occurred over the last decades.

No consistent global inventory exists, however, for quantifying industrial developments and their linkages to rural landscapes. Global snapshots of nighttime lights detected by satellite illustrate the rapid changes in both urban extent and electrification of the cities and their surroundings, but it is not clear yet how they can be further developed for land-cover change analysis beyond illustration purposes.

Lifestyle-Driven Changes and Their Impacts

With globally rising economic activities and living standards, the structure of consumption has changed, opening a wide-ranging potential for effects on land-use expansion (see Sect. 3.3.2). Generally, with increasing incomes, the demand for food initially increases and then stabilizes. This is accompanied by an overall decline in the agricultural sector in terms of the size of the labor force and the revenues to the state. This is particular to developed countries, but the process of deagrarianization is also underway in newly industrializing and even less developed countries of Africa (Bryceson 1996). In addition, overall consumption becomes diverted towards industrial goods and services and more diverse food expenditures, with increasing share of non-food crops consumed (Dicken 2003).

Only national statistical data, such as those provided by the FAO, WTO, USDA and IFPRI are available to quantify life-style and consumption-change-driven land-cover changes. For example, clothing fashions since the 17th century have driven the expansion of cotton plantations worldwide. The same is true for the fashion of consuming stimulants such as coffee, tea and tobacco (Heilig 1994). Tobacco, for example, is grown in more than 100 countries, thus being the world's most widespread non-food crop. Between 1982 and 1996, land under tobacco globally expanded at an average annual rate of 2%, which was slightly below the overall expansion rate of arable land in the same time period. FAO data show that the bulk of land under tobacco is located in low-income countries of the (sub)tropical zones. Tobacco land expansion there outpaced global tobacco expansion by a factor of 5. In countries such as Pakistan, Philippines, China, Zimbabwe, Zambia, Uganda and Malawi, land under tobacco increased at rates up to 10 times higher than for arable land. Most importantly, three quarters of the tobacco grown in the developing world are artificially cured varieties using heat from external sources such as wood and coal to dry leaves on the farm for the production of American blend type cigarettes. Wood from natural ecosystems, rather than from plantations, is most commonly used by African tobacco producers. Based on crop-specific wood consumption rates, deforestation related to tobacco curing in the developing word can be estimated to contribute to roughly 5% of total net losses of forest cover there in the 1990–1995 period (Geist 1999b, 2000).

In sum, no global inventory exists to link shifts in production and consumption and life-style changes to land-cover outcomes in situ. It remains a major caveat that social process-specific approaches to the study of land-cover change (such as remittances, agro-industrialization, and contract farming) as well as crop-specific approaches (such as the boom in non-food crops discussed here) cannot be fully substantiated in quantitative terms, and proxy indicators need to be used instead (Sack 1992; Heilig 1994; Geist 1999b). This results in a weak understanding of production-consumption relationships; for example, in the example outlined above, we had to resort to national-level statistical data to highlight the declining economic importance of food cropping.

2.4.2   Characterizing the Complexity of Changes

Conversion Versus Modification of Land Cover

Land cover is defined by the attributes of the Earth's land surface and immediate subsurface, including biota, soil, topography, surface and groundwater, and human structures. Land-cover data sets represent the land surface by a set of spatial units, each associated with attributes. These attributes can be either a single land-cover category (i.e., leading to a discrete or Boolean representation of land cover) (Loveland et al. 2000) or a set of continuous values of biophysical variables (i.e., leading to a continuous representation of land cover) (DeFries et al. 1995a). A discrete land-cover data set has the advantages of conciseness and clarity, but it has led to an overemphasis of land-cover conversions and a neglect of land-cover modifications. Land-cover conversions are defined as the complete replacement of one land-cover type by another (e.g., agricultural expansion, deforestation, or change in urban extent). Land-cover modifications are more subtle changes that affect the character of the land cover without changing its overall classification.

Recently, there has been increased recognition of the importance of the processes of land-cover modification. For example, agricultural intensification – defined as higher levels of inputs (including use of high-yielding crop varieties, fertilization, irrigation, and pesticides) and increased output of cultivated or reared products per unit area and time – permitted an increase in the world's food production over the last decades, outpacing human population growth (Matson et al. 1997; Tilman 1999). In the Brazilian Amazon, every year forest impoverishment caused by selective logging and fires affects an area at least as large as that affected by forest-cover conversion (Nepstad et al. 1999). Woody encroachment on the western United States grasslands, following fire suppression and overgrazing, may have contributed to a large carbon sink (Houghton et al. 1999; Pacala et al. 2001; Asner et al. 2003) (see Sect. 4.4.4). Declines in tree density and species richness in the last half of the twentieth century were observed in a region of Senegal in the West African Sahel, potentially indicative of desertification (Gonzalez 2001). Another study in western Sudan, a region that was allegedly affected by desertification, however, did not find any decline in the abundance of trees despite several decades of droughts (Schlesinger and Gramenopoulos 1996).

The monitoring of land-cover conversion can be performed by a simple comparison of successive land-cover maps. In contrast, the detection of land-cover modifications requires a continuous representation of land cover, where the surface attributes vary continuously in space and time, at the seasonal and interannual scales (DeFries et al. 1995a; Lambin et al. 1999). This allows detection of, for example, changes in tree density, in net primary productivity, or in the length of the growing season. Earth observation from satellites provides repetitive and spatially explicit measurements of biophysical surface attributes, such as vegetation cover, biomass, vegetation community structure, surface moisture, superficial soil organic matter content, and landscape heterogeneity. Analyses of multi-year time series of these attributes, their fine-scale spatial pattern, and their seasonal evolution have led to a broader view of land-cover change. In particular, data from wide-field-of-view satellite sensors reveal patterns of seasonal and interannual variations in land-surface attributes that are driven not by land-use change, but rather by climatic variability. These variations include the impact on vegetation and surface moisture of the El Niño Southern Oscillation (ENSO) phenomena (Eastman and Fulk 1993; Plisnier et al. 2000; Behrenfeld et al. 2001), natural disasters such as floods and droughts (Lambin and Ehrlich 1997b; Lupo et al. 2001), changes in the length of the growing season in boreal regions (Myneni et al. 1997), and fluctuations in the southern margins of the Sahara driven by rainfall fluctuations (Tucker et al. 1991).

Progressive Versus Episodic Land-Cover Changes

Time series of remote sensing data reveal that land-cover changes do not always occur in a progressive and gradual way, but rather often show periods of rapid and abrupt change followed either by a quick recovery of ecosystems or by a nonequilibrium trajectory. Such short-term changes, often caused by the interaction of climatic and land-use factors, have an important impact on ecosystem processes. For example, droughts in the African Sahel and their effects on vegetation are reinforced at the decadal timescale through a feedback mechanism that involves land-surface changes caused by the initial decrease in rainfall (Zeng 1999; Wang and Eltahir 2000), although this mechanism is still disputed for the Sahel (Giannini et al. 2003; also see review by Xue et al. 2004). Grazing and conversion of semiarid grasslands to row-crop agriculture are the sources of another positive desertification feedback by increasing heterogeneity of soil resources in space and time (Schlesinger et al. 1990). The role of the Amazonian forest as a carbon sink (in natural forests) and source (from land-use changes and fires) varies from year to year as a result of interactive effects between deforestation, abandonment of agricultural land reverting to forests, fires, and interannual climatic variability (Tian et al. 1998; Houghton et al. 2000). In Indonesia, periodic El Niño-driven droughts lead to an increase in the forest's susceptibility to fires. Accidental fires are more likely under these conditions and lead to the devastation of large tracts of forests (Siegert et al. 2001), and to the release of huge amounts of carbon from peatland fires (Page et al. 2002). Large landholders also seize the opportunity of drought conditions to burn large tracts of forest to convert them to plantations. Forests that have been affected by forest fragmentation, selective logging, or a first fire subsequently become even more vulnerable to fires as these factors interact synergistically with drought (Cochrane 2001; Siegert et al. 2001) (see Sect. 3.3.1).

In summary, both land-cover modifications and rapid land-cover changes need to be better accounted for in land-cover change studies. Climate-driven land-cover modifications do interact with land-use changes. Slow and localized land-cover conversion takes place against a background of high temporal frequency regional-scale fluctuations in land-cover conditions caused by climatic variability, and it is often linked through positive feedbacks with land-cover modifications. These multiple spatial and temporal scales of change, with interactions between climate-driven and anthropogenic changes, are a significant source of complexity in the assessment of land-cover changes. It is not surprising that the land-cover changes for which the best data exist – deforestation, changes in the extent of cultivated lands, and urbanization – are processes of conversion that are not strongly affected by interannual climatic variability. By contrast, few quantitative data exist at the global scale for processes of land-cover modification that are heavily influenced by interannual climatic fluctuations, e.g., desertification, forest degradation and rangeland modifications.

Box 2.7. Fire on the landscape

Fire plays a significant, yet complex, role in relation to land-use/cover change in many parts of the world, including boreal forests, Mediterranean ecosystems, tropical savannas and dry (or seasonally dry) forests (see Box 4.8). It may be perceived as a "natural" part of an ecosystem, as part of a "disturbance regime", or as an integrated part of natural resource management.

While fires do occur naturally in many ecosystems, humans have modified the fire regime significantly, both through increased burning in some places, and more recently, through fire suppression. For example, when humans arrived in Australia roughly 40 000 years ago, the fire frequency increased greatly (on the basis of fossil charcoal evidence, e.g., Singh and Geissler 1985), leading to extensive change in land cover. Nowadays, fires are a common tool in natural resource management, including crop production (especially in shifting cultivation or slash-and-burn agriculture), rangeland management, forestry and hunting.

While a clear correlation between fire occurrence and land-cover change may exist, the relationship between fires and land-cover change is complex. Clearing a closed canopy forest may create conditions that allow the development of a herbaceous layer. This may, in turn, help to sustain fire, and keep out the woody vegetation. In other cases fires may be the cause of land-cover change at a time scale of decades (see Sect. 3.3.1), yet at longer time scales it may be perceived as being part of the ecosystem. Thus, individual species and whole ecosystems have co-evolved with the prevailing fire regime, and over the millennia of human induced fires. The structure and functioning of these ecosystems are shaped by the fire regime, and it may not be appropriate to talk about fire as a "disturbance regime" in this context.

2.5   Power and Limitations of Remote Sensing

2.5.1   Remote Sensing of Global Land Cover

Recently, satellite-based observations of the Earth have provided a spatially and temporally consistent picture of the state of global land cover. Earlier efforts were a painstaking compilation of different maps from different periods in time, and that were often inconsistent in terms of the land-cover classes used. Satellite-based remote sensing began in 1959, with the first space photograph taken by the Explorer 6 satellite. NASA launched Landsat 1 in 1972 to monitor the Earth's natural resources. A series of Landsat satellites followed, with the most recent, Landsat 7, launched in 1999, making it the longest running space-based remote-sensing program.

Landsat, with 30 m spatial resolution multispectral data, has become the workhorse for land-cover change studies, and has been used extensively to study land-cover change around the world. The first large-scale deforestation assessment for the Brazilian Amazon was made by Tardin and colleagues (Tardin et al. 1980), who pioneered the use of satellite remote-sensing imagery to map deforestation over a 5 million km^2 area for years 1974 and 1978. Nearly a decade later, new remote sensing surveys of deforestation were repeated for the Amazon (Tardin and Cunha 1989; Skole and Tucker 1993) and extended for much of the tropics (Chomentowski et al. 1994) through the NASA Pathfinder Humid Tropical Deforestation project. The Pathfinder deforestation project also pioneered in the dissemination of remote sensing imagery and land-cover change maps for the tropics. More recently, Achard et al. (2002), through the TREES project associated with the Joint Research Center (JRC) of the European Commission, used Landsat data to estimate deforestation rates for the humid tropics. Such studies are making it possible to monitor the impacts of human land-use activities.

In addition to Landsat, there are other, similar sensors that are being used to monitor land cover. In early 1978, the France launched the SPOT (Système Pour l'Observation de la Terre) program. The series of SPOT-1, -2, -3 and -4 satellites have provided 20-m resolution multispectral data. The Indian Remote Sensing (IRS) program was launched in 1998, with the capability of sensing land cover at 23 m resolution in three different wavelength bands. Russia, China, and Japan have also launched satellites.

While such data cover the globe, the high spatial resolution of the data make it resource intensive to compile and classify over the entire globe. Moreover, cloud cover and lack of temporal data indicating vegetation phenology limit the usefulness of Landsat for global land-cover mapping. With the use of moderate-resolution satellite sensors (~1 km resolution) acquired throughout the year, it has recently become possible to characterize land cover

Table 2.3. Global land cover data sets from Earth observation data

Name of product	Creator	Sensor used	Year of data	Spatial resolution	Reference
UMD 1 degree land cover	University of Maryland	AVHRR	1987	1 deg	DeFries and Townshend (1994)
UMD 8 km global land cover	University of Maryland	AVHRR	1984	8 km	DeFries et al. (1998)
UMD 1 km global land cover	University of Maryland	AVHRR	1992–1993	1 km	Hansen et al. (2000)
DISCover	United States Geological Survey's EROS Data Center, the University of Nebraska-Lincoln, and the Joint Research Centre, European Commission	AVHRR	1992	1 km	Loveland et al. (2000)
MOD12Q1	Boston University	MODIS	2001	1 km	Friedl et al. (2002)
GLC2000	Joint Research Centre, European Commission	SPOT VEGETATION	2000	1 km	Bartholomé and Belward (2005)
MODIS VCF (Vegetation Continuous Fields)	University of Maryland	MODIS	2000	500 m	Hansen et al. (2003)
GLOBCOVER	European Space Agency and EC Joint Research Center	ENVISAT MERIS	2005	300 m	GOFC-GOLD (2005)

globally. Several efforts have emerged in the last decade to develop global land-cover data sets (see Table 2.3). These efforts have either classified the global land cover into ~13–22 different land-cover classes, or in the case of the MODIS Vegetation Continuous Fields (VCF) product, characterized the landscape using a continuous description of the landscape (percentage tree cover, herbaceous and bare ground, as well as leaf type and phenology). These global data sets have provided a comprehensive global view of the Earth's land surface for the first time. The next generation of global land-cover information will build upon these experiences and improve the spatial and thematic detail for land characterization. One prominent example, the GLOBCOVER project of the European Space Agency (ESA), will produce a global land-cover data set for the year 2005 using 300-m resolution ENVISAT MERIS data.

In recent years, in addition to these global data sets, several commercial satellites have been launched that provide very high-resolution imagery. IKONOS, launched by Space Imaging, is designed to provide 4-m resolution data in four multispectral bands. QuickBird, launched by EarthWatch, Inc., is very similar. While these sensors can provide a very detailed picture of land cover, they are too expensive, and it would be an onerous task to compile data from them at the global scale. Therefore, they have mostly been used to calibrate or validate land-cover data derived from Landsat or the other high-to-moderate resolution sensors.

2.5.2 The Challenge of Monitoring

Observations of land-cover change face numerous challenges resulting from disagreements in definitions of land cover and the processes causing land-cover change, dif-

ficulties in characterizing the complexity of land-cover changes, and methodological difficulties (Sect. 2.4). In this section, we describe the various remote-sensing challenges in observing land-cover change.

Wall-to-Wall or Sampling?

Numerous studies have focussed on measuring the extent of tropical deforestation. Studies conducted at local or national levels were based on the analysis of wall-to-wall coverage of fine spatial resolution satellite images (e.g., Skole and Tucker 1993) or aerial photographs. For scaling up the remote sensing estimates to the global scale, two main methods have been tested:

- Measuring change by extrapolating from a sample of fine resolution (30 m) satellite imagery (the FAO Remote Sensing Survey (FAO 2001b) and the TREES project (Achard et al. 2002). This approach requires a sampling strategy designed to take into account the spatial variability of the phenomena to be estimated (Richards et al. 2000).
- Measuring change using wall-to-wall coarse resolution (>250 m) satellite imagery (DeFries and Achard 2002; Hansen et al. 2003). This approach, which is targeted towards estimating changes in tree cover percentage, must be carefully calibrated with local studies.

Sampling in natural resource assessment is a standard technique to provide an estimate at a feasible cost. The cost of carrying out a 100% survey of very large regions such as the tropics with fine spatial resolution (10–30 m) satellite imagery, has until now been prohibitively high in terms of image acquisition, data management, image interpretation and extraction of results.

There has been much debate surrounding the issue of whether wall-to-wall coverage is required to estimate deforestation. Some researchers have argued that wall-to-wall coverage is required because deforestation occurs in clusters, along roads and rivers, and is not randomly distributed across the landscape (Tucker and Townshend 2000). Tucker and Townshend (2000) suggested that ~80% of the total area needs to be covered to estimate deforestation rates accurately. On the other hand, Czaplewski (2003) demonstrated that estimation accuracy is less related to the percentage of samples, and rather a question of having sufficient number of samples, and that a robust sampling strategy can provide accurate estimates at global to continental levels. Furthermore, the efficiency of the sample can be improved by using stratified sampling; indeed this technique has been used to estimate forest change in the humid tropics (Achard et al. 2002). In this particular case, the stratification was created from the delineation of "deforestation hot spot areas" (Achard et al. 1998). However, it should be noted that the TREES (Achard et al. 2002) and FAO (FAO 2001a,b) Landsat-derived estimates can only be reported at continental or global levels as they were derived from a global-targeted sampling scheme. A statistical survey aiming at producing national estimates would require a dedicated sampling scheme with a larger number of samples within individual countries as compared to the TREES and FAO FRA-2000 surveys. FAO is planning such a sampling strategy for the remote sensing survey of their next FRA-2010 exercise, allowing the provision of national estimates for most countries (Mayaux et al. 2005).

In summary, wall-to-wall coverage of fine resolution satellite images would be desirable to improve accuracy in estimates of deforestation. The 6.5% sample area of the TREES exercise and the 10% sample area of the FAO remote sensing survey provided estimates with 13% and 15% standard error, respectively (Achard et al. 2004). However, the cost and effort involved currently in making wall-to-wall estimates (10 to 15 times more than the FAO or TREES exercises), the marginal gain in accuracy, and problems with persistent cloud cover in many parts of the world (Asner 2001), suggest that sampling strategies or coarse-resolution imagery will continue to be used for a long time until methodologies improve and remote sensing imagery becomes cheaper.

Frequent Monitoring or Snapshots?

Deforestation is most often estimated by developing percentage tree cover maps for the end points of the time period under consideration, and subtracting the two images to estimate changes (DeFries et al. 2002b), or by developing forest-cover change maps directly from the comparison of two images (Achard et al. 2002). For example, Achard et al. (2002) derived estimates of deforestation for the humid tropics by producing forest-cover change maps by overlaying the forest-cover map for the year 1990 with the satellite image of the year 1997. Pixels that showed a decrease in forest cover in visual interpretation of the two layers (image and map in 1990 *versus* satellite image in 1997) were classified as deforested, while pixels that showed an increase in forest cover were classified as regrowth. However, such interpretations of land-cover change from two (or more) snapshots may disguise the full land-cover dynamics that occur within those two snapshots. For example, if the snapshots are sufficiently separated in time, it is likely that some pixels could be deforested and regain a full canopy cover between the two snapshots, and show no change. This is especially the case with short fallow systems, where land that is cleared for agriculture is abandoned and allowed to re-grow 2 to 3 years after the initial clearing. In the humid tropics, the spectral signatures of regrowths may be confused with forest spectral signatures after as little as 10 to 20 years. However, the ecological and physical properties will be different. For example, biomass and soil carbon will take a long time to recover to the original state. Moreover, the biodiversity of secondary forests is often lower than in a primary forest. Therefore, to adequately address issues such as the global carbon cycle and biodiversity loss (see Chap. 4), it is important to measure the interannual variability of land-cover change, preferably monitoring at least every 2 to 3 years, if not every year.

Furthermore, while satellite imagery has been used mainly to identify initial clearing associated with land-use change, additional information is needed on the fate of the cleared land following the initial change. Is the land being cultivated or grazed? Is the land abandoned, and if so how long is it in agricultural use before abandonment? Do subsequent conversions or changes in land management occur following the initial use? Such information is difficult to discern at regional or global scales and remains a challenge for remote sensing.

Spatial Resolution

As described in Sect. 2.5.1, high-spatial-resolution data (e.g., 30 m resolution from Landsat) has become the standard for monitoring land-cover change. However, such high-spatial resolution data are difficult to use at continental-to-global scales because of the prohibitive cost of the data, the difficulty in manually classifying images, problems with cloud cover, as well as the effort involved in classifying thousands of images (57 784 images to cover the world). Moreover, because of all these issues, only one snapshot during the year is normally classified when high-spatial-resolution data are used, and classifying the land cover is difficult unless images are chosen during the right season, and accuracies are low even then. For example, cultivated lands can look very similar to the surrounding landscape during the non-growing season. Coarse-spatial resolution data have the obvious disadvantage of being poor at identifying land-cover change features that often occur at spatial scales that are finer than that of the sensor. On the other hand, coarse-resolution data are less expensive (free to the user in many cases), are easier to handle, and have the full phenological information during the year to distinguish between different land uses (e.g., DeFries et al. 1995b; Loveland et al. 2000). They are also the only data sets that have been used to obtain a globally complete characterization of land cover.

Harmonizing Land-Cover Classification Systems

A land-cover classification is a systematic framework to describe the situation in the field using well-defined diagnostic criteria (or classifiers) (Di Gregorio and Jansen 2000b). A classification system provides names of classes, the criteria used to distinguish them, and the relationship between classes. Researchers have developed numerous classification systems to characterize landscapes around the world. Some of the systems are *a priori* defined, and the observations are made to fit into the classification system, while others allow the data, *a posteriori*, to determine the classification system; similarly, while some systems are hierarchical, others are not (Di Gregorio and Jansen 2000b). The profusion of land-cover change research has also resulted in an overabundance of land-cover classification systems. Unfortunately, this has resulted in the inability to compare land-cover maps made by different groups for different locations, and even maps made by different groups for the same location at different points in time. Consequently, it has become impossible to scale the numerous local-scale land-cover mapping efforts to the global scale. This has hampered global scale synthesis efforts to identify rates, locations, and patterns of rapid land-cover change around the world (Lepers et al. 2005).

There is no internationally accepted land-cover classification system today (Di Gregorio and Jansen 2000b). Recognizing the need to meet this challenge, the FAO promoted the development of LCCS (Land Cover Classification System), a new hierarchical, *a priori*, classification system, which is flexible, but also has systematic and clear class boundary definitions (see Box 2.8). The LCCS has already been adopted by several regional and global land-cover mapping efforts, including AFRICOVER (http://www.africover.org), as well as the Global Land-Cover classification for the year 2000 (GLC2000) global land-cover mapping effort. The LUCC project recommended the adoption of LCCS to its members in 2000 (McConnell and Moran 2001), as did the Global Observation of Forest and Land Cover Dynamics (GOFC-GOLD) group, a panel of GTOS (Global Terrestrial Observing System) (Herold and Schmullius 2004). LCCS is en-

dorsed and promoted to all actors involved in land observations and mapping. Ongoing developments include capacity building and LCCS application (e.g., UN Global Land Cover Network (GLCN) – http://www.glcn-lccs.org), harmonization of case studies, and the bridging of harmonization and validation efforts towards operational terrestrial observation (Herold et al. 2006; Strahler et al. 2006).

Integrating Stakeholders in Monitoring: How Can We Make Our Efforts Useful?

One purpose of monitoring land-use and land-cover change is to create credible information about the state of our landscapes that might be useful to those who manage them (see Chap. 7). This will allow land managers to make fundamental decisions about different futures for their landscapes so that they can evaluate the trade-offs of those decisions (see Box 2.1). The information on monitoring described in the previous sections goes one step toward creating that credible database, but if this information is to be used locally, there are some fundamental challenges to be addressed. Effectively integrating stakeholders in monitoring requires several steps, each of which alone is not sufficient to allow credible monitoring information to be relevant and legitimate to different actors. Despite this, small improvements in the relevance, credibility, and legitimacy of the process and practice of monitoring have the potential to reap large benefits in linking science to action (e.g., Clark et al. 2002; Cash 2003).

Potentially our most important challenge is that scientific interests drive many of the monitoring analyses we do today, often resulting in information that has no connection or relevance to the users (stakeholders), either in content or structure. This requires a turnabout in that scientists need to listen to the questions that stakeholders ask, and design monitoring that sheds light on those questions (Tomich et al. 2004a). Much of our credible science is left inaccessible to land managers in technical publications. "Translator" organizations, like assessment institutions or Non-Governmental Organizations (NGOs), can help by linking scientists and different actors together, and translating information into a mutual language and crossing institutional boundaries (Cash 2003).

Another disconnect is that of scale. While remote sensing provides a coarse-scale understanding of land changes, many of the important, fine-scale changes in land use and land cover are invisible from all of our current remote sensing data platforms. For example, subtle changes in land use, like increased grazing pressure or fence building, which have big impacts on land management, are often impossible to see from satellite or aircraft-borne sensor. Thus, if global and regional-scale monitoring is to be relevant to local action, we need to make strong links across scales. Global and regional analyses can identify hot spots of change, which can be translated into landscape-level analyses by local teams. In turn, the local analyses can confirm or revise coarse-scale analyses, explain why changes are happening and anticipate the consequence of those changes. If these local scientific efforts are then linked to local land managers and policy makers, scientific information could inform local decisions. This has the side benefit of linking international and national scientific efforts and strengthening the capacity of both of these groups to monitor change and link to stakeholders.

One final need is to incorporate knowledge from different sources, to strengthen the credibility of the final monitoring assessments produced. For example, agricultural scientists are finding that their efforts are more relevant, credible and legitimate if they integrate the long-term knowledge of farmers and herders with the insights and methods developed by modern science (Humphries et al. 2000). Integration of information from different sources will improve monitoring efforts, make products more relevant to local problems, but also open the channels of communication between scientists and stakeholders, as they jointly evaluate shared information (see Chap. 5, 6 and 7).

Summary and Conclusions

Remote sensing of land cover continues to face several challenges including the need to attain both high spatial and temporal resolution (30 m or less, and every 2–3 years, and multiple seasons within a year), while at the same time attaining large spatial coverage (continental-to-global scales) over long time periods (multiple decades). It appears that wall-to-wall spatial coverage with high resolution imagery can be attained for smaller regions, while understanding land-cover change over larger regions will come from moderate-to-coarse resolution data or through the use of stratified sampling methods with higher resolution data. Furthermore, it is important to recognize that the appropriate resolution for remote sensing varies among regions depending on the varying spatial scale and patterns of land cover in different parts of the world. It is also important to understand the full land-cover change dynamics using remote sensing, including the initial clearing, the subsequent land uses, and abandonment and regrowth if any. A recent review concluded that the "creative use of remote sensing inputs as well as ancillary data sources will improve the mapping of land cover more than further development of classifiers and algorithms" (Woodcock and Ozdogan 2004). Standardized land-cover and land-use classification systems also need to be developed and adopted to ensure compatibility across different study regions.

To enable the continuous monitoring of land-cover change and its consequences, critical developments are needed in the global extension of monitoring programs,

---

Box 2.8. The Land-Cover Classification Systems LCCS 1 and 2

The Land-Cover Classification System (LCCS) was developed and implemented by FAO and UNEP, to describe different land-cover features in a standardized way (Di Gregorio and Jansen 2000a; McConnell and Moran 2001). LCCS provides a comprehensive methodology for the description, characterization, classification and comparison of land-cover data worldwide.

LCCS is an *a priori* classification system, but uniquely provides a scale-independent, hierarchical method for classifying land cover. The approach uses a set of universally valid diagnostic classification criteria that uniquely identify the land-cover classes worldwide and enables a comparison of land-cover classes regardless of data source, sector or country. LCCS is designed to operate in two phases. Eight major land-cover types are defined in the initial step. At the second level, land-cover classes are created by combining sets of pre-defined classifiers, which have been carefully defined to describe land-cover variations present within each major land-cover type and also to avoid inappropriate combinations of classifiers. The system is highly flexible, while providing internal consistency by allowing land-cover classes to be clearly characterized. The methodology is applicable at a variety of mapping scales and to any geographic location. The diagnostic criteria used allow correlation with existing classifications and legends, and can therefore serve as an internationally acceptable reference standard for land cover.

A software program, now in its second version, has been developed to assist in land-cover interpretation. Despite the large number of classes that can be created, the user deals with only one classification at a time. A land-cover class is built up by a stepwise selection in which a number of classifiers are aggregated to derive the class. Updates to LCCS can be followed at http://www.africover.org/LCCS.htm.

---

enhanced capacity building especially in developing countries, and the design of data collection, quality control and statistical analysis programs (Balmford et al. 2005). Remote sensing scientists need to pay more attention to the particular needs of the stakeholders for whom the land-cover change information is being developed, and incorporate them in the process from the beginning.

2.6 Conclusions

In the last decade, significant advances have been made toward estimating the rates and patterns of historical land-cover change at global scales. For example, new global databases of land-cover changes during the last 300 years have been developed, among many others. Collectively, these indicate that land-cover change due to human land-use practices has occurred for millennia, and is not a recent phenomenon. The pace of change today is unprecedented; however, while rapid land-cover changes in the tropics capture much public attention today, it is important to note that rapid and/or extensive changes occur in the extratropical zones of the world as well, and have certainly been the focus of rapid changes in the past.

Results from the BIOME 300 project indicate that, between 1700 and 1990, global cropland area increased from ~3–4 million km^2 in 1700 to ~15–18 million km^2; while grazing land (the definition of which is problematic) increased from ~500 million km^2 in 1700 to 3 100 million km^2; forests decreased from ~53 million km^2 to ~43–44 million km^2; and savannas and grasslands decreased from 30–32 million km^2 to 12–23 million km^2. Much of the expansion of croplands came at the expense of forests, while much of today's grazing land was formerly grassland. However, there are notable exceptions to these trends – for example, the North American Prairies were lost mainly to croplands, and many Latin American forests are being cleared for ranching.

The 20th century has witnessed an acceleration of the pace and intensity of land-cover change. Since the 1960s, spurred in part by the Green Revolution, a shift in land-use practices toward agricultural intensification has been observed. Indeed, between 1961 and 2002, while cropland areas increased by only 15%, irrigated areas doubled, world fertilizer consumption increased 4.5 times, and the number of tractors used in agriculture increased 2.4 times. These factors causing land-cover change during the 20th century have been modulated by rapid and pervasive globalization. By removing regional and national barriers to global trade, globalization has enabled formerly economically isolated places to be rapidly connected to global markets, often resulting in substantial land-use and land-cover changes, as in the case of Borneo. Globalization and transportation infrastructure has also enabled agricultural commodities grown in distant locations to support urban centers, resulting in agricultural abandonment in proximity to these centers.

A recent assessment identified regions of the world that underwent the most rapid land-cover changes in the last 20 years. The study found that the rapid land-cover changes that have been observed are not randomly or uniformly distributed but clustered in particular locations – for example, along the edges of forests and along roads. More spatially-diffuse changes involving land-cover modifications (i.e., subtle changes that affect the character of the land cover without changing its overall classification; e.g., forest degradation from logging) have been more difficult to observe than land-cover conversions (i.e., the complete replacement of one cover type by another). The study also found that different processes of land-cover change have taken place in different parts of the world over the last two decades; for example, a decrease in cropland area in temperate regions and an increase in the tropics.

The advances in synthesizing land-cover change studies to a global scale also revealed significant gaps in our knowledge. While the development of remote sensing techniques have allowed us to quantify land-cover change more accurately, most of the recent studies of this change have focussed on the more-easily observed process of deforestation in the humid tropics. Sig-

nificant uncertainties persist about the rates and patterns of change in the dry tropical forests; about forest-change dynamics due to fires, logging, and insect damage; alteration of wetlands; soil erosion; dryland degradation; expansion of built-up areas; and lifestyle-driven changes. Even with advances in remote sensing, significant challenges remain including the need to monitor change with sufficient spatial and temporal resolution, the balancing of expensive wall-to-wall analysis against the cheaper but potentially more inaccurate sampling schemes, and the inability to compare across studies using different land-cover classification systems.

The lessons learned from the LUCC global land-cover change synthesis activities suggest future directions for research. It is important to use historical and paleo-data to understand past changes in order to place current changes in the proper perspective. For example, many ecosystems today are still responding to (recovering from) past land-cover change (e.g., regrowing forests of the eastern United States of America). A long-term perspective will also help understand non-linear behavior (i.e., hysteresis or that exhibiting thresholds) resulting from the interaction between fast and slowly changing components of the same system (e.g., climate variability interacting with shifts in vegetation distribution and changes in soil properties). Moreover, a long-term perspective allows the study of land-use change as a process undergoing predictable transitions with economic development (see Sect. 2.5.3).

Future monitoring methods must also maintain a global perspective, even if particular attention is paid to hot spots of land-cover change, lest some important processes are missed in non-hot spot regions. Another priority is to move beyond a focus on humid forest systems, to study changes in agricultural lands, drylands, built-up areas, fires, and wetlands, among others. The use of coarse-resolution satellite data to identify locations of rapid change, and then high-resolution data within those hot spots to estimate rates and patterns of change accurately seems like a promising idea. However, the challenge of connecting global land-cover change estimates to narratives about land-use change from particular places and times remains unresolved and needs significant thought.

Land-change monitoring should also aim to characterize the full land-cover dynamics or land-use transitions, and these dynamics and transitions will require superior detail in the land-cover classes analyzed. The integration of remote sensing data with other sources of information including, for example, household surveys, census data, land economic surveys, is again a promising direction for future research (see Chap. 8).

Finally, more attention needs to be paid to integrating stakeholders in monitoring systems, in order to improve the relevance, credibility, and legitimacy of monitoring.

Chapter 3

Causes and Trajectories of Land-Use/Cover Change

Helmut Geist · William McConnell · Eric F. Lambin · Emilio Moran · Diogenes Alves · Thomas Rudel

3.1 Introduction

One of the key activities of the Land-Use/Cover Change (LUCC) project has been to stimulate the syntheses of knowledge of land-use/cover change processes, and in particular to advance understanding of the causes of land change (see Chap. 1). Such efforts have generally followed one of two approaches: broad scale cross-sectional analyses (cross-national statistical comparisons, mainly); and detailed case studies at the local scale. The LUCC project applied a middle path that combines the richness of in-depth case studies with the power of generalization gained from larger samples, thus drawing upon the strengths of both approaches. In particular, systematic comparative analyses of published case studies on land-use dynamics have helped to improve our knowledge about causes of land-use change. Principally, two methods exist for comparative analyses of case studies. These methods are sufficiently broad geographically to support generalization, but at a scale fine enough to capture complexity and variability across space and time.

A first method is to organize *a priori* a set of standardized case studies, wherein a common set of variables is collected at a representative sample of locales, according to common protocols that can support inferential statistical modeling. These case studies are required to use a common structure and address a pre-set collection of factors or hypothesized causal mechanisms. This approach has been successfully applied to land-change questions aimed at exploring the relationship between population growth and agricultural change (Turner et al. 1993a), identifying regions at risk of environmental change (Kasperson et al. 1995), testing the relationship between population and urban as well as rural land-use dynamics (Tri-Academy Panel 2001), and examining broad types of forest ecosystems for their relationship with institutional arrangements, mainly (Turner et al. 2004; Sader et al. 2004; Moran and Ostrom 2005). Although comparative research has been widely touted as an important goal of research (Ragin 1987; Moran 1995), there are just a handful of synthesis efforts involving the systematic collection of data *in situ* across a variety of national boundaries using common data protocols, mainly because it

requires a large investment to coordinate comparative research. Actually, there is no widely accepted protocol for carrying out field studies about land-use/cover change, despite long-standing calls for standardization. Researchers opposing standardization argue that each study site is unique and that results therefore cannot be extrapolated. Some view the human-environment processes under study as simply being too complex to support robust generalization. For example, some authors assert that desertification owes to multiple causative factors that are specific to each locality and time period, revealing no distinct patterns (e.g., Warren 2002; Dregne 2002). Likewise, proponents of complexity state that correlations between tropical deforestation and multiple causative factors are contextual, many and varied, again not exhibiting any distinct pattern (e.g., Bawa and Dayanandan 1997). Undeniably context matters, yet a systematic comparison of multiple case studies often reveals a limited and recurrent set of variables associated with major land-change processes.

A second method is the *a posteriori* comparison of case studies already published in the literature (Cook et al. 1992; Matarazzo and Nijkamp 1997), preferably at the sub-national scale. It can illuminate the factors that have been found important in case studies from different parts of the world but that share the same outcome (e.g., deforestation, agricultural intensification, desertification). It also identifies how theses factors have been studied at different times, in different regions, and from the perspective of different disciplines. This provides key information for the design of future research that will be even more amenable to comparative analysis (e.g., Guo and Gifford 2002; Parmesan and Yohe 2003; Root et al. 2003; Nijkamp et al. 2004; Misselhorn 2005). The bulk of the findings presented in this chapter are synthesized from three recent meta-analyses drawing upon case studies published in peer-reviewed literature, including reviews of tropical deforestation case studies (Geist and Lambin 2001, 2002), cases of dryland degradation (Geist and Lambin 2004; Geist 2005) and a review of agricultural intensification (McConnell and Keys 2005; Keys and McConnell 2005). Other comparative studies dealt with forest-cover change (Unruh et al. 2005), agricultural change (Wiggins 2000), and urbanization (Seto et al.

2004; Elvidge et al. 2004). All these studies produced insights into the causes of land-use change and their mode of interaction (Rudel and Roper 1996; Angelsen and Kaimowitz 1999; Petschel-Held et al. 1999). In total, the first three meta-analyses concerned approximately 400 cases at the sub-national scale, mainly in the tropics. In order to ensure a basic standard of quality, the cases were identified primarily from the Web of Science of the Institute for Scientific Information (ISI); in the case of agricultural intensification, some supplementary cases were drawn from other indexes such as JSTOR (http://www.jstor.org/, subscription required) and AGRICOLA (http://agricola.nal.usda.gov/), and books. Each of the cases were coded into databases recording the presence in the case study of each of a suite of social and biophysical factors found to be associated with the outcome of interest (e.g., deforestation). These databases were then analyzed to detect patterns of co-occurrence of causal and contextual conditions, using multiple cross-tabulation (Geist 2006a).

These meta-analyses have identified sets of underlying causes of land-use/cover change at a time scale of around 300 years from now (see Chap. 2). They are detailed in the following, for changes in tropical forests, drylands, areas of intensive agricultural production, and urban zones worldwide. They are clustered in terms of biophysical (Sect. 3.3.1), economic and technological (Sect. 3.3.2), demographic (Sect. 3.3.3), institutional (Sect. 3.3.4) and cultural factors (Sect. 3.3.5). These various groups of drivers are strongly interlinked across two or several levels of organization of human-environment systems. They were found to interact directly via feedbacks, and thus often have synergetic effects (Lambin et al. 2003; Steffen et al. 2004). It has also been argued that the many processes of globalization cross-cut the local and national pathways of land-use/cover change, amplifying or attenuating the driving forces by removing regional barriers, weakening national connections, and increasing the interdependency among people and nations (Lambin et al. 2001, 2002). Likewise, an integration of diverse, causal factors across temporal and spatial scales has been promoted by the concept of land-use transition or, more narrowly, forest transition (Mather et al. 1998, 1999; Rudel et al. 2000, 2002b; Mather 2004; Rudel et al. 2005).

Other important concepts are those of pathways or trajectories of land-use change, also referred to as spirals or "syndromes" (Moran et al. 2002; McCracken et al. 2002; Lambin et al. 2003; Mustard et al. 2004; Geist et al. 2006). Over the last decade, both place-based research and comparative analyses of land-use change studies identified some dominant pathways leading to specific outcomes. They are presented in this chapter as typical successions or dominant "stories" of causes and events of, for example, tropical deforestation. They vary substantially between major geographical entities and over time. Finally, from summarizing a large number of case studies, an attempt is presented to arrive at a limited number of fundamental, high-level causes of land-use/cover change (Lambin et al. 2003).

3.2   Explaining Land Dynamics

There are two fundamental steps in any study of land change, i.e., detecting change in the landscape, and ascribing that change to some set of causal factors. Establishing the change in the dependent variable is by no means simple, but advances in the acquisition, processing and interpretation of remotely sensed imagery over the past decade have made it much easier (see Chap. 2). This task pales in comparison, however, to that of explaining the observed change, i.e., identifying and assigning causal power to candidate factors. The research approach of detecting change in land cover and elaborating the causal and contextual factors responsible for that change bears little resemblance to classical experimentation, as understood and practiced in many other realms of global change research. Some study designs, however, may pretend to quasi-experimentation, for example in the case of so-called "natural experiments". In natural experiments, one identifies real-world situations that allow controlling for as many potential causal factors as possible, while looking for variation in one key factor that distinguishes the cases from one another. Transboundary situations, for example, permit comparative analysis of the implications for land use of contrasting macro-economic policies or land-tenure systems. In order to properly address the causes of land dynamics, it is important, first, to be clear on the distinction between land cover and land use (see Chap. 1), and, second, to broadly distinguish between proximate *versus* underlying causes.

3.2.1   Proximate Versus Underlying Causes

Identifying the causative factors requires an understanding of how people make land-use decisions and how various factors (including the biophysical setting and changes therein) interact in specific contexts at the local, regional, or global scale to influence land-use decision-making. The links between human activities and land-use/cover change, as adopted by the LUCC project, have been conceptualized by Turner et al. (1993a), and Ojima et al. (1994), among others – see Fig. 3.1. An important distinction is between proximate and underlying causes of land-use change (Turner et al. 1993a, 1996; Lambin et al. 2001). This framework has been widely applied (e.g., Nielsen and Zöbisch 2001; Xu and Wilkes 2004; Geist 2005; Misselhorn 2005).

Fig. 3.1.
Links between human activities
and land use and land cover.
Source: Ojima et al. (1994)

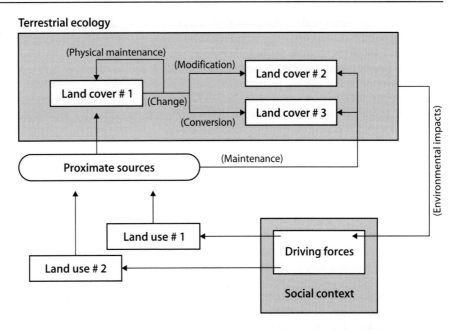

Terrestrial ecology

Land use is the sum of the proximate causes of land-cover change, i.e., human activities or immediate actions that originate from the intended manipulation of land cover (see Chap. 1). Proximate (or direct) causes involve a physical action on land cover and are usually limited to a recurrent set of activities such as agriculture (or agricultural expansion), forestry (or wood extraction), and infrastructure construction (or the extension of built-up structure). Proximate causes generally operate at the local level, for example, of individual farms, households or communities (Lambin et al. 2003; Mather 2006a). These are considered "direct drivers" of ecosystem change, along with other proximate factors such as species introduction or removal (see Chap. 4).

Underlying (or root, or indirect) causes are fundamental forces that underpin the more proximate circumstances. They operate more diffusely (i.e., from a distance), often by altering one or more proximate causes. Underlying causes are formed by a complex of social, political, economic, demographic, technological, cultural and biophysical variables (Brookfield 1999) that constitute structural (or systemic) conditions in human-environment relations. In contrast to proximate causes, underlying driving forces may originate from the regional (districts, provinces, or national), or even global levels, with complex interactions among levels of organization (Mather 2006b). A limited set of about half a dozen broad fundamental forces or root causes is consistently used in global environmental change research, i.e., technological, economic, political, institutional, demographic, and (socio)cultural factors. At the global scale, these fundamental forces influence the level of production and consumption of ecosystem services and collectively control the trajectory of (non)sustainable land or resource use

(U.S. National Research Council 1999; Millennium Ecosystem Assessment 2003, 2005). Changes in any of these indirect drivers usually result in changes in one or more of the proximate factors, thus triggering land-use/cover changes. Especially in tropical zones, underlying causes are often exogenous to the local communities managing land and are thus difficult to control by these communities. Only some local-scale factors are endogenous to decision makers (Lambin et al. 2003).

In explaining land change, a web of factors thus needs to be considered that links the proximate and underlying levels (Kaimowitz and Angelsen 1998). Note that the proximate/underlying distinction depends on the spatial and temporal scales of analysis. Land-use decisions are made at a variety of scales (individual, household, community, nation and international environmental/trade agreements), and understanding is sought all the way from the very local to the global scale. Factors that appear quite distal and therefore exogenous for the purposes of a local case study (such as a government credit scheme) may be entirely endogenous to a national study aimed at assessing the effectiveness of that very policy.

One of the best examples where both the distinction between and the interaction among proximate and underlying causes can clearly be seen is tropical deforestation (Walker 2004). Based on the works of Ledec (1985), Angelsen and Kaimowitz (1999) and Contreras-Hermosilla (2000), among others, a meta-analytical framework was applied to identify the broad categories of proximate causes and underlying driving forces which were further subdivided into specific variables as found in a wide array of case studies from various regions of the world. At the level of proximate causation, the broad category of agricultural expansion, for example, falls into cropping and livestock activities with further subdivisions such as shifting cultivation and sedentary cropping, to be further subdivided into large-scale *versus* smallholder farming. Likewise, the broad category of wood extraction falls into commercial timber logging, fuelwood and polewood extraction for domestic uses, and charcoal production, with further subdivisions possible between clear-cutting, selective logging, state-run *versus* private company activities, etc. (Geist and Lambin 2001, 2002). At the level of underlying causation, the most prominent causal clusters are made up of economic factors, institutions, and national policies, with subsequent subdivisions (Geist and Lambin 2001, 2002).

3.2.2 The Context of Land Change: Slow Versus Fast, and the Role of Mediating Factors

It is useful to recognize that both anthropogenic and biophysical processes can be gradual, slow-moving and/or delayed, with long turnover times (e.g., the domestication of wild plants, tectonic forces), or they can work quite rapidly and be immediate, as trigger forces of land-cover change (e.g., violent conflict leading to mass movement of people, extreme weather events). Slow-intervening factors with long turnover times usually determine the boundaries of sustainability and collectively govern a land-use trajectory (such as the spread of salinity in irrigation schemes or declining infant mortality). However, fast variables or trigger events drive land-use changes as well. Generally, land-use dynamics are driven by a combination of factors or processes that work gradually and factors that happen intermittently (Lambin et al. 2001; Stafford-Smith and Reynolds 2002). Also, a random element can be important in several land-change situations, as discussed for Sudano-Sahelian land-use systems by Reenberg (2001).

The interplay between underlying and proximate causes may be shaped or modified by a number of mediating factors. In particular, underlying factors do not operate individually; rather they are themselves shaped by other factors. For example, population increase in a given area – often considered an underlying cause of land change – may be amplified or modulated by existing or changing social norms, and by fertility or resettlement programs, which may in turn be influenced by changes in knowledge and policy at national and international levels. It is helpful to recognize that some factors concern the motivation to change behavior, while others function in contextual ways, often filtering the effects of other factors (Turner 1989; Moran 2005).

A mediating factor – sometimes also labeled intermediate, filter or context variable – constitutes a biophysical or socio-economic causative factor which shapes, modifies or intervenes into the interplay between under-

lying driving forces and proximate causes. Often cited examples of mediating factors are gender, ethnic affiliation, class or wealth status (and thus power relations), and institutional arrangements regulating the access to land (e.g., privately-held, communally-held, and federal- and state-held forests), but also include biophysical properties (Turner 1989; Agrawal and Yadama 1997; Young 2002a, 2003; Tole 2004; Moran 2005). Researchers have found that demographic and economic factors in particular do not work in an unmediated fashion. For example, in the Mayan zone of the Yucatán peninsula, the presence of male population increases the probability of deforestation in a statistically significant manner, while the presence of female population decreases the same probability (Sader et al. 2004). This begs the question of the effects of mediating sociocultural and institutional factors.

Biophysical factors conditioning land use include the properties of the landscape – its soils, terrain, climate, hydrology, as well as native flora and fauna, and location relative to human settlement, thus contributing to various degrees to land quality, in particular the condition of land relative to the requirements of a given land use (Pieri et al. 1995; Stone 1996; Dumanski and Pieri 2000). While these factors generally constitute the context within which land use takes place in the sense of initial conditions (or pre-disposing environmental factors), their dynamics – soil degradation and aridification, for example – can assume causal power. Thus, dramatic biophysical changes, such as increased aridity or drought, may be considered proximate causes, while they may be seen as contextual factors when operating gradually, shaping both natural (potential) land cover as well as land-use dynamics (Brookfield 1999). Similar is the dual character or role of institutions in causing and/or mediating land-use/cover change (Young 2002a, 2003).

At the proximate level, the conversion of tropical forests into agricultural uses, for example, is often found to be mediated by the unequal relations between large-scale farmers or corporate agricultural enterprises and smallholders eking out a living, thus creating "entrepreneurial" *versus* "populist" agricultural frontiers with rather distinct land uses (Turner 1920; Schneider 1995; Walker et al. 2000; Pacheco 2006a,b). Likewise, all categories of what has been called "agrodiversity" (Brookfield 2001) in settled agricultural zones – i.e., biophysical diversity, management diversity, agro-biodiversity, and organizational diversity – are shaped by factors that play out differently at various time and spatial scales. For example, crop choice and type of conservation practices often differ between poor and rich farmers, thus affecting the pattern of management diversity, and feeding back to enlarge differences in natural land quality (Brookfield et al. 2003; Xu and Mikesell 2003).

At the underlying level, mediating factors may severely alter the impact of similar demographic forces, shaping the trajectory of land change towards degradation or restoration. Whether or not increasing population is damaging or beneficial, for example, depends upon a variety of institutional, ecological, or technological factors. This implies that population growth can cause land degradation in the short term, but it can also spur innovation and agricultural intensification as well as the adoption of conservation techniques (Boserup 1965, 1975, 1981, 2002; Mortimore 1993a; Mortimore and Tiffen 1994; Tiffen and Mortimore 1994; Tiffen et al. 1994a).

Furthermore, mediating factors are crucial for the response of land managers to external forces, i.e., feedbacks are strongly mediated by local factors such as access to land, gender, education and institutional arrangements. In particular, institutions need to be considered at various scales to identify those local mediating factors that, together with peoples' adaptive strategies or responses to changing market opportunities, shape land-use change (Agrawal and Yadama 1997). Local participation in natural resource conservation, for example, is strongly mediated by a community's interactions with non-local actors such as national governments, transnational corporations, and international non-governmental organizations (Sundberg 2003). Seen together with other examples, these "conservation encounters" can shape landscapes and livelihoods in rather contradictory ways. In the Mayan zone, for example, local evidence of high deforestation can be found close to locations where exceptionally low rates of deforestation have occurred, with intervening institutional factors making the difference (Bray et al. 2004; Klepeis and Chowdhury 2004).

It is important to understand that, as land use is conditioned on the biophysical and social milieu, its effects cascade through the human-environment system, altering that milieu, and thereby changing the perception by land managers of the conditions for future land-use decisions. Thus, neither the social nor biophysical contexts are static. Quite the contrary, they reflexively shape, and are shaped by the collective actions of land managers (Lambin et al. 2003; Steffen et al. 2004). Therefore, a significant obstacle to the synopsis presented below is that factors that are crucially important in explaining change in one place may be irrelevant in other nearby places, and therefore not mentioned in a study of that other place. By the same token, a given factor (e.g., improved market access) may be implicated in opposite land-cover outcomes (e.g., increase/decrease in woody biomass). This happens for two reasons. First such factors are never identical from one instance to another (e.g., the particular incentives provided by a market to which access has been improved). In addition, even when the factor in question is quite similar, its effects will depend on the biophysical and socio-cultural context within which each land manager experiences it.

3.3 Synopsis of Broad Factors Affecting Land Change

3.3.1 Biophysical Factors

General Remarks

Biophysical factors – whether gradual processes, trigger events or filter variables – define the natural capacity or predisposing environmental conditions for land-use change, with the set of abiotic and biotic factors – climate, soils, lithology, topography, relief, hydrology, and vegetation – varying among localities and regions and across time (Lambin et al. 2001). The variability in biophysical factors and natural environmental changes interact with the human causes of land change. For example, biophysical limitations such as steep slopes and difficulty of access can provide considerable but not necessarily sufficient protection for a forest. From a wide array of case studies, it appears that institutional factors (see Sect. 3.3.4), in combination with biophysical limitations, play a major role in protecting limited forest areas from deforestation and erosion (Moran 2005).

Highly variable ecosystem conditions driven by climatic variations amplify the pressures arising from high demands on land resources, especially under dry to subhumid climatic conditions, whereas the role of climatic influences, for example, in temperate and humid zones is less pronounced (Lambin et al. 2003). Natural and socioeconomic changes may operate as synchronous but independent events. In the Iberian Peninsula during the 16th and 17th centuries, for example, the peak of the Little Ice Age occurred almost simultaneously with large-scale clearing for cultivated land following the consolidation of Christian rule over the region, which triggered changes in surface hydrology and significant soil erosion (Puigdefábregas 1998). In part because of human activities, the Earth's climate system has changed since the re-industrialize era, and is projected to continue to change throughout the 21st century – in terms of warmer temperatures and spatial and temporal changes in precipitation patterns, among others (see Chap. 4).

Natural variability may also lead to socioeconomic unsustainability, for example when unusually wet conditions alter the perception of drought risks and generate overstocking on rangelands. When drier conditions return, the livestock management practices are ill adapted and cause land degradation. This overstocking happened several times in Australia and, in the 1970s, in the African Sahel (Geist 2005). Land-use change, such as cropland expansion in drylands, may also increase the vulnerability of human-environment systems to climatic fluctuations and thereby trigger land degradation (Okin 2002).

Forest Change

In tropical forest zones, land characteristics or features of the biophysical environment – e.g., soil quality, low lying zones, flat and gently sloping areas, high density of marketable woods, and closeness to water – were found to be among causative factors of deforestation (in 14% of the cases reviewed) – see Table 3.1. Soil-related features clearly dominated in cases of forest-pasture conversion in Latin America (less so in cases of forest-cropland conversion) (Geist and Lambin 2001). However, this phenomenon can be related to both forests on fertile soils located on flat ground (i.e., most of the soil-related cases) and to forests on poor soils (i.e., in some of the cases), since meager soil endowment sets the context for accelerated clearing to put more land into cultivation (Hecht 1993). In addition, biophysical triggers – such as soil fertility collapse, drought, weed intrusion and forest fires – appeared in 18% of the deforestation cases (Geist and Lambin 2001). The impact of mostly natural fires on land cover in boreal regions has been well documented (Kasischke et al. 2002), mainly using remote sensing data (which is also true for the mostly anthropogenic fires in tropical regions; e.g., Pereira et al. 1999). In contrast to drylands, where increased aridity is a widespread factor in desertification, drought-induced forest fires are important so far only in the Amazon Basin or Indonesia. In Indonesia, for example, periodic El Niño-related droughts in the late 1990s lead to an increase in the forest's susceptibility to fires, with accidental fires becoming more likely under such conditions, leading to the devastation of large tracts of forests (Siegert et al. 2001). Forests that have been affected by forest fragmentation, selective logging, or a first fire subsequently become even more vulnerable to fires as these factors interact synergistically with drought (Siegert et al. 2001; Cochrane 2001; Csiszar et al. 2004). In general, fires as causative factors of land-use/cover change result from a combination of climatic factors (which determine fuel availability, fuel flammability, and ignition by lightning), and factors related to land-use/cover change that control fire propagation in the landscape and human ignition (Lavorel et al. 2005).

Dryland Change

In dryland zones of the world, soil conditions constitute key criteria in assessing the presence and severity of land degradation there, and in particular climatic factors are of overriding importance (in 86% of the cases reviewed) (Geist and Lambin 2004) – see Table 3.2. As underlying driving forces leading to increased aridity at the proximate level, climate factors can affect land cover in the form of prolonged droughts (Nicholson et al. 1998). Likewise, rainfall trends at meteorological stations in north-

Table 3.1.
Biophysical factors associated with forest conversion and modification in the tropics

	All	Asia	Africa	Latin America
Land characteristics (biophysical features)				
Soil quality	12	1	0	11
Slope, topography	7	0	2	5
Watercourse/water body	6	0	0	6
Vegetation	4	1	0	3
Biophysical triggers				
Soil-related	7	4	2	1
Water, climate	10	3	2	5
Vegetation-related	18	9	4	5

Total number of cases is 152; multiple counts possible. *Source:* Geist and Lambin (2001), p. 31.

Table 3.2.
Biophysical factors causing land degradation in drylands

	All	Asia	Africa	Latin America
Indirect impact (climatic variability)				
Higher rainfall deficit	37	3	24	1
Warmer, drier	34	23	5	0
Direct impact (on land cover/surface vegetation)				
Changes in fire regime	16	0	3	0
More oscillations of dry/wet conditions	12	0	0	6
Simple occurrence of droughts	42	17	20	2

Total number of cases is 132; multiple counts possible; not included are data for Europe, Australia and North America. *Source:* Geist (2005), p. 100.

western Senegal show a negative slope until the 1990s which is in congruence with all data from the West African Sahel (Gonzalez 2001). Climatic factors also operate indirectly, through changes in land use resulting from variation in rainfall (Nicholson 2002); rainfall changes at the landscape level, for example, can trigger significant shifts in soil type priorities (Reenberg 1994; Reenberg et al. 1998). Estimating from a wide array of case studies, the most widespread mode of causation by biophysical factors in drylands is reported to be climatic conditions operating concomitantly or synergistically with socio-economic driving forces such as technological changes.

Cropland Change

In zones of intensified agricultural production, biophysical factors figure prominently, namely precipitation, topography, presence and proximity of water bodies, and soil conditions (in almost 40% of the cases reviewed) (McConnell and Keys 2005) – see Table 3.3. Frequently, soil factors – mainly declining fertility, but also erosion – affected the specific location of different agricultural practices, for example when farmers adopt new practices to exploit micro-environments (e.g., bottom lands) (Kasfir 1993), often as a result of a change in access to land. Likewise, it has been shown that soil erosion on the Greek island of Lesvos was an important factor in the abandonment and reallocation of cereals in intense, mechanized agricultural systems during the 1886–1996 period (Bakker et al. 2005). Alternatively, choices about which lands to continue cultivating and which to let revert to forest regrowth have been observed to change over time, as settler communities learn about local environmental conditions in agricultural frontiers (Moran et al. 2002). Climatic factors, primarily changes in precipitation, were found in just over a quarter of the cases of agricultural intensification in croplands (McConnell and Keys 2005; Keys and McConnell 2005).

3.3.2 Economic and Technological Factors

General Remarks

Economic factors appear to play a strong role. This should not come as a surprise since global economic activity increased nearly sevenfold between 1950 and 2000 (while global population doubled in roughly the past 40 years), thus increasing the demand for many ecosystem goods and services (Millennium Ecosystem Assessment 2005). Available case studies highlight that, at the timescale of a couple of decades or less, land-use changes mostly result from individual and social responses to economic conditions, which are mediated by institutional factors (Agrawal and Yadama 1997; Lambin et al. 2001). Opportunities and constraints for new land uses are created by markets and policies and are increasingly influenced by global factors (see Sect. 3.4.3).

Economic factors (and related policies) encompass a number of distinct processes that require individual treatment. They define a range of variables that have a direct impact on the decision making by land managers, e.g., input and output prices, taxes, subsidies, production and transportation costs, capital flows and investments, credit access, trade, and technology (Barbier 1997). In particular, taxes and subsidies are important driving forces of land-use dynamics and related land cover and ecosystem changes. Currently, many subsidies substantially increase rates of resource consumption and negative externalities. It has been estimated that currently about 2 000 billion U.S.$ are spent in the form of "perverse subsidies" (Myers and Kent 2001) each year, which equals the annual income of the most impoverished 1.3 billion people on Earth (including agriculture, especially irrigation farming, and forestry, but also fishery, transport, and energy production). The 2001–2003 average subsidies, for example, paid to the agricultural sectors of member states of the Organization for Economic Cooperation and Development (OECD) were over U.S.$324 billion annually, encouraging greater food production and associated water consumption and nutrient and pesticide release (see Chap. 4). At the same time, many developing countries also have significant agricultural production subsidies. On the other hand, fertilizer taxes or taxes on excess nutrients, for example, provide an incentive to increase the efficiency of the use of fertilizer applied to crops and thereby reduce negative externalities (Millennium Ecosystem Assessment 2005).

Consumption ranks high among economic factors (Myers 1997; Kates 2000). The market demand for forest products and for agricultural output, including livestock-based products, not only encompasses basic needs (i.e., food crops for human and animal diets, fiber crops for clothing, timber for shelter), but also derived or relative needs (Keynes 1936; Maslow 1943) which go beyond the

Table 3.3.
Biophysical factors associated with land-use dynamics in croplands

	Important	Not important	Absent
Precipitation variation	30	30	48
Watercourse/water body	23	35	50
Soil properties	43	27	38

Total number of cases is 108; multiple counts possible. *Source:* McConnell and Keys (2005), p. 334.

immediate satisfaction of fundamental livelihood requirements (e.g., exotic tropical timber and fruits, counterseasonal fresh agricultural produce) (see Sect. 3.3.5). Increasing demand affects both the expansion of cropland and pastures into forests (e.g., cattle, soya) and drylands (e.g., cotton, rice, vegetables), as well as various forms of intensification of existing farmland, including the planting of trees (e.g., coffee, fruit trees) (Geist and Lambin 2002; McConnell and Keys 2005; Geist 2005).

Available case studies highlight that the effects of local consumption on land-use patterns often is decreasingly important relative to external consumption (Tri-Academy Panel 2001). Be it in core agricultural lands or at tropical forest and dryland margins, much of the demand originates from nearby urban areas as well as from very distant (global) markets (McConnell and Keys 2005; Geist et al. 2006). Market demand exerts a "pull" on rural producers to engage in land-use practices beyond subsistence production. The possibility (or necessity) of purchasing goods or services constitutes a "push" factor. Of course, very few people are completely disengaged from markets, and even before subsistence demands are satisfied, rural producers are often prompted to commercialize at least some portion of their production. As they gain access to a wider range of products and services, and to information about lifestyles in other parts of their country, or the world, consumer aspirations rise (see Sect. 3.3.5). At the same time, government policies contribute to push factors as well, with market access remaining largely conditioned by state investments in transportation and other infrastructure (in fact there are few, if any, market factors that are free of the influence of the state).

Related to this demand-driven pattern are two global observations. First, subsistence croplands are decreasing in extent, while land under crops for markets is increasing, with a parallel increase in agricultural intensity, strongly driven by agro-technological measures of the Green Revolution since about the 1960s (see Chap. 2). And, second, local consumption has changed in response, with a shift in diets from traditional grains or starchy staples (such as rice, wheat, and potatoes) to diets including more fat (such as meat, dairy products, and fish) but also more fruits and vegetables (Tri-Academy Panel 2001; Millennium Ecosystem Assessment 2005). The former is true even for world regions with strong religious taboos on nonvegetarian food – see Box 3.1.

Forest Change

In tropical forest zones, economic and technological factors are prominent underlying driving forces, found in a preponderance of the cases reviewed (Geist and Lambin 2001, 2002) – see Table 3.4. Among the economic factors, commercialization and the growth of national and international timber markets as well as market failures are frequently reported to drive deforestation. Economic variables such as low domestic costs (for land, labor, fuel, or timber), product price increases (mostly for cash crops), and the demands of remote urban-industrial centers underpin about one-third of the cases, whereas the requirement to generate foreign exchange earnings at a national level intervenes in a quarter of the cases. With few exceptions, factors related to economic development through a growing cash economy show little regional variation and, thus, constitute a robust underlying force of deforestation. Likewise, technological factors such as agrotechnological change – i.e., land-use intensification

---

Box 3.1. Land-use change in Haryana, India, during the 2nd half of the 20th century

Haryana, located in an arid to semi-arid environment in the northwestern part of India, comprises part of the wheat-growing "breadbasket" of the country, together with its northern neighbor state of Punjab. Major transformations of land cover into rice-wheat rotations – as in the rest of the Indo-Gangetic Plains – coincided with the introduction of Green Revolution technologies. Mainly initiated in the period 1967–1978, major aspects of the Green Revolution were the expansion of the cropland, the adoption of double-cropping systems (i.e., two crop seasons per year) and seeds that had been improved genetically (i.e., high-yielding varieties (HYVs) of wheat, rice, corn, and millet), and high inputs of fertilizer and water for irrigation. This attracted not only large numbers of migrants from other parts of India, but also made it possible that increased agricultural productivity kept pace with population growth, the rate of which is among the highest in India. During the state's high-growth period over the past three decades, the sectoral change in land use was rather small – i.e., the area devoted to croplands has remained fairly constant since 1971 at 81% of the total area (indicating that the potential for expansion of cultivation was already exhausted then) –, but farmers moved away from the production and consumption of traditional staple crops (such as maize, barley, gram, mil-

let/bajra and pulses) and modified their farming systems toward income-producing cash crops (such as rice, wheat, and cotton). State policies amplified this trend by favoring the semiarid Green Revolution areas with infrastructural projects at the expense of the more arid western parts of Haryana. For example, public priority is given to large-scale investments (expansion of canals and pumping of groundwater for irrigation), state subsidies are provided for electricity (tube wells), credits and marketing facilities, price policies stabilize output prices (wheat) and favor cotton as well as oilseeds, and most of the (wheat) production surplus is procured by government agencies for sale through public distribution system networks in India. With continuation of the price support system for wheat and rice throughout the 1970s and beyond, rice-wheat crop rotations became a lucrative proposition for the farmers, and Haryana continues to be an important supplier of food for the country. In terms of food consumption, however, the share of cereal grains declined significantly in rural as well as urban areas from 1972 to 1993: milk and dairy products have replaced cereal grains as the now most important component of food expenditure, also fulfilling a sizable portion of the demand for livestock products in Delhi and several other urban centers (Vashishta et al. 2001).

Table 3.4.
Economic and technological factors of forest conversion and modification in the tropics

	All	Asia	Africa	Latin America
Market growth, commercialization	103	30	15	58
Sectoral market growth	(78)	(23)	(13)	(42)
Demand/consumption	(69)	(24)	(13)	(32)
Market failures	52	22	6	24
Urban-industrial growth	58	23	5	30
Foreign exchange	38	16	5	17
Special variables (low cost, price change)	48	9	5	34
Agrotechnological change	70	28	8	34
Poor timber/wood extraction techniques	69	39	8	22

Total number of cases is 152; multiple counts possible. *Source:* Geist and Lambin (2002), p. 148.

Table 3.5.
Economic and technological factors causing land degradation in drylands

	All	Asia	Africa	Latin America
Market growth, commercialization	64	37	8	9
Economic depression, impoverishment	48	22	9	8
Innovative developments, introductions	73	35	12	7
Deficiencies of technical applications	67	31	17	3

Total number of cases is 132; multiple counts possible; not included are data for Europe, Australia and North America. *Source:* Geist and Lambin (2004), p. 823.

as well as agricultural expansion –, and poor technological applications in the wood sector (leading to wasteful logging practices) have no distinct impact regionally (Geist and Lambin 2002).

Perhaps most striking in the analysis of economic and technological factors in forest zones is their multiple and sometimes contradictory effects (see Sect. 3.5.3). The most general pattern of economic effects follows directly from differences in the local abundance of forest resources. Forest-rich regions like the Amazon, insular Southeast Asia, and central Africa become the focus for large-scale logging and agricultural expansion, driven in part by a desire to capitalize on the store of natural resource value on the land, so in these instances economic incentives accelerate deforestation. Forest-poor regions like South Asia and peri-urban places in East Africa see very different trends. In these places, increases in the prices of scarce forest products induce afforestation, and both smallholders and the state respond to economic incentives by planting trees where there were none (Rudel 2005; Unruh et al. 2005).

Improved agricultural technology – while providing secure land tenure and giving farmers better access to credit and markets –, can potentially encourage more deforestation rather than relieving pressure on the forests. The differing impact of agricultural development on forest conversion depends on how the new technologies affect the labor market and migration, whether the crops are sold locally or globally, how profitable farming is at the forest frontier, as well as depending on the capital and labor intensity of the new technologies (Angelsen and Kaimowitz 2001b).

Dryland Change

In dryland zones, economic and technological factors were prominent underlying driving forces in about two-thirds of the cases of land degradation (or desertification) reviewed (Geist and Lambin 2004) – see Table 3.5. Economic factors are reported to underlie desertification in the form of a mixture of "boom" and "bust" factors, though with considerable regional variations. Boom factors relate to market growth and commercialization, mainly entailing export-oriented market production, industrialization, and urbanization. Farmers respond to market signals reflecting high external demands for cotton, beef, and grain, with mostly native grassland increasingly put under rain-fed or irrigated production. Bust factors relate to the overuse of land because of land scarcity, low investments, low labor availability, indebtedness, lack of employment in the formal nonagrarian sector, or poverty (Geist et al. 2006). In dryland zones of Asia, cases of desertification are mainly driven by remote influences such as urbanization and commercialization. For example, among the many drivers of land change in various regions of Syria, most prominent are those which are the result of individual decisions made for economic opportunity, supported by state planning (Hole and Smith 2004) – see Box 3.2. In many cases from Australia and Latin America, local farmers' response to an unfavorable economic situation, coupled with cycles of low rainfall, is reported to underlie desertification: declining prices in the export-oriented sheep sector, for example, cause farmers to go into debt when their farms are no longer economically

---

Box 3.2. Economic factors of steppe conversion in Syria during the 2nd half of the 20th century

Since the founding of the Syrian state in 1946 at the end of the French mandate, the socialist government created a series of Five Year Plans for overall economic and social development. From the standpoint of land use, the most important of these was to increase agricultural productivity to accommodate a rapidly growing population and increasingly affluent society. These plans gave rise to land reform, creation of agricultural cooperatives, economic incentives and subsidies for production, the building of reservoir and canal systems, grain silos and a first class road system. While production has never met the ambitious goals set in these plans, most of the potentially productive steppe land has now been transformed (Hole and Smith 2004).

---

viable, inducing the overuse of scarce natural resources, especially during droughts (Geist 2005).

Different from the regional variations in economic factors driving desertification, most of the technological factors are pervasive driving forces. Most strikingly, technological innovations are reported to be associated with desertification (but also deficiencies of technological applications). Innovations mainly comprise improvements in land and water management through motor pumps and boreholes (at the village level) or through the construction of hydrotechnical installations such as dams, reservoirs, canals, collectors, and artificial drainage networks (for large-scale irrigation schemes) – see Boxes 3.1 and 3.2. When applied, these developments are often coupled with high water losses due to poor maintenance of the infrastructure, especially in the Asian studies. In addition, they induce fundamental and often irreversible changes to the natural hydrographic network, altering hydrological cycles in most cases. The disaster of the Aral Sea is an extreme case of such perturbation (see Sect. 4.7). Technological applications associated with desertification also include transport and earthmoving techniques (trucks, tractors, carterpillar-tracked vehicles) and new processing and storage facilities (refrigeration containers on ships and trucks). These innovations can trigger rapid increases in production at remote sites (e.g., greater numbers of irrigated garden products or herds of sheep, both destined for distant markets). It should be noted though that some research, especially in Asia, is devoted to technologies that might be used to stabilize the sand that is threatening expensive highway, railroad, and irrigation infrastructure. Thus, technology may also make it possible to mitigate some of the adverse impacts of desertification (Geist and Lambin 2004; Geist 2005).

Cropland Change

In zones of high-intensity agriculture, market demand was reported in the case studies more often than population as a causal variable (McConnell and Keys 2005) –

see Table 3.6. Improved market access, as a separate variable, was found to be important less frequently than market demand but showed regional variations and usually occurred concomitantly (cases where market access did not occur concomitantly with demand imply that there was improved access to a largely unchanged market in terms of demand). A related variable, standard of living, was important less often than market access but when present occurred almost always in conjunction with market access. A possible linkage was also discovered between market access and the availability of off-farm employment, which was judged important in less than one-third of all cases. Technological factors – such as agrotechnical change or the provision of water-related infrastructure – rank lower in core agricultural zones, implying that there was sufficiently developed infrastructure no longer triggering agricultural change (McConnell and Keys 2005).

In Sub-Saharan Africa, case studies from the 1970s to the 1990s, done at the village or district levels, confirm that demand from and access to a market is essential for agricultural development (which has been the single biggest idea in the policy reforms of the 1980s), but they also underline the importance of the detail of policy, i.e., in remedying failures in public investment in technology and in product, capital and insurance markets (Wiggins 2000). From an array of other, partly overlapping cases, explored by McConnell and Keys (2005), it could be seen that land-use intensification involved a change of cultivars and livestock without any explicit change in water management. Gains in productivity were seen to be coming from more frequent use of the land, that is, reduction in length of fallows. Land-use changes largely consisted of three dynamics. Farmers used farmland more frequently (decreasing fallow time); shifted from mainly consumption-oriented production of staple foods toward the adoption of cash crops like peanuts and cotton, and tree crops such as coffee, tea, palms, and vanilla; and switched from rain-fed production to small-scale irrigation, in the form of urban and kitchen gardening (Eder 1991; Mortimore 1993b; Drescher 1996). The adoption of high-yield varieties, particularly maize, was seen in several cases and resulted in increased output. Finally, changes in livestock practices, including replacement of grazing with cropping and intensive stabling (zero grazing), also were seen (Benjaminsen 2001; Bernard 1993; Carney 1993; Conelly and Chaiken 2001; Ford 1993; Goldman 1993; Gray and Kevane 2001; Kasfir 1993; Kull 1998; Laney 2002; Netting et al. 1993; Okoth-Ogendo and Oucho 1993; Tiffen et al. 1994b).

In Latin America, market-driven agricultural extension efforts were credited with the adoption of new crops, such as cocoa, rubber, coconut and improved pasture, as well as mechanical technology (e.g., tractors), credit (e.g., marketing cooperatives and soft loans), and infrastructure (e.g., roads and small-scale irrigation) (McConnell

Table 3.6.
Economic and technological
factors causing land dynamics
in croplands

	Important	Not important	Absent
Market access	58	18	32
Standard of living	48	32	28
Market demand	69	11	28
Off-farm employment	30	28	50
Water provision program	16	0	92
Infrastructure program	33	10	65

Total number of cases is 108; multiple counts possible. *Source:* McConnell and Keys (2005), p. 333.

and Keys 2005). Specific government policies included fines for leaving fields fallow, such as in Peru (Wiegers et al. 1999; Coomes et al. 2000), and nature conservation and import controls (Taussig 1978). Likewise, nongovernmental organizations were credited with the provision of capital and knowledge (e.g., in limiting erosion on hillsides and green manure application). Changes in labor input play an important role, with reference, in some cases, to increased labor requirements associated with aging fields, and, in more cases, to new labor-intensive tasks such as those associated with terracing (McConnell and Keys 2005). In particular, there appears to be an issue of labor bottlenecks created in the adoption of news crops (e.g., chilies), or green manure application (Keys 2004). These arrangements are particularly problematic as Latin America is known for a variety of new, more labor-intensive crops soon to be widely introduced in the region (e.g., soybeans) (Hecht 2005; Jepson 2005). They may however, be foregoing the opportunity cost of their home-based labor for the perception of a much greater income in other locales (Schelhas 1996).

In Asia, which has the longest record of continuous large-scale irrigated agriculture, water management is an intricate part of the process of land-use/cover change (Brown and Podolefsky 1976; Abrol et al. 2002; McConnell and Keys 2005). While increased frequency of cultivation appears to be as strong as in other tropical regions, changes in cultivars seem much less frequent than in Latin America. In particular, the adoption of high-yielding rice varieties has often been accompanied by increased use of chemical inputs, demonstrating a most dramatic input of green revolution technologies (Leaf 1987; Turner and Ali 1995; Vashishta et al. 2001) – see Box 3.1. Other crops mentioned in Asian cases of agricultural change include beans, cotton, okra, Job's tears, maize, manioc, millet, mustard, peanuts, sesame, soybeans, squash, sweet potatoes, and taro. Notably, the intensification of forest-product collection and the adoption of agroforestry practices have been rather high, including bananas, cashews, coconuts, coffee, pepper, and rubber. Asian farmers, generally having secure land tenure and access to markets, manage non-timber forest products like crops, i.e., they grow them in plantations or manage them intensively in forests, and the families –

usually not the poorest ones – specialize in a particular product and, indeed, get most of their income from it (Ruiz-Pérez et al. 2004). When information on market access and demand is present, access to nearby markets and changing urban market tastes has spurred notable changes in the types of crops farmed and the land-cover intensity of these crops (Leaf 1987; Eder 1991; Shidong et al. 2001b). Economic factors and related policies include direct agricultural policies such as import quotas, rice reserve requirements, and rice premiums, and the encouragement of soybean production, subsidies for market vegetables, and irrigation credits, but also non-governmental organization programs as well as broad national or government policies such as China's Open Door policy, or tax policies favoring (agro)industrialization, market intervention, and even tax policy favoring coconuts and rubber over rice (George and Chattopadhyah 2001; Shidong et al. 2001a).

Urban Change

In major urban or peri-urban zones, economic changes together with technological and also demographic changes (e.g., growth of urban aspirations and urban-rural population distribution) have led to a greater integration of rural and urban economies. Farmers within city boundaries or in peri-urban lands have, in particular, been intensifying land use on sites which are themselves often in demand for residential or industrial development, mainly through adjusting crop types to satisfy urban food demand (e.g., Eder 1991; Guyer and Lambin 1993; Kasfir 1993; Gumbo and Ndiripo 1996; Godoy et al. 1997; Alves 2002a). As an example of one of the above-mentioned remote influences, urbanization affects land change elsewhere through the transformation of urban-rural linkages. Urban commodity demands, and, especially, the impact of rapidly growing cities, have been triggering considerable land-use/cover change (Tri-Academy Panel 2001), also affecting ecosystems goods and services, or the flow of natural resources in urban zones and well beyond in remote hinterland or watershed areas (Fox et al. 1995; Humphries 1998; Indrabudi et al. 1998; Mertens et al. 2000) (see Chap. 4).

Residential preferences for private houses in a "green" environment, and economic incentives provided by private land developers and/or the state to achieve this, drive the extension of peri-urban settlements primarily in but not limited to the developed world, fragmenting the landscapes of such large areas that various ecosystem processes are threatened. In turn, however, excessive urban sprawl (and, thus, ecosystem fragmentation) may be offset by urban-led demands for conservation and recreational land uses (Lambin et al. 2001). Economically and politically powerful urban consumers tend to be disconnected from the realities of resource production, largely inattentive to the impacts of their consumption on distant locales (Sack 1990, 1992; Heilig 1994) (see Sect. 3.3.5). For example, urban inhabitants within the Baltic Sea drainage depend on forest, agriculture, wetland, lake and marine systems that constitute an area about 1 000 times larger than that of the urban area proper (Folke et al. 1997) (see Chap. 2 and 4 for the related notion of ecological footprint).

In China, and to a lesser degree in some other developing or newly industrializing countries, urbanization usually outstrips all other uses for land adjacent to the city, including prime croplands (Shidong et al. 2001b; Seto et al. 2004). In many cases, prior occupants such as farmers or herders have been displaced into marginal dry land sites, resulting in land degradation (Geist 2005). However, cities also attract a significant proportion of the rural population by way of permanent or circulatory migration, and, given the fact that many new urban dwellers in developing countries still own rural landholdings, urban remittances to the countryside have contributed to economic growth and landscape changes in both close and distant regions (Browder and Godfrey 1997; Lambin et al. 2001). These changes often run counter to the effects of remote urban consumers in that urban remittances have relieved pressures on local natural resources. It has been shown, for example, that in a small island of Micronesia, international migration, foreign aid, and monetary remittances from family members living overseas in urban agglomerations have removed the pressures of economic crowding on mangrove forests, despite an increase in population and a decline in local government jobs (Naylor et al. 2002). Likewise, some regions in the tropics currently show signs of signification reforestation which can at least partly be traced back to urban remittances (Rudel et al. 2000). Perhaps most importantly, this urbanization changes ways of life fundamentally, associated with demographic transitions, increasing expectation about consumption and potentially a weakened understanding of production-consumption relationships which has so far been mainly noted in the developed world (Lambin et al. 2001).

For thousands of years, China was mainly rural but is becoming increasingly urban, with land-use changes there dominated by an urban transformation unprecedented in human history (nearly one quarter of the 488 major urban centers in the world are located in China; see Chap. 2). In the Pearl River Delta, which is one of the most economically vibrant regions in China, nearly all land-use changes can be attributed to an array of economic factors associated with remarkable growth and linked to respective policies supporting economic growth (as well as population mobility). For example, the establishment of three special economic zones (SEZs) in the 1980s (Shantou, Shenzen, and Zhuhai), and the formation of the Pearl River Delta (PRD) Economic Open Region in 1985, helped the area to attract foreign investment and transform itself into an export-oriented region. As a consequence, entrepreneurs from Hong Kong – due to geographic proximity and cultural ties – moved their operations into the area (accounting for almost 75% of foreign direct investments in 1996). Their overseas ventures have exerted a considerable impact on the pace and structure of economic and urban development in the PRD due to large investment flows, access to technological innovations, and managerial acumen (Seto et al. 2004).

Industrial Change

Industrialization – i.e., the transition, made possible by large-scale technological changes (coal, steam power, electrification), from agricultural society to an economy based on large-scale, machine-assisted production of goods by a concentrated, usually urban labor force (Krausmann 2006) – has driven – or gone hand-in-hand with – urbanization since the middle of the 18th century. The process has been accompanied by a surge in labor productivity in both industry and agriculture with fundamental implications for land use, expressed in terms such as those of an agrarian or "agricultural revolution" in today's developed countries (Jeleček 1995, 2006), and a "green revolution" in today's newly industrializing, less or least developed countries especially during the 1960s and 1970s (Ewert 2006).

In forest zones of the tropics, for example, more than a quarter of deforestation cases reviewed reported the growth of wood- and mineral-related industries as an underlying driving force steering economic demands stemming from the build-up of basic, heavy steel and iron industries in today's newly industrializing countries (Geist and Lambin 2002). This had also been true for historic processes of industrialization in Europe and in the eastern United States of America (Williams 1994, 2003). In drylands of the world, especially in Asia, industrialization is one of the remote influences – together with urbanization and commercialization (i.e., export orientation, market competition) – which combines with local factors such as agricultural intensification and crop choices in favor of agricultural cash produce to drive land-use changes and perhaps even degradation (Geist 2005).

Through large-scale processes of spatial specialization and concentration of population and production, industrialization has affected practically every region of the world, especially after World War II. It constitutes a global and still ongoing process which exerts effects not only on the overall economic and social structure, but is also related to land use and major biophysical transformation processes. In particular, the linkage of agriculture with the agricultural industry (e.g., sugar, tobacco, distilling, milk, and brewing industries) and with agricultural engineering (biotechnology) introduced an industrial character into agriculture in terms of the global agro-industry (see Sect. 3.4.3). It is also considered, in conjunction with social, political, and demographic changes, to be the major factor behind forest transitions worldwide (see Sect. 3.5.3).

3.3.3 Demographic Factors

General Remarks

At least since the classic essay by Malthus (1798), population growth and the pressure it puts on land use (and agricultural practices, in particular) have been central to thinking about the human-environment condition. A general agreement has developed, however, that not the sheer number of people but aspects of population composition and distribution, namely changes in urbanization and in household size, have become the most important characteristics of population aspects, acknowledging the importance of indirect or consumptive demands on the land by an increasingly urbanized population (Lambin et al. 2001). Also, it has long been recognized (but frequently overlooked) that it is "population in context" (Rindfuss et al. 2004a) that matters (see Sect. 3.2.2), i.e., any effect of population change – be it fertility, mortality, in- or out-migration – likely interacts with other factors as diverse as social organization (e.g., networks, institutional arrangements), technology (e.g., level of agricultural yields), lifestyle (e.g., income, diet pattern) and consumption patterns (e.g., staple food *versus* non-food crops) (Ehrlich and Holdren 1971; Jolly and Torry 1993; Heilig 1994). Usually, there is a complex of factors that determines the direction and extent to which population growth will lead, for example, to forests being converted to cropland, or *vice versa* (Waggoner and Ausubel 2001). The expansion of forest land between 1935 and 1975 across the southeastern part of the United States of America, for example, related to urbanization, industrialization and increased agricultural yields elsewhere (Rudel 2001). With global population having doubled in the past 40 years and increased by 2 billion people in the last 25 years (reaching 6 billion in 2000), demographic variables, and in particular, population growth must be expected to play a major role in explanations of land

change (Millennium Ecosystem Assessment 2005). By and large, population growth rates in tropical countries have been – and continue to be – strongly positive, while European and North American populations approach stability or tend to be on a decline. However, there is an unprecedented diversity of demographic patterns across regions and countries, which does not allow for sweeping generalizations. For example, some high-income countries such as the United States of America are still experiencing high rates of population growth (mainly due to immigration), while some developing or newly industrializing countries such as China, Thailand, and North and South Korea have very low rates (Millennium Ecosystem Assessment 2005).

While population growth may underlie many land-cover changes (Bilsborrow and Okoth-Ogendo 1992; Cropper and Griffiths 1994), its effects are frequently manifest through migration (including temporary and/or circulatory migration) or displacement of groups of people, either spontaneously or with direct government support (Tri-Academy Panel 2001). At a given location under study, migration in its various forms clearly is the most important demographic factor causing land dynamics at timescales of a couple of decades (Geist and Lambin 2004; Angelsen and Kaimowitz 1999). Undeniably, high fertility in the areas of origin may be implicated, and it is also true that, once on the frontier, migrant families usually exhibit high fertility rates (Carr 2004). Nonetheless, migration operates as a significant factor with other nondemographic factors, such as government policies, changes in consumption patterns, economic integration, and globalization. Some policies resulting in land-use change either provoke, or are intricately linked with increased migration. From a wide array of case studies, some form of relocation was found in well over a third of deforestation cases (Geist and Lambin 2002), and in a quarter or more of desertification (Geist and Lambin 2004) and agricultural intensification cases (McConnell and Keys 2005). While spontaneous movements may often occur within a context of high density in the source region, in many instances specific triggers, such as drought, conflict, or major government (re)settlement programs were identified. Government programs to encourage settlement in the Brazilian Amazon (e.g., Moran 1981) and Indonesia (e.g., Fearnside 1997) are well-known, and other, smaller instances were seen where market demand and government incentives for the establishment of plantations also lead to relocation and subsequent land change in areas as different as Costa Rica (e.g., Schelhas 1996), Sumatra (e.g., Imbernon 1999a), and Zambia (Petit et al. 2001). In other cases, residents returning to a region after long absences initiated changes in local land use (e.g., Boyd 2001; Tiffen et al. 1994b). The creation of infrastructure, especially roads, is a crucial step in facilitating settlement and triggering land-use intensification in a region (e.g., Conelly 1992), and much road construction can be construed in this sense (see Sect. 3.3.4).

Thus, while population growth is clearly associated with a great deal of land change, there are always other factors that shape the expression of that growth: the location, timing and nature of the change, as well as who undertakes it, and who benefits from it. The treatment of demographic factors in land-change research is becoming increasingly sophisticated, and a population analysis of great nuance is required. For example, demographic factors go well beyond growth rates, density, or the shift from high to low rates of fertility and mortality (as suggested by the demographic transition) to include age and sex structure of the population, the characteristics of migration cohorts, and the demographic composition of households, among others (Moran and Brondizio 1998; Walker et al. 2000, 2002; Geist 2003a; Lambin 2003; Moran et al. 2003; Carr 2004). These life-cycle features arise from and affect rural as well as urban environments. They result from households' strategic responses to both economic opportunities (for example, market signals indicating higher crop profitability) and constraints (due to economic crisis conditions, for example). They shape the trajectory of land-use change, which itself affects the household's economic status. The longitudinal research of the Carolina Population Center in the United States and its partners, among others, is exemplary in its consideration of seasonal and permanent migration and the evolution of settlement patterns in shaping land trajectories in Nang Rong, Thailand (e.g., Entwisle et al. 1998; Rindfuss et al. 2003; Walsh et al. 2003).

Several concerted efforts have been undertaken to examine specifically the role of population growth in land-change processes. A set of commissioned case studies in high density areas of Africa, for example, was largely able to confirm the Boserupian hypothesis (Boserup 1965, 1975, 1981, 2002) linking population pressure on land to the transformation of agriculture (Turner et al. 1977, 1993a). Looking beyond tropical Africa, a set of case studies commissioned by a consortium of the National Academies of India, China and the United States of America described regions in those countries – i.e., Pearl River Delta and Jitai Basin in China, Kerala and Haryana Provinces in India, and southern Florida and Chicago in the United States of America – where agricultural production was increased without major detriment to the environment. This comparative analysis highlights the importance of economic and policy variables in shaping land-use practices, although initially it was assumed that population growth alone could be a significant driver of land-use change in many of the regions (Tri-Academy Panel 2001).

Another comparative study also wanted to address the role of population, seeking to examine a number of cases in three major types of forest ecosystems worldwide (i.e., temperate, tropical humid and tropical dry forests), and along a variety of institutional arrangements (i.e., privately held, communally held, and federal- and state-held forests), wherein could be tested the degree to which population density or its distribution is associated with loss of forest, or its recovery (Moran 2005). It has been found that the role of population not only varies by scale but is also often counterintuitive (Geist 2003a; Unruh et al. 2005), as in the case of the forest transition (see Sect. 3.5.3).

Forest Change

Case study evidence on land-use dynamics in forested tropical zones largely confirms the expectation that population plays a major, though complex role in the explanation of land change, with demographic factors implicated in almost two thirds of deforestation cases reviewed (Geist and Lambin 2001, 2002) – see Table 3.7. Among these factors, only in-migration of colonizing settlers into sparsely populated forest areas, with the consequence of increasing population density there, shows a notable influence on deforestation. This pattern tends to feature African and Latin American rather than Asian cases. While not denying a role of population growth in tropical deforestation (e.g., Allen and Barnes 1985; Amelung and Diehl 1992; Bilsborrow and Geores 1994; Pichón 1997a,b; Ehrhardt-Martinez 1998; Cropper et al. 1999; Carr 2005), most case studies fail to confirm the simplification "more people, less forest" in lieu of other more important, if complex forces (e.g., Anderson 1996; Rudel and Roper 1996; Barraclough and Ghimire 1996; Fairhead and Leach 1998) – see Box 3.3. Historical experience and current comparative research would suggest that there is no permanent, rigid or deterministic rule linking population and forest trends, but the role of population is located in a wider context, including agricultural and wider development trends, and concentrating on its role is perhaps to focus on the symptom rather than on the underlying condition or context (Mather and Needle 2000; Lambin et al. 2003; Geist 2003a). It has further been found that population does indeed show an association with deforestation at aggregate scales, but at local to regional scales it does not (Rindfuss et al. 2004a). Moreover, some of the most successful cases of forest management

Table 3.7.
Demographic factors of forest conversion and modification in the tropics

	All	Asia	Africa	Latin America
In-migration	58	12	9	37
Growing population density	38	12	6	20

Total number of cases is 152; multiple counts possible. *Source:* Geist and Lambin (2002), p. 148.

occur at the highest population densities (Tri-Academy Panel 2001; Moran 2005; Unruh et al. 2005) (see Sect. 3.5.3).

Population increase due to high fertility rates is not a primary driver of deforestation at a local scale and over a time period of a few decades. There is no single common effect of fertility on land use, nor is one expected. The relationship between land-use change and fertility flows in both directions, and, as a review of the literature shows, the effect of fertility on land use varies from place to place and over time (Rindfuss et al. 2004a). Fertility intervenes in only 8% of the reviewed cases of land change (Geist 2003a; Geist et al. 2006), it is never a sole factor, but always combined with other, at least equally important factors (Angelsen and Kaimowitz 1999), and though it is significantly associated with deforestation at the global and regional scales, evidence for population links to deforestation at micro-scales – where people are actually clearing forests – is scant. For example, where

---

Box 3.3. Misreading West African forest landscapes

Many influential analyses of West Africa take it for granted, that old-growth forest cover has progressively been converted and savannized during the 20th century by growing populations. By testing these assumptions against historical evidence, exemplified in case studies from the forest-savanna transition zones of Ghana, Guinea and Ivory Coast, it has been shown that these neo-Malthusian deforestation narratives badly misrepresent people-forest relationships. They obscure important non-linear dynamics, as well as widespread anthropogenic forest expansion and landscape enrichment. These processes are better captured, in broad terms, by a neo-Boserupian perspective on population-forest dynamics. However, comprehending variations in locale-specific trajectories of change requires fuller appreciation of social differences in environmental and resource values, of how diverse institutions shape resource access and control, and of ecological variability and path dependency in how landscapes respond to use (Fairhead and Leach 1996; Leach and Fairhead 2000).

Box 3.4. Household dynamics and forest-cover modification in the Amazon

In humid forest frontiers in South America, the internal dynamics of traditional and colonist families, which are mainly related to households' capital and labor constraints, explain the microlevel dynamics of land-cover modification by forest types (Coomes et al. 2000), land quality (Marquette 1998), and gender division, as well as the changing social context of deforestation in the Amazon Basin (Pichón 1997a,b; Sierra and Stallings 1998; Perz 2002). Forest clearing is caused by a variety of actors, with differing effects (Rudel 2005): recent immigrants practice slash-and-burn agriculture, and their children's families shift to fallow agriculture, while long-settled families practize diversified production; small families have crop/livestock combinations (associated with high rates of forest losses), while large families employ perennial production modes (associated with low rates of forest losses); and small ranchers, large ranchers, or upland croppers are displaced by lowland ranchers (Humphries 1998; McCracken et al. 1999; Walker et al. 2000). As a rule, microlevel dynamics shape the trajectory of land-use change, in turn affecting the household's economic status (Walker et al. 1996; Sunderlin et al. 2001).

---

tropical deforestation is linked to the increased presence of shifting cultivators, triggering mechanisms invariably involve changes in frontier development and policies by national governments that pull and push migrants into sparsely occupied areas (Rudel 1993, 2005; Mertens et al. 2000; Carr 2005). In some cases, these "shifted" agriculturalists (Bryant et al. 1993; Bryant and Bailey 1997) exacerbate deforestation because of unfamiliarity with their new environment; in other cases, they may bring new skills and understandings that have the opposite impact (Lambin et al. 2001) – see Box 3.4. This is not to deny empirical evidence that the link between high fertility and high deforestation can be shown at local scales for certain stages in the demographic cycle of settler households (e.g., Pichón 1997a,b; Carr 2005).

Dryland Change

As in other types of land change, case study evidence largely confirms the expectation that population plays a major role in the explanation of dryland change, with demographic factors implicated in over half of the cases of land degradation – see Table 3.8. However, and thus repeating the pattern found for forest zones, closer inspection reveals that even when population growth is an important explanatory factor, the archetypal process of a burgeoning population expanding into virgin lands is rare in the case study literature. For example, it has been found that population increase due to high fertility rates among impoverished rural groups, at a local scale and over a time period of a few decades, is not a primary driver of dersertification, appearing in just 3% of the cases reviewed (Geist and Lambin 2004). More important are family or life-cycle features that relate mainly to labor availability at the level of households, which is linked to migration, urbanization, and the breakdown of extended families into several nuclear families. As an example of the latter phenomenon, the splintering of family herds in the West African Sudan-Sahel zone over the past 25 years (due to increases in nuclear households and the transfer of livestock wealth from herding families to merchants, agriculturalists, and government officials) led to increased investment in crop production, reduced labor availability among pastoral households, lower energy and skills applied to livestock husbandry, and reduced livestock mobility, which increased the risk of land degradation (Turner 1999, 2002, 2003). Fuelwood demand by households in Africa differs between nuclear family units and larger consuming units; the latter are generally more energy efficient. Small consuming units thus cause more forest degradation, especially in peri-urban environments (Cline-Cole et al. 1990).

Demographic factors in dryland degradation show distinct regional clusters, with Asian and African cases of desertification most commonly cited as reflecting human population dynamics – see Table 3.8. Most widespread are cases in which (remote) population growth, overpopulation or population pressure is reported as a driver. The growth or increased economic influence of urban population often triggers migration of poor cultivators or herders from high-potential, peri-urban zones into marginal dryland sites. Consequently, the sometimes rapid increases in the size of local human populations in drylands are often linked to the in-migration of cultivators into rangelands or regions with large-scale irrigation schemes, or of herders into hitherto unused, marginal sites, resulting in rising population densities there (Geist 2005). Prominent examples of migration-driven desertification stem from ancient or historical irrigation (oasis) sites in Central Asia, such as the Tarim and Hei River Basins or the Aral Sea region. Until recently, traditional irrigation farming practices in these regions had a relatively small impact on dryland ecosystems. Only in the second half of the 20th century did advances in hydrotechnical infrastructure combine with population influx from remote zones, likewise driven by outside economic demands and related policies, i.e., attaining self-sufficiency in food and clothing, so that cotton monocultures and irrigated food crops became key crops in areas of rapid settlement. In the period 1949 to 1985 alone, population in the Hei River Basin of northern China almost doubled, from 55 million to 105 million people, with the total irrigated area tripling from 8 to 24 million ha and the number of reservoirs increasing from 2 to 95 in the same period of time (Sheehy 1992; Genxu and Guodong 1999; Yang 2001; Feng et al. 2001; Lin et al. 2001).

Cropland Change

Case study evidence also confirms the expectation that population plays a major role in the explanation of land dynamics in agricultural intensification zones, with demographic factors implicated in almost two thirds of the reviewed cases, though not working in a universal, or unmediated fashion – see Table 3.9 (McConnell and Keys

2005). As for other land-change classes, it has been found that population, usually together with national economic policy, plays an important role in regional studies as explanatory variable of change. However, at the village level, it may become clear that features of the household life cycle are more important (Vance and Geoghegan 2004). For example, it has been shown that the effects of population change in northeastern Thailand, when expressed as a change in household size, had a larger impact on the conversion of land for use in upland crops (e.g., cassava, corn, sugar cane) than when expressed as counts of individuals (Rindfuss et al. 2003). Likewise, historical demography is a powerful way to bring attention to the fact that a complete explanation of ecosystem change in agricultural core zones must include the actual sequence and timing of events that produce an observed structure or function. The age-gender structure of human populations is a summation of their historical experience and can provide powerful ways to examine land change in light of the changing structure of households (Netting 1986; Butzer 1990; Batterbury and Bebbington 1999; Redman 1999).

Urban Change

Today, about half the people in the world live in urban areas, up from less than 15% at the start of the 20th century. High-income countries typically have populations that are 70 to 80% urban. Some developing-country regions (e.g., parts of Asia) are still largely rural, but Latin America (at 75% urban) is indistinguishable from high-income countries in this regard (Millennium Ecosystem Assessment 2005).

Urban populations are not randomly scattered across the globe, but are commonly located at transportation break points and places of opportunity, with highest population densities at low coastal elevations and in topographic basins adjacent to mountain ranges (see Chap. 2). Rural-urban migration stories are not simple, and they involve both pull (facilitating) and push factors. There is an important life cycle aspect to how households use land, and timing of fertility is an important

Table 3.8.
Demographic factors associated with land degradation in drylands

	All	Asia	Africa	Latin America
Population growth, increases in size	42	31	4	2
In-migration, rising population densities	33	15	14	0

Total number of cases is 132; multiple counts possible; not included are data for Europe, Australia and North America. *Source:* Geist and Lambin (2004), p. 823.

Table 3.9.
Demographic factors associated with land dynamics in croplands

	Important	Not important	Absent
Population numbers, density	70	22	16
Population composition	8	12	88
Settlement, migration	34	41	33

Total number of cases is 108; multiple counts possible. *Source:* McConnell and Keys (2005), p. 333.

aspect. For urban migrants, push factors at the place of origin historically often include population pressure as a legacy of prior fertility levels. Rural-urban migration will transfer part of the impact of rural fertility to urban places and play a role in the conversion of land to urban uses. The longer-term effect would involve increased rates of household formation, and, although fertility has declined in most parts of the world (especially in urban zones), a legacy of high levels in the past is a continuing growth in the numbers of young people coming of age, forming their own households, and using land for dwelling units and for some type of productive activity. Thus, even though reduced fertility leads to diminished growth of the base (ages 0–4), the legacy of past fertility leads to substantial increases in the numbers of men and women entering their 20s and 30s many years after the decline in fertility, known as "population momentum" (Rindfuss et al. 2004a). On the other hand, turnarounds in forest-cover change have been widely associated with urbanization and industrialization, and the processes facilitating reforestation likely includes urbanization (Rudel 1998; Mather and Needle 1998) (see Sect. 3.5.3).

With the rising affluence commonly associated with the transition from agricultural to urban-industrial societies, a shift has occurred to smaller household sizes, i.e., the number of individuals living in a household, for which there is a variety of reasons (McKellar et al. 1995). Other things being equal, declines in fertility will lead to smaller household sizes. Also, many countries have experienced increases in divorce, especially in urban zones, and this often turns one household into two. And, in some countries of the industrialized world, a stage in the life course has emerged wherein children leave the parental household but have not yet formed their own family, frequently resulting in the creation of an additional household. Likewise, when rising affluence permits mobility from multi-generational households (or extended families), splitting into smaller units is typical. In sum, declining household size affects urban land use through a variety of mechanisms. There is demand for more housing units, and typically these units will spread horizontally across the landscape, contributing to urban sprawl. More dwelling units usually leads to more demand for building materials, etc., and smaller household size commonly also translates into less efficient use of various resources (Rindfuss et al. 2004a).

3.3.4 Institutional Factors

General Remarks

The preceding presentation of demographic, economic and technological factors makes it clear that it is also important to understand institutions (political, legal, economic, and traditional) and their interactions with individual decision-making (Agrawal and Yadama 1997; Ostrom et al. 1999; Young 2002a, 2003). In particular, government policy plays a ubiquitous role in land change, either directly causative or in mediating fashion (see Sect. 3.3.2). In the last case discussed in the preceding section, for example, governments intervene to reduce fertility and encourage transmigration (thus influencing demographic factors), while in the economic realm they control prices, subsidize inputs, provide credit, promote industrialization and export, and provide and maintain infrastructure. Throughout history and throughout most major regions of the world, the expansion of agricultural land has often served as a tool of population redistribution and has also played a key role in the formation and consolidation of nation states (Richards 1990; DeKoninck and Dery 1997). In the latter case, access to land, labor, capital, technology, and information is structured (and is frequently constrained) by local and national policies and institutions (Batterbury and Bebbington 1999). Also, crucial issues of property rights lie clearly in the institutional domain, and land managers have varying capabilities to participate in and to define these institutions. Relevant nonmarket institutions are, for example, property rights regimes, decision making systems for resource management (e.g., decentralization, democratization, and the role of the public, of civil society, and of local communities in decision making), information systems related to environmental indicators as they determine perception of changes in ecosystems, social networks representing specific interests related to resource management, conflict resolution systems concerning access to resources, and institutions that govern the distribution of resources and thus control economic differentiation (Lambin et al. 2003).

Probably the most closely scrutinized realm of policy influence on land dynamics is economic policy. National governments exert a huge influence on land-use decisions through economic and finance policy. Broad policy factors, often associated with structural adjustment (e.g., market liberalization, privatization, currency devaluation), were cited in all types of land-use change reviewed (Kaimowitz et al. 1999; Mertens et al. 2000; Sunderlin et al. 2001). Specific policies, including the provision of credits, price supports and subsidies, as well as the imposition of tariffs and taxes, were detected in a third of the cases of agricultural intensification, where subsidized inputs and price supports enabled farmers to profitably adopt new crops (McConnell and Keys 2005). More examples of policies that influence land-use change are state policies to attain self-sufficiency in food (Xu et al. 1999), decentralization (Becker 1999), (low) investments in monitoring and formally guarding natural resources (Agrawal and Yadama 1997), resource commodification (Remigio 1993; Deininger and Minton 1999; Sohn et al. 1999; Tri-Academy Panel 2001; Keys 2004), land consolidation (Imbernon 1999b; Pfaff 1999), and nationalization or collectivization (Xu et al. 1999; Tri-Academy Panel 2001) as well as privatization (Watts 1989, 1994, 1996). Credits and subsidies for the forest sector played strong roles in over a quarter of the deforestation cases reviewed (Barbier 1993; Pichón 1997a,b; McCracken et al. 1999; Deininger and Minton 1999; Hecht 1993, 2005), while such factors appear to be somewhat less important in cases of desertification (Geist 2005).

Forest, Dryland and Cropland Change

As mentioned above, the linkage between infrastructure expansion and deforestation has long been recognized and debated, and the meta-analyses of land-use dynamics bear this out. Overall, government-sponsored migration (resettlement) schemes exert an overwhelming influence in deforestation in certain cases, such as the Brazilian Amazon and Indonesia (Geist and Lambin 2002). While the most frequently cited form of infrastructure facilitating forest settlement is transportation, this was much less prevalent in prompting agricultural intensification, occurring in barely one quarter of the cases (McConnell and Keys 2005), and even more rarely associated with desertification (Geist and Lambin 2004). By contrast, the provision of water resource infrastructure (reservoirs, dams, canals, levies, boreholes and pump stations) was seen as an important causal factor in over a third of the desertification cases, and played a crucial role in agricultural intensification involving irrigation (Johnson 1986; Hopkins 1987; Ewell and Merill-Sands 1987; Carney 1993; Shively 2001). In their studies on land-use change in Punjab and Haryana, the Indian heartlands of green revolution applications, Leaf (1987) and Vashishta et al. (2001) both find that the two most crucial public policies were regionally biased infrastructure development (roads as well as irrigation infrastructure) and the pricing of crop inputs and outputs by the state, which is widely supported by other case study evidence (e.g., Deininger and Minton 1999; Tri-Academy Panel 2001) – see Box 3.1.

Direct government participation in extractive industries, such as agricultural or forestry plantations, can have locally powerful consequences. Likewise, the state's encouragement of energy and mineral resources development has led to pressure on water resources, triggering desertification. As a general rule, it appears as if land degradation is more prominent when macropolicies, either capitalist or socialist, undermine local adaptation strategies (Geist and Lambin 2003). In particular, "perverse subsidies" for road construction, agricultural production, forestry, and so forth are thought to be one of the biggest impediments to environmental sustainability (Myers and Kent 2001) (see Sect. 3.3.2).

The flip side of the influence of government policy is its failure, i.e., ill-defined policies and weak institutional enforcement. This can involve the lack of access to government services by particular groups (e.g., highlanders, ethnic minorities), as well as more widespread inability, for example, to provide extension services, or to enforce land-use regulations. In some instances, such failure is seen to result from simple lack of resources, while in others, authors assert that clientelism and other forms of corruption are to blame. In Indonesia, for example, widespread illegal logging is linked to corruption and to the devolving of forest-management responsibilities to the district level (Jepson et al. 2001). In the Brazilian Amazon, significant examples of policy failure are the widespread disrespect of the limits to clear cut determined by the federal Forest Code and the difficulties of implementing prescribed land-zoning programs (Alves et al. 2003; Mahar 2002). On the other hand, recovery or restoration of land is also possible with appropriate land-use policies (Tri-Academy Panel 2001; Mather 2006c). Also, war, insurgency, and violent conflicts over land lead to the disruption of land management, thus triggering dryland degradation, for example (Geist 2005).

Clearly one of the most important sets of factors influencing people's actions on the landscape is their rights to use, alter and extract resources from the land. In much of the tropics, property rights have been quite dynamic over the past few decades, as traditional community tenure systems cede to increasingly private, individualized regimes, generally in the context of colonial and post-colonial influences. In fact, the delineation of colonial territory by the European powers was often purposely designed to subdivide the territory of ethnic groups, and the legacy of this continues to be cited as an important factor shaping land dynamics. These shifts in access to and control over land resources have of course been experienced differently by different groups within any country, and even within localities (see Sect. 3.2.2). At the same time, states have exerted – and have sometimes subsequently relaxed – ownership of all or part of their national territory (e.g., forest lands). An important recent manifestation of this is the creation of biodiversity conservation areas, which entails denying or restricting access to lands considered crucial to existing livelihoods. At the same time, consolidation of land resources in the hands of few has been an important process, and the redressment of this through land reform (redistribution) has had major consequences (Bebbington 2000; Coomes et al. 2000).

Not surprisingly, then, property rights issues emerge as important factors in almost half of the deforestation cases reviewed (Geist and Lambin 2002) – see Table 3.10. Of particular relevance in this domain are logging concessions, liberalization of land markets, easy transfer of public lands for private use, state regulations favoring large land holdings, tenure insecurity, and malfunctioning customary tenure regimes. Though much discussed as a robust cause of deforestation (e.g., Deacon 1994, 1995,

Table 3.10.
Institutional factors associated with forest conversion and modification in the tropics

	All	Asia	Africa	Latin America
Formal deforestation policies	105	46	7	52
on land development	60	28	5	27
on economic growth	51	22	5	24
on credits/subsidies	39	11	1	27
Property rights issues	67	33	5	29
Policy failures	64	31	1	32
Mismanagement	38	13	1	24

Total number of cases is 152; multiple counts possible. *Source:* Geist and Lambin (2002), p. 148.

Table 3.11.
Institutional factors associated with land dynamics in croplands

	Important	Not important	Absent
Property regime	65	34	9
Government/NGO policy	55	24	29
Income affecting program	36	14	58
Infrastructure program	33	10	65

Total number of cases is 108; multiple counts possible. *Source:* McConnell and Keys (2005), p. 333.

Table 3.12.
Institutional factors associated with land degradation in drylands

	All	Asia	Africa	Latin America
Malfunctioning common property regulation	42	21	8	6
New land tenure, land zoning measures	37	19	11	1
Agricultural development policies	35	27	7	0

Total number of cases is 132; multiple counts possible; not included are data for Europe, Australia and North America. *Source:* Geist and Lambin (2004), p. 823.

1999; Mendelsohn 1994; Mendelsohn and Balick 1995), it appears as if property rights issues are mainly a characteristic of Asian cases and tend to have ambiguous effects upon forest cover, i.e., insecure ownership, quasi-open access conditions, maladjusted customary rights, as well as the legalization of land titles, are all reported to influence deforestation in a similar manner (Geist and Lambin 2002). Virtually all of the agricultural intensification cases reported some information on property regimes. The information was part of a still larger set of nonmarket institutional variables that emerged as frequently as other important causes, with policies and programs of the government or non-governmental organizations somewhat less frequently reported than property regimes. The latter were particularly important in those cases involving the adoption of tree crops, which often imbues the owner with a greater degree of control over the land (McConnell and Keys 2005) – see Table 3.11. Among the institutional and policy factors that underlie about two-thirds of reported cases of desertification, modern policies and institutions are as much involved as are traditional institutions (or, in other words: the failure of traditional land-tenure regimes under circumstances of other pressures such as aridification or market integration). It appears that the failure of institutional aspects of traditional land tenure (e.g., equal sharing of land and splintering of herds because of traditional inher-

itance law) are as important in driving desertification as are growth-oriented agricultural policies (including measures such as land distribution and redistribution), agrarian reforms, modern sector development projects, and market liberalization policies. Both traditional and modern institutions and policies thus reduce flexibility in management and increase the pressure on constant land units. The introduction of new land-tenure systems, whether under private (individual) or state (collective) management, is another factor associated with land degradation in drylands (Geist 2005) – see Table 3.12.

Underlying the institutional arrangements for land management and property rights regimes are broad sociopolitical factors that encompass, among others, the amount of public participation in decision-making, the groups participating in public decision-making, the mechanisms of dispute resolution, and the role of the state relative to the private sector. Over the past 50 years, there have been significant changes in these forces. The changes include, among others, a declining trend in centralized authoritarian government (but also in the importance of the state relative to the private sector), an increased involvement of non-governmental and grassroots organizations in decision-making processes (expressed, for example, in the worldwide recognition by the Norwegian Nobel Prize Committee of Wangari Maathai and the Green Belt movement in Africa, linking women's rights, democracy, ecological restoration, and grassroots activism in favor of sustainable development), and an increase in multilateral environmental agreements such as the United Nations Convention to Combat Desertification (UNCCD) (Millennium Ecosystem Assessment 2005). With increasingly interconnected market forces and the rise of international conventions, the impact of institutional drivers moves from the local to the global level (Taylor et al. 2002a). It can be expected that many of the rules used for making land-related policies will continue to be relevant factors. This will be important because in the history of human-environment relations there has often been a widespread mismatch between environmental signals reaching local populations and conventional macrolevel institutions (Redman 1999; Tri-Academy Panel 2001), and any changes should help to ensure that local users are able to better influence resource-management institutions (Poteete and Ostrom 2004). These institutions need to be (re)considered at various scales, to identify the local mediating factors and adaptive strategies and to understand their interactions with national- and international-level institutions (Klepeis and Chowdhury 2004; Mather 2006c).

Urban Change

A prime example of economic and related policies associated with the growth of urban zones is China. On the one hand, beginning in the late 1970s, urban regions benefited from national reform policies in the agricultural sector (price reform, elimination of collective farming), which triggered increased crop yields and a surplus of agricultural workers available for urban economic sectors. On the other hand, decentralization policies allowed provincial and local city governments more autonomy to devise and implement their growth-oriented development strategies (e.g., incentives to stimulate investment, economic development and conversion into urban-industrial zones) (Seto et al. 2004). A land reform in 1988 further allowed the transfer of land-use rights through negotiation, auction, or bid, with the consequence that both individuals and collectives can rent or lease their land to local and foreign ventures (Sharkawy et al. 1995). Movement to cities was made possible through reforms, which have relaxed the so-called *hukou* and reduced the importance of the *danwei* systems, both limiting population mobility, especially from rural to urban areas. *Hukou* has been a household registration system which determined the residency status of an individual, while the work unit, *danwei*, was an important provider of basic goods and services such as housing, health care, food ration tickets, and education, with both systems controlling internal migration and urbanization before 1978 (Mallee 1996; Smart and Smart 2001). At the national

scale, again beginning in the late 1970s, the central government initiated a series of sweeping reforms that included the promotion of township and village enterprises (TVEs) which had originally been agricultural collectives. Urban TVEs in China turned into veritable pillars of economic growth, since they were built upon low labor costs due to rural surplus labor and relative freedom from state or bureaucratic control, thus becoming attractive partners for foreign investments (Putterman 1997).

3.3.5   Cultural Factors

General Remarks

Numerous cultural factors also influence decision making on land use, and it is important not to divorce these cultural conditions and trends from underlying political and economic conditions, including political and economic inequalities such as the status of women, ethnic minorities and resource-poor households, that affect resource access and land use (see Sect. 3.2.2). The ways in which people frame land-use choices represent an important set of proximate factors that influence decision-making, but these framing practices in turn influence and are influenced by the other driving forces discussed in this chapter. Land managers have various motivations, collective memories, and personal histories, and it is their attitudes, values, beliefs, and individual perceptions which affect land-use decisions, for example, through their perception of and attitude toward risk (U.S. National Research Council et al. 1999). Understanding the mental models (i.e., cognition, volition, will, etc.) of various actors may thus help explain the management of resources, adaptive strategies, compliance with or resistance to policies, or social learning, and therefore social response in the face of land-use change (Lambin et al. 2003).

Forest Change

In tropical forest zones, cultural factors are reported to underlie mainly economic and policy forces in the form of attitudes of public unconcern towards forest environments, and these factors also shape the rent-seeking behavior of individual agents causing deforestation (e.g., Deininger and Binswanger 1995) – see Table 3.13. Most notably the so-called cattle complex, or the high status accorded cattle ranching in Latin America, explains some important variations in regional patterns of land use, i.e., pasture creation for cattle ranching as a striking cause of deforestation reported almost exclusively for humid lowland cases from mainland Latin America (Geist and Lambin 2002). The cultural preference for cattle ranching stems from colonial Iberian experiences in the

17th and 18th century in the Americas. This common cultural legacy explains in part why cattle ranching is so prevalent in land poor Central America as well as in land rich South America. When penetration roads were built through these regions during the 1960s and 1970s, this cultural preference catapulted cattle ranching into one of the key driving forces behind tropical deforestation in the Western Hemisphere (Shane 1986; Hecht 1993). These cultural preferences also have spillover effects, spreading from majority to minority groups in a society. Some of the most populous and acculturated indigenous peoples in Latin America became cattle ranchers during the 1970s in an effort to secure titles to what had been forested land. Some indigenous peoples reverted to more culturally familiar patterns of shifting cultivation after they obtained formal land tenure, but others remained cattle ranchers (Rudel et al. 2002a).

et al. 2001; Lin et al. 2001; Jiang 2002). Such land-use change is very often linked to the belief that water is a "free good" and that grazing is "inefficient" when compared with grain production. In particular, water has always been regarded as a common good to be used freely, and there is usually little incentive to conserve when the cost of irrigation from individual wells is only the cost of extraction, and when costs for water drawn from canals is a low annual fee independent of volume and frequency of use (Hole and Smith 2004). Contrasting with this pattern are the Latin American cases, in which desertification seems to be predominantly driven by the individual responses or motivations of ranchers, and the Australian cases, in which a frontier mentality is not explicitly promoted by the state but seems to reflect a private attitude (Geist 2005). In Africa, ethnicity can have a strong bearing on adaptive land-use strategies (Reenberg and Paarup-Laursen 1997).

Dryland Change

In drylands affected by land degradation, public attitudes, values and beliefs are as frequently associated with cases of desertification as are individual or household behavior, but there are regional variations – see Table 3.14. In Asia, land-use change leading to desertification is sometimes driven by public encouragement of a frontier mentality and by efforts to improve living standards and attain self-sufficiency in food. An example of the former cultural complex is the official support for land consolidation in the northern and, especially, northwestern territories of China (Jiang et al. 1995; Genxu and Guodong 1999; Feng

Cropland Change

In agricultural core areas with settled farming practices and pronounced land use intensification, religion, ethnicity and education have been identified as factors shaping land use decisions. These include strong preferences for staple crops, or for particular cropping practices (McConnell and Keys 2005) – Table 3.15. In addition, cultural and religious factors often shape restrictions (i.e., taboo) on the use of certain parts of the landscape, for example, reserving the hillsides surrounding family tombs for cultivation only in special circumstances in respect of the founders of the village (McConnell 2002).

Table 3.13.
Cultural factors associated with forest conversion and modification in the tropics

	All	Asia	Africa	Latin America
Public attitudes, values and beliefs	96	45	5	46
Public unconcern	66	25	3	38
Missing basic values	55	33	2	20
Individual/household behavior	80	38	6	36
Situation-specific (esp. rent-seeking)	74	36	5	33
Unconcern by individuals	48	20	4	24

Total number of cases is 152; multiple counts possible. *Source:* Geist and Lambin (2002), p. 148.

Table 3.14.
Cultural factors associated with land degradation in drylands

	All	Asia	Africa	Latin America
Public attitudes, values and beliefs	52	30	10	5
Violent land conflicts, war	10	3	6	0
Perceptional issues	12	7	0	0
Individual and household behavior	53	18	10	9
Indifference	13	4	5	3
Perceptional issues	9	4	0	0

Total number of cases is 132; multiple counts possible; not included are data for Europe, Australia and North America. *Source:* Geist and Lambin (2004), p. 823.

Table 3.15.
Cultural factors associated with land dynamics in croplands

	Important	Not important	Absent
Religion/ethnicity	7	41	60
Education	21	11	76

Total number of cases is 108; multiple counts possible. *Source:* McConnell and Keys (2005), p. 333.

Urban Change

The influence of cultural preferences includes landscapes of consumption as well as landscapes of production. Urbanization, for example, very likely changes ways of life fundamentally, with increasing expectations about raised consumption and potentially a weakened understanding of production-consumption relationships (see Sect. 3.3.3). Demands originating from urban-industrial zones often exert remote influences on rural and marginal sites, and urban entrepreneurs are often cited as being responsible for what has been called "speculative cultivation" outside the built-up zones, affecting property rights regimes there. In the humid forest zones of mainland Latin America, for example, pasture creation by large ranchers and absentee landlords is often reported as an unproductive, profit-seeking activity to add value to land, thus raising the value of land for speculation purposes and driving "speculative deforestation" (Hecht 1993; Walker et al. 2000). Likewise, it has often been reported that in the wake of rising prices of irrigation key crops (such as cotton and rice) urban entrepreneurs start investing in land, tractors and combine harvesters to cultivate large tracts of what had previously been rangelands (Geist 2005). As in the case of felling old-growth forest trees for pasture, steppe could be claimed by plowing it. Enormous areas of marginal land were thereby brought under speculative cropping, mostly funded by urban investors, such as in the semi-arid Syrian Khabur Region between the Tigris and Euphrates Rivers (Hole and Smith 2004). Cultural factors also shape land-use dynamics within urban zones. The English preference, for example, for lawns contributed to suburban sprawl in North America after World War II, and more recently, the preference for suburban landscapes of consumption has spread from North America to South America and shows signs of spreading to disparate other world regions (Leichenko and Solecki 2005).

3.4   Causation Revisited

3.4.1   Factor Interaction and Conjunctural Causation

This presentation of causal factors highlights several issues. First, any given factor can have multiple and often contradictory effects, depending on its specific nature, and on the context in which it occurs. For example, an increase in world coffee prices may cause farmers in Central America to clear forest land to make way for coffee groves, while at the same time in East Africa, land may be converted from field crops to coffee groves (Goldman 1993; Kasfir 1993; Okoth-Ogendo and Oucho 1993). The net effect on woody biomass at the two sites will be quite different. Likewise, the effects of globalization (see Sect. 3.4.3), in the sense of the geographical expansion of free trade, have had dramatically different effects in different regions: increasing pressure on forest resources in forest-rich regions like the Amazon, while reducing pressure in forest-depleted regions like West Africa or South Asia (Rudel 2005).

In addition to the ambiguous effects of a given causal factor, as noted above, no objective framework exists for the classification of factors into broad groups; rather the framework applied depends on the analytical lens of the researcher. The construction of roads, for example, can be analyzed according to the resulting difference in farm gate prices, or as part of a government policy to encourage transmigration, which itself may be seen as an outcome of rapid population growth. In fact, it has been argued that roads can only facilitate land change, but are themselves insufficient in the absence of price incentives, and that inputs must also be in place (Angelsen and Kaimowitz 1999) – see Box 3.5 and Fig. 3.2.

The strongest finding emerging from the meta-analyses of case studies is a resounding rejection of single-cause explanations of land-use change. No factor ever works in isolation. While some factors, such as population growth, may be very widely implicated in land change around the world and through time, their effects depend not only on their particular nature, but also on the specific biophysical and social contexts in which they occur. Given the impossibility of carrying out classical experimentation, isolation of the "independent" effects of any factor is fruitless. Thus, the focus should be causal synergies or the interaction of factors, rather than the individual factors or groups of factors (sectors). For example, a recurrent combination of interacting factors associated with desertification entails a change in precipitation combined with government policy promoting growth in the agricultural sector, along with the introduction of new technology, in the context of an inflexible tenure regime ill-suited to these new circumstances (Geist and Lambin 2004).

Different patterns or modes may represent the interactions between the various causes of land change (Young 2002a; Lambin et al. 2003). First, while no key factor operates in isolation, one cause may completely dominate the other cause, assuming that land use in a given locality is influenced by whatever factor exerts the greatest

Box 3.5. Debating the role of roads in deforestation

As illustrated by the case of roads and deforestation, the direction of causality may be difficult to establish, even at short timescales. For example, 81% of the deforestation in the Brazilian Amazon between 1991 and 1996 occurred within 50 km of four major road networks (Lele et al. 2000; Alves 2002a). Is it the national demand for land and the (high) agricultural suitability of some forest areas that lead to policy decisions to expand the road network in these areas, which then gives access to the forest for migrants who clear land? Or is it the expansion of local logging or agricultural activities in some forest areas that then justifies the construction of new roads to link these active production areas to existing markets? Or does the construction of a road for reasons unrelated to land use in the forest (e.g., to connect major cities) induce new deforestation by its mere presence, through a spatial redistribution of population and activities? Or, in the latter case, does the road simply attract to a given location a preexisting demand for land that would have led to deforestation elsewhere if the road had not been built? In this case, are there other intervening factors like the creation of forest reservations or a more strict enforcement of existing land appropriation regulations? In other words, is a road an endogenous or exogenous factor in deforestation and does it affect just the location or also the quantity of deforestation in a given country? The likely answer to these questions is that, in most cases, national demand for land, policies to develop the forest frontier, capital investments in logging and agricultural activities, population movements, commodification of the economy, the development of urban markets, and infrastructure expansion are highly interdependent and co-evolve in close interaction as part of a general transformation of society and of its interaction with its natural environment (Lambin et al. 2003).

constraints. Second, factors driving land-use/cover dynamics can be connected as causal chains, i.e., interconnected in such a way that one or several variables (underlying causes, mainly) drive one or several other causes (proximate causes, mainly). Third, different factors can intervene in concomitant occurrence which describes the independent but synchronous operation of individual factors leading to land change. Finally, and the modes of interaction might not be exhausted herewith, different factors may also intervene in synergetic factor combinations, i.e., several mutually interacting variables driving land-use change and producing an enhanced or increased effect due to reciprocal action and feedbacks between causes (see Sect. 3.4.2). In meta-analyses of case studies of tropical deforestation (Geist and Lambin 2002) and dryland degradation or desertification (Geist and Lambin 2004), the proportion of cases in which dominant, single, or key factors operate at either the proximate or underlying level was low (ca. 5 to 8%); concomitant occurrence of causes was more widespread (ca. 25%); and the most common type of factor interaction was found to be synergetic factor combinations (in ca. 70 to 90% of the case studies reviewed).

Quantitative social science has long recognized the implausibility of the assumption of complete independence among so-called "independent" variables, and a great number of sophisticated techniques have been proffered to accommodate – that is, to remove the effects of –

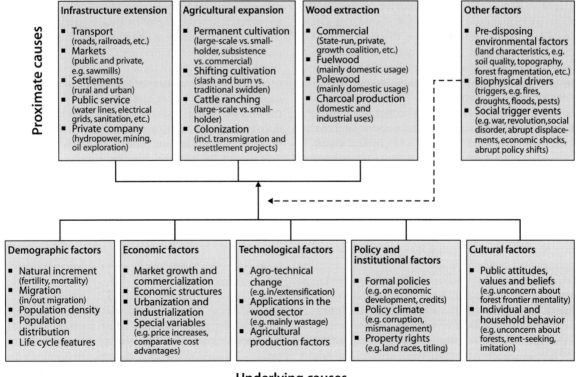

Underlying causes

Fig. 3.2. Proximate causes and underlying driving forces of forest decline. *Source:* Geist and Lambin (2002), p. 144

such interaction (e.g., York et al. 2003). A different approach is to reject the notion that individual variables have independent effects, and can be substituted one for another in causing an outcome, and instead embrace these interactions as the heart of explanation (Ragin 1987). This approach seeks to identify how suites of interacting factors work in conjunction with one another, and to identify typical, or recurring causal clusters. Case studies of land-cover change underline the importance of meso-level variations in land-cover change in which different rain forest regions exhibit distinct clusters of causes that together drive land-use change. For example, Geist and Lambin (2002) identify distinct continental clusters surrounding cattle ranching in Latin America, smallholder agriculture in Africa, and a logging-smallholder tandem in Southeast Asia. Rudel (2005) finds variations in causal clusters between regions with large forests (Amazon, Central Africa, and Southeast Asia) and regions with small forests (Central America, West Africa, East Africa, and South Asia): well financed landowners and corporations drive deforestation in regions with large forests, while villages and smallholders are important actors in places with small forests.

3.4.2　Feedbacks, Thresholds, Endogeneity, and Co-Evolution

The patterns of causation discussed above are in most cases simplifications that are useful for communicating about particular environmental issues or for modeling (see Chap. 5). In reality, however, there are functional interdependencies in reality between all the causes of land change, both at each organizational level ("horizontal interplay"), and between levels of organization ("vertical interplay") (Young 2002b). Thus, the relationship between causes and outcomes is neither linear nor unidirectional (Mather 2006b). Anthropogenic land change invariably alters all aspects of the biophysical system to some degree (and extent), and as those alterations become sufficiently great, they are detected by the land users (or by their neighbors or regulating bodies), and this detection eventually leads to a change in behavior. When the reaction exacerbates perceived negative consequences on the system, the result is degradation; when the reaction moderates such negative effects, the result may be remediation or rehabilitation. In other words, amplifying mechanisms (or "positive" feedback loops) lead to an acceleration of change, in some cases triggering a rapid degradation of ecosystems and the impoverishment or even collapse of the societies using these ecosystems (Diamond 2005). By contrast, attenuating mechanisms dampen the human impact on the environment, in some cases leading – in the form of institutional and technological innovations, for example – to "negative" feedback loops associated with a decrease in the rate of change or even a rever-

sal of the land-cover change trend (see Sect. 3.5.3). Thus, feedback is an important system property associated with changes in land use that can impact the speed, but also intensity and mode of land change (Lambin et al. 2003).

Adding to the system properties of land-use/cover dynamics are thresholds (hidden points or "break points"), that steer fundamental, but reversible changes. Sudden, abrupt and irreversible shifts from one land use into another (or into collapse) can occur at control (or switch and choke) points (Steffen et al. 2004). Often, biophysical and socioeconomic factors tend to operate in what could be called "multiple thresholds", governing the trajectory towards degradation or remediation in conjunction with feedback mechanisms, occasionally in an event-driven manner (Reenberg 2001). In dryland areas, for example, common examples of multiple thresholds are dry climate conditions (limiting water provision for cropping and determining germination conditions), critical minimum soil depths, the regenerative capability of vegetation to develop back to dense growth, and the degree of flexibility among rural societies for informal arrangements to cope with these factors (Geist 2005).

As a matter of co-evolution, many factors driving land-use/cover change – such as new economic policies or technological developments in agriculture – appear to be exogenous forces (thus largely out of control by local land managers), but as the timescale of analysis expands, all causes – from demographic changes to technological innovations (including new environmental policies) – become endogenous to the human-environment system and are affected in some degree by land dynamics. Actually, the changes in ecosystem goods and services that result from land-use/cover change (see Chap. 4) lead to important feedback on the drivers of land dynamics. These changes affect the availability and quality of some of the natural resources that are essential to sustain livelihoods, create opportunities and constraints for new land uses, induce institutional changes at the local to global levels in response to perceived and anticipated resource degradation, modify the adaptive capacity of land managers (by affecting their health, for example), and give rise to social changes in the form of income differentiation (when there are winners and losers in environmental change) or increased social complexity (e.g., by increasing interactions between urban and rural systems) (Lambin et al. 2003).

3.4.3　Globalization

Globalization – i.e., the worldwide interconnectedness of places and people through global markets, information, capital flows and international conventions, for example – is a process that underlies the driving forces discussed above. Global markets, in particular, increase complexity and uncertainty, raising concerns about risk from the global-local interplay of driving forces. Examples include forces of globalization that underlie processes of tropical deforestation (e.g., through an expansion and liberalization of the markets for forest products), rangeland modifications (e.g., by the application to dryland regions of inappropriate land-management systems designed elsewhere), agricultural intensification (e.g., through domestic and international capital flows leading to agricultural specialization), and urbanization (e.g., by the diffusion of urban culture and the increasing disconnection of the sources of demand from the location of production) (Lambin et al. 2001). For example, the depletion of accessible stocks of tropical hardwoods in Southeast Asia has prompted Asian buyers and companies to investigate and begin purchasing old growth timber from other continents, most recently from the Central African and Amazon-Orinoco forests; at the same time, these firms closed down their operations in already depleted areas like peninsular Malaysia or Thailand (Rudel 2005).

The various processes of globalization accelerate or dampen the impact of drivers of land change, i.e., they cross-cut the local and national pathways of land-use/cover change, and they therefore attenuate or amplify the driving forces by removing regional barriers, weakening connections within nations, and increasing the interdependency among people and between nations. Throughout the history of land transformation, rapid land-use changes often coincide with the incorporation of a region into an expanding world economy such as in the process of European colonization of the New World (Richards 1990). In an increasing manner, global forces replace or rearrange the local factors determining land use, building new, global cause-connection patterns in their place, such as contract farming schemes and global agro-food chains (Watts 1996; Goodman and Watts 1997; Fold and Pritchard 2005). On the other hand, globalization also affects land change indirectly. Examples are eco-labeling and global organic food networks, information technologies leading to better forecasts on weather or market prices for farm management, or land monitoring using earth observation satellites that provide control and global sanctioning such as in the case of forest fires in Indonesia (in 1998). In particular, international institutions – be they organizations within the United Nations system or nongovernmental organizations – can be instrumental in promoting and funding policies aimed at combating environmental degradation, setting political agendas, building consensus, and creating constraints and incentives for sustainable land management (Lambin et al. 2002).

It appears as if globalization, in the sense of trade liberalization and the spread of neo-liberal macroeconomic policies, is particularly important in countries or areas with fragile ecosystems (e.g., semiarid lands and mangrove forests). In Ghana and Mexico, for example, land-use/cover changes during the 1980s and 1990s were identified as the immediate and principal impact stemming from economic liberalization and globalization, mostly trade liberalization and reforms to open up the agro-industrial sector. Increased agricultural productivity directly triggered forest conversion and increased land degradation from unsustainable production methods, and, indirectly, agro-industrial development displaced the landless and rural poor, who were then pushed to marginal agricultural lands or to the forest frontier (Barbier 2000a).

Globalization also has a cultural component that most visibly affects consumption landscapes in expanding urban areas. The spread of recreational norms embodied in games like golf leads to the construction of golf courses and second homes in seemingly unlikely settings in newly industrializing, prosperous nations (Leichenko and Solecki 2005). Although the MacDonaldization thesis refers to a form of bureaucratic rationality within enterprises (Ritzer 1998), one could appropriate the term and use it to describe the common element that makes emerging urban landscapes in the more affluent and newly affluent parts of the world look so similar.

3.5 Syndromes, Pathways, and Transitions

3.5.1 Syndromes of Land Change

Case study comparisons revealed that not all causes of land change and all levels of organization are equally important. This prompted an attempt to reduce the complexity of the analysis of causes by identifying a limited suite of processes and variables which makes the problem tractable at aparticular scale. For any given human-environment system, a limited number of causes are essential to predict the general trend in land-use/cover change (Stafford-Smith and Reynolds 2002; Reynolds et al. 2006). This is the basis, for example, for the syndrome approach, which describes archetypal, dynamic, co-evolutionary patterns of human-environment interactions (Petschel-Held et al. 1999; Petschel-Held 2004). A taxonomy of syndromes links processes of degradation to both changes over time and status of state variables. The approach is applied at the intermediate functional scales that reflect processes taking place from the household level up to the international level. For example, the "over-exploitation syndrome" represents the natural and social processes governing the extraction of biological resources through unsustainable industrial logging activities or other forms of resource use. Policy failure is one of the essential underlying driving forces of this syndrome (e.g., lobbyism, corruption, and weak or no law enforcement) (Petschel-Held et al. 1999). The typology of syndromes reflects expert opinion based on local case examples, and the overall approach aims at a high level of generality in the description of mechanisms of environmental degradation.

Summarizing from a large number of case studies (Geist and Lambin 2002, 2004), the authors found that land change is driven by a combination of the following fundamental high-level causes (or "syndromes"), making a difference between "slow" and "fast" variables – see Table 3.16 (Lambin et al. 2003):

- resource scarcity leading to an increase in the pressure of production on resources,
- changing opportunities created by markets,
- outside policy intervention,
- loss of adaptive capacity and increased vulnerability, and
- changes in social organization, in resource access, and in attitudes.

Some of these fundamental causes are experienced as constraints. They force local land managers into degradation, innovation, or displacement pathways. The other causes are associated with the seizure of new opportunities by land managers who seek to realize their diverse aspirations. Each of these high-level causes can occur as slow evolutionary processes that change incrementally at the timescale of decades or more, or as fast changes

that are abrupt and occur as perturbations that affect the land system suddenly. As may be seen from the cases collected by Puigdefábregas (1998), only a combination of several causes, with synergetic interactions, is likely to drive a region into a critical trajectory. In short,

$$\text{land use} = f \text{ (pressures, opportunities, policies, vulnerability, and social organization)}$$

with

pressures = f (population of resource users, labor availability, quantity of resources, and sensitivity of resources);
opportunities = f (market price, production costs, transportation costs, and technology);
policies = f (subsidies, taxes, property rights, infrastructure, and governance);
vulnerability = f (exposure to external perturbations, sensitivity, and coping capacity); and
social organization = f (resource access, income distribution, household features, and urban-rural interactions),

with the functions f having forms that account for strong interactions between the causes of land change.

Table 3.16. Typology of the causes of land change

	Resource scarcity causing pressure of production on resources	Changing opportunities created by markets	Outside policy intervention	Loss of adaptive capacity and increased vulnerability	Changes in social organization, in resource access, and in attitudes
Slow	Natural population growth and division of land parcels Domestic life cycles that lead to changes in labour availability Loss of land productivity on sensitive areas following excessive or inappropriate use Failure to restore or to maintain protective works of environmental resources Heavy surplus extraction away from the land manager	Increase in commercialization and agro-industrialization Improvement in accessibility through road construction Changes in market prices for inputs or outputs (e.g., erosion of prices of primary production, unfavourable global or urban-rural terms of trade) Off-farm wages and employment opportunities	Economic development programs Perverse subsidies, policy-induced price distortions and fiscal incentives Frontier development (e.g., for geopolitical reasons or to promote interest groups) Poor governance and corruption Insecurity in land tenure	Impoverishment (e.g., creeping household debts, no access to credit, lack of alternative income sources, and weak buffering capacity) Breakdown of informal social security networks Dependence on external resources or on assistance Social discrimination (ethnic minorities, women, lower class people, or caste members)	Changes in institutions governing access to resources by different land managers (e.g., shift from communal to private rights, tenure, holdings, and titles) Growth of urban aspirations Breakdown of extended family Growth of individualism and materialism Lack of public education and poor information flow on the environment
Fast	Spontaneous migration, forced population displacement, refugees Decrease in land availability due to encroachment by other land uses (e.g., natural reserves or the tragedy of enclosure)	Capital investments Changes in national or global macro-economic and trade conditions that lead to changes in prices (e.g., surge in energy prices or global financial crisis) New technologies for intensification of resource use	Rapid policy changes (e.g., devaluation) Government instability War	Internal conflicts Illness (e.g., HIV) Risks associated with natural hazards (e.g., leading to a crop failure, loss of resource, or loss of productive capacity)	Loss of entitlements to environmental resources (e.g., expropriation for large-scale agriculture, large dams, forestry projects, tourism and wildlife conservation), which leads to an ecological marginalization of the poor

Source: Lambin et al. (2003), p. 224.

Some of the fundamental causes triggering land change are mainly endogenous (such as resource scarcity, increased vulnerability and changes in social organization), even though they may be influenced by exogenous factors as well. The other high-level causes (such as changing market opportunities and policy intervention) are mainly exogenous, even though the response of land managers to these external forces is strongly mediated by local factors (see Sect. 3.2.2).

3.5.2 Typical Pathways of Land-Use/Cover Change

The various drivers of land change discussed above are strongly linked within and between levels of organization. They interact directly, are linked via feedback, and thus often have synergetic effects. Any land manager also constantly makes trade-offs between different land-use opportunities and constraints associated with a variety of external factors (Geist et al. 2006) (see Chap. 7). Moreover, various human-environment conditions react to and reshape the impacts of drivers differently, which leads to specific pathways of land dynamics (Lambin et al. 2001). As noted above, despite of the large diversity of causes and situations (or contexts) leading to land change, the complexity of causative factors giving rise to land dynamics can be greatly reduced. Thus, the critical challenge is to identify dominant pathways or trajectories, which also illuminate associated risk factors for each trajectory (Lambin et al. 2003).

This is the basis, for example, of the approach to study "regions at risk" and environmental criticality by Kasperson et al. (1999). Several case studies of regions under environmental degradation were described qualitatively by their histories. These qualitative trajectories were represented in terms of development of the wealth of the inhabitants and the state of the environment. A "critical environment" was defined as one in which the extent or the rate of environmental degradation precludes the maintenance of current resource-use systems or levels of human well-being, given feasible adaptation and the community's ability to mount a response (Kasperson et al. 1995). Different typical time courses of these variables were identified and interpreted with respect to more or less problematic future development of the regions. The Aral Sea, for example, was unquestionably a critical region after a few decades of Soviet-sponsored, ill-conceived large-scale irrigation schemes (Glazovsky 1995). Assigning a particular case (e.g., the present situation and the history in a specified region) to one of these classes should allow for a restricted prognosis of its possible future development, which is a prerequisite for mitigation or adaptation (Kasperson et al. 1995).

In summary, and drawing the information from Table 3.16, there are some generalizable patterns of change that result from recurrent interactions between driving forces, following specific sequences of events. Even though, at the detailed level, these sequences may play out differently in specific situations, their identification may confer some predictive power by analogy with similar pathways in comparable regional and historical contexts (Lambin et al. 2003).

Trajectories of dryland degradation (or desertification), for example, are quite distinct on different continents (Geist and Lambin 2004; Geist 2005). In Central Asia, two central pathways of partly irreversible desertification are the expansion of grain farming into steppe grazing land, triggering soil degradation and overstocking, and the invasion of large-scale hydraulic agro-industries into desert ecosystems that historically supported only localized, traditional oasis farming. The most spectacular outcome, notably in low-lying sea region basins (such as the Aral Sea) and northern China, is a widespread increase in desert-like sand cover, which is linked to the exceptionally strong impact of socioeconomic driving forces such as centrally planned frontier colonization and (sometimes forced) population movements. In contrast, a typically African pathway of desertification involves the spatial concentration of farmers and pastoralists, very often as a result of national sedentarization policies, around infrastructure nuclei and water resources. This local, sometimes forced concentration of population results in overgrazing, intensive fuelwood collection, and high cropping intensities, ultimately leading to degraded vegetation and declining soil productivity during periods of drought. "Beefing up" of drylands, with little or no involvement of cropping, frequently characterizes the desertification pathways of Australia and of North and South America. Historically, these rangeland zones typically shared common patterns of land use, such as the rapid introduction by European settlers of exotic livestock species and commercial pastoralism into ecosystems that had not undergone these uses before. Since about the 1950s, however, these trajectories diverged. In Australia, the livestock industry and its complex of related infrastructure developed sufficient flexibility to counterbalance droughts and avoid spectacular desertification, and in the U.S. Southwest, principal land uses shifted away from cattle ranching to meet urban-driven aspirations. In contrast, Patagonia and northern Mexico suffered from a lack of advanced technologies and alternative land uses or diversification options to deal with the vagaries of oscillating natural resource productivity. Local farmers find themselves with no viable alternative but to continue raising livestock, often under conditions of impoverishment and deprivation. Consequently, dryland degradation in these areas is not just a historical phenomenon, but continues to advance (Geist and Lambin 2004; Geist 2005).

Likewise, some typical pathways can be identified for tropical humid forest regions, and deforestation notably (Rudel and Roper 1997; Lambin and Geist 2003b). In some frontier regions, however, determining prevailing land-use/cover change pathways may be difficult due to complex, rapidly changing dynamics over time. In the case of the Brazilian Amazon, for example, unsustainable cattle ranching appears to have evolved to market chains to satisfy local and national demand for cattle-based products (Hecht 1993; Faminow 1997; Veiga et al. 2004). Thus, a trajectory of land-cover change for the Amazon may start with rubber extraction for the world market (from end of 19th to mid-20th century), which was followed by integration of forested regions into national economic development, mainly through pasture creation (2nd half of 20th century). More recently, cattle ranching that depended heavily on subsidies and land speculation in the 1970s and 1980s evolved into intensified land uses for (semi)urban markets, relying upon well-developed transport and other infrastructure to satisfy local as well as national demand for cattle-based products (Alves et al. 2003). More recently, there are indications that global market demands regain power in local land-use decisions to convert forests for soybean (increasingly) and beef (again). Thus, what appears to be a typically homogenous agricultural frontier pathway in the land-use history of forested mainland South America, related to individual colonists' land-use decisions, is indeed driven by local urban as well as remote economic influences, with strong oscillations and overlaps between poverty- and capital-driven land-use dynamics (Perz 2002; Pacheco 2006a,b).

What has been lacking so far is the development of an integrative framework that would provide a unifying theory for the insights on causes and these pathways of land change, as well as a more process-oriented understanding of how multiple macrostructural variables interact to affect micro agency with respect to land (Lambin et al. 2003). The concept of land-use transition represents a first step in this direction.

3.5.3   Land-Use Transitions

Land-use dynamics have been construed as constituting about a dozen processes. In particular in tropical zones, which are the focus of this chapter, these processes are:

- urbanization (or the increase of built-up areas),
- conversion of forest to cropland (classic expansion, but virtually always intensification),
- conversion of grassland to cropland (classic expansion, but virtually always intensification),
- change of crop on existing cropland (will always entail change in intensity),
- more intensive use of croplands (decreased fallow – up to and beyond double cropping –, change of cultivar, terracing, irrigation, use of chemical and mechanical technology),

- incorporation of trees into cropland (usually considered intensification, when it is an economic species such as coffee, tea, cocoa, or vanilla),
- conversion of cropland to forest (considered disintensification, if abandonment; or intensification if for economic gain),
- conversion of forest to pasture (often cropland as an intermediate step),
- conversion of cropland to pasture (may appear less intensive, but yield higher rewards),
- more intensive use of pasture (usually through increased inputs),
- incorporation of livestock into cropland, and, finally,
- conversion of pasture to cropland.

In the following, we do not provide an integrative or unifying framework for all these land-change processes, but attempt to detail some of the aspects only as laid out above. Even considering just a small number of broad land-use/cover states, a large number of land-change processes are possible. This is illustrated in a very simplistic form in Fig. 3.3. The figure considers just two broad natural land-cover types (forest and grassland), and two broad land-use types (cropland and pasture). Changes among these four classes yield a minimum of twelve possible transitions (only some shown for simplicity). Quite different processes, however, may account for a given transition, yielding a much greater array of land-change processes. For example, cropland may begin to look more like forest because of forest succession due to fallow or farm abandonment, or because farmers replace field crops with arboreal species, i.e., practice agroforestry.

Through a series of transitions, land-use change is associated with other societal and biophysical changes (Raskin et al. 2002; Mustard et al. 2004). A transition can be defined as a process of societal change in which the structural character of society (or a complex subsystem of society) transforms. It results from a set of connected changes, which reinforce each other but take place in several different components of the system. Multiple causality and co-evolution of different sectors of society caused by interacting developments are central to the concept of transition. Transitions in land use must be viewed as multiple and reversible dynamics. They are not set in advance, and there is substantial variability in specific causes and situations (or contexts). There is thus a strong notion of instability and indeterminacy in land-use transitions (Lambin et al. 2003). Transitions should be viewed as possible development paths where the direction, size, and speed can be influenced through policy and specific circumstances (Martens and Rotmans 2002).

The concept of transition has been applied in land-change studies at different spatial and temporal scales. In the early 1990s Alexander Mather began using the term "forest transition" as a shorthand way of summarizing the historical changes in forest cover that occurred in

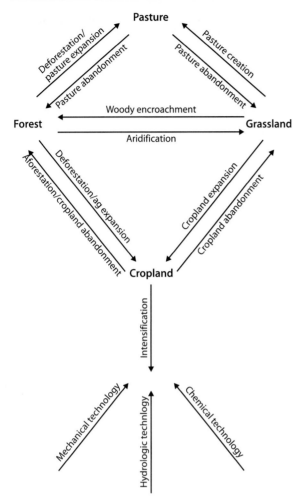

Fig. 3.3. Potential transitions between two land-use/cover states (pasture/cropland, forest/grassland)

Europe during the past two hundred years as European societies underwent industrialization and urbanization (Mather 1992; Mather and Needle 1998; Mather 2001). He saw a series of mainly northern European societies undergo deforestation as rural populations grew during the 18th and 19th centuries. Beginning in the 19th century, the creation of industrial jobs and amenities in cities induced widespread rural-to-urban migration. The departure of rural residents for cities led to the abandonment of the most marginal agricultural lands, and some of these lands reverted to forests. As the extent of abandoned lands grew, a transition in forest cover trends occurred, with net forestation rather than net deforestation coming to characterize these countries (Mather 2006d).

Analyses of forest cover trends during the 1990s suggest that forest-cover transitions take two somewhat different forms. In the more affluent European and American societies labor scarcities in agriculture continue to drive land abandonment and aforestation. In the poorer Asian and African contexts forest product scarcities brought about by the widespread destruction of natural

forests induce landholders to plant trees and, on a larger scale, plantations. The increase in the extent of these re-planted areas largely explains why forestation rather than deforestation now characterizes forest-cover trends in these countries (Rudel et al. 2005).

3.6   Conclusions

This chapter presents a synthesis of the suite of social and biophysical factors that have been associated with land-use change and, thus, land-cover dynamics. At first glance, there seems a universe of land-change studies that presents an effectively unlimited number of land-cover changes, and of associated human and biophysical factors. In general, synthesis of these factors is inherently a process of simplification, and of establishing some order among these factors (e.g., Brookfield 1962; Turner et al. 1977; Petschel-Held et al. 1999). Further examination, mainly by reviewing meta-analytical studies, reveals a limited suite of recurrent core variables of land change or variable configurations, which are detailed above. As a result, the richness of explanations has greatly increased over the last decade, but this has often happened at the expense of the generality of explanations, and no general land-change theory is yet in sight (see Chap. 1).

Nonetheless, over the last decade, research on the causative factors (or causal clusters) has largely dispelled simplifications or "myths" such as that only the growth of the local population, aggregated to a global level, and, to a lesser extent, its increase in consumption were thought to drive the changes in land conditions. Thus, our understanding of the causes of land change has moved from simplistic representations of two or three driving forces to a much more profound understanding that involves situation-specific interactions among a large number of factors at different spatial as well as temporal scales (Lambin et al. 2001, 2003). Concerning the latter, it is well known that explanations for processes vary by the scale at which they are studied. Thus, specificity of scale is essential, but also, ideally, the results of each causative factor analysis should be scaleable, both up and down, from the original scale of analysis (see Chap. 5 and 6). Such improved understanding also helps to account for the growing human capacity to transform vast areas of the land surface through agriculture, the building of roads and dams, and the rise of cities with vast impervious areas (see Chap. 4). For example, for the monasteries in Western Europe it took several centuries to deforest a substantial portion of the landscape in the early to late Middle Ages. By the 19th century, in contrast, it was possible for homesteading farmers to move across the forested lands of North America and cut down most of the existing forests in less than one century. Today, comparable deforestation is possible in a matter of decades – because of much greater technological capacity, favorable government policies, and much larger populations acting simultaneously to make forests into agropastoral and urban areas (Moran 2005). Likewise, one thousand years ago, a postulated combination of factors including population growth, political instability and warfare, environmental degradation and climate change may have led to the collapse of the ancient Mayan Civilization (Diamond 1997, 2005; Turner et al. 2004), a situation threatening to repeat itself in today's forests in southern Mexico, Belize and northern Guatemala (the largest contiguous tropical moist forest remaining in Central America) with its current inhabitants, in spite of a much lower population and a much shorter time frame (Sever 1998; Sader et al. 2004).

The synopsis presented in this chapter relies upon case study material, and while the breadth and depth of that literature is to be celebrated, its idiosyncrasy is a major impediment. The meta-analyses by necessity depend on *ex post* operationalization of variables, which will be inherently unsatisfactory. Greater success may be expected from case studies undertaken with comparative analysis in mind from the outset. Now that a more coherent set of relevant factors has been codified, this should be increasingly likely.

While the causes and trajectories of certain land-change processes are commonly analyzed (e.g., deforestation), there is no consensus on specific definitions. These depend upon the observational perspective used, which in turn depends on the observer's analytical purpose. Likewise, the optimum organization of causal and contextual factors depends on their intended use. A researcher whose objective is a critique of existing land-related policy will likely call upon a different theoretical framework than one interested in generating a model capable of predicting spatio-temporal trends in net primary productivity. In part, these are issues of differing spatial and temporal scales of analysis, but it is important to remember that land change in and of itself is generally an intermediate analytical outcome. Since changes in land cover reverberate throughout the ecosystem, the impacts are many (see Chap. 4), and different causal and contextual factors are likely relevant. It must be recognized that with multiple stakeholders come multiple sets of values. Different stakeholders have values that are often not part of how scientists study land change, and even individual stakeholders may have internally inconsistent values (see Chap. 5 and 6).

Chapter 4

Multiple Impacts of Land-Use/Cover Change

Abha Chhabra · Helmut Geist · Richard A. Houghton · Helmut Haberl · Ademola K. Braimoh · Paul L. G. Vlek
Jonathan Patz · Jianchu Xu · Navin Ramankutty · Oliver Coomes · Eric F. Lambin

4.1 Introduction

Local changes in land-use/cover are so pervasive that, when aggregated globally, they may significantly affect central aspects of the Earth System functioning and thus life support functions and human livelihoods. Estimates of the areal extent, spatial expression or likewise quantitative estimate of the impact of land change more or less converge, while estimates driven by notions of "imprint", "impress", "footprint" or "carrying capacity" are larger, thus appearing more dramatic – see Box 4.1 (and Chap. 2).

Six observations seem noteworthy, though. First, understanding of land-use transitions is crucial since the most profound impacts usually occur during periods of transitions between different land-use states (Mustard et al. 2004). Second, the rate of expansion of land-use systems across the world over the past 300 years has not been uniform, but has followed a path which was mostly determined by European economic and political control or colonization (Richards 1990), cross-cut at present by the various modes of globalization (Lambin et al. 2001).

Third, impacts are scale-dependent in that some affect the local environment (e.g., local water quality), while other impacts extend far beyond the location where they arise (e.g., carbon cycle, climate change) (Mustard et al. 2004). Fourth, not all land changes have global impacts, and, fifth, not all land changes are irreversible. Finally, multiple impacts may overlap and reinforce each other, with some mitigating and even canceling each other. This implies the existence of various multi-directional impacts on both ecosystems and people, with biodiversity loss and soil degradation possibly being the sole truly irreversible global environmental change impacts.

As a matter of fact, many if not most impacts are associated with "positive" influences such as continuing increases in food and fiber production, resource use efficiency, wealth, livelihood security, welfare and human well-being (Lambin et al. 2003). Indeed, land-use/cover change as a mega-trend or "forcing function in global environmental change" (Turner 2006) conforms generally with the development of human societies and civilizations – at least over the last 300 years. However, when looking back at historical to ancient impacts, some land-use/cover changes

---

Box 4.1. Rough estimates of the spatial magnitude of land-use/cover change impacts

About 50% of the ice-free surface of the Earth is considered to be modified by human action (Vitousek et al. 1997). The estimate conforms with a recent assessment stating that roughly about half of the world's land surface is still covered by "wilderness" areas (Mittermeier et al. 2003). Other estimates of the transformation of natural ecosystems for the cultivation of food, feed and fiber come to the conclusion that 37 million ha, or 34% of the global land surface, is directly occupied by cultivated systems, which represents the greatest single use of terrestrial ecosystems by humans, with about three quarters of the world's population living within the boundaries of those managed ecosystems. In a non-trivial manner, cultivated systems overlap with other ecosystems or biomes such as forests, mountains and drylands (Cassman et al. 2005).

Likewise, estimates of the amount of net primary production (NPP) "appropriated" by humans – i.e., either harvested and consumed by humans and their domesticated animals or diminished through ecosystem degradation, soil sealing, etc. – demonstrate human domination of ecosystems through land-use/cover change. According to Vitousek et al. (1986, 1997) up to 40% of terrestrial NPP are directly or indirectly used by humans or foregone due to changes in productivity caused by current or past land use or land-cover change. A recent study revealed that human consumption of NPP is around 11.5 Pg C or 20.3% (uncer-

tainty range: 14.1–26.1%) of current terrestrial NPP (Imhoff et al. 2004) (note, however, that this study did not include NPP foregone due to land-use/cover change). Spatially explicit studies on a national scale have shown that the human appropriation of NPP (HANPP) is considerably higher in densely populated, intensively used industrialized countries such as Austria. While aboveground HANPP for Austria as a whole was around 50% in the late 1990s, HANPP reached over 90% in intensively used regions (Haberl et al. 2001). However, an uncertainty in detailed estimates or a wide range in estimates remains (Rojstaczer et al. 2001).

Most of the fertile lands of the world are already under cultivation with relatively little scope for further expansion – if not into humid forest ecosystems or into drylands (Young 1999; Döös 2002). It is likely that on the populated surface of the Earth, there is practically no place where cultivation of trees, crops, livestock and fisheries does not occur (Cassman et al. 2005). Today, virtually no land surface remains untouched in some way by humankind (Turner 2002), especially if one considers that humans have indirectly influenced global ecosystems by changing the chemistry of the atmosphere. A recent study by the World Conservation Service, including urban-industrial uses and taking into account remote urban influences beyond the agricultural sector, estimates that the wider "human footprint" covers 83% of the global land surface (Sanderson et al. 2002).

triggered the decline and collapse of whole societies (see Sect. 2.2.1). In addition, some contemporary influences on climate and ecosystem services and conditions can clearly be associated with undesirable or "negative" influences. Altering ecosystem services – i.e., the benefits people obtain from ecosystems such as provisioning services (e.g., food, water), regulating services (e.g., flood and disease control), cultural services (e.g., spiritual and recreational benefits), and supportings services (e.g., nutrient cycling) that maintain the conditions for life on Earth (Millennium Assessment 2003) –, affects the ability of biological systems to support human needs (Odum 1989; Ojima et al. 1994; Vitousek et al. 1997; Cassman et al. 2005).

For land-use patterns to be sustainable, the legitimate concerns over the losses of certain ecosystem functions of global importance (such as carbon stocks and biodiversity) need to be balanced with the equally legitimate interests of national economic development and sustaining local livelihoods. Therefore, trade-offs need to considered between what is to be sustained and what is to be developed, taking into account that synergetic effects and complementarities may exist or arise. This implies that a consideration of the impact of land-use/cover change in situ, i.e., to reconcile ecosystems and human well-being at the local level, needs to be complemented by looking at external impacts upon other people and ecosystems as well. Thus, trade-offs, complementarities and externalities may lead to some important land changes such as urbanization, industrialization, tropical deforestation, desertification and agricultural intensification (Asner et al. 2004; DeFries et al. 2004a; Stoorvogel et al. 2004; Geist et al. 2006).

In this chapter, we adopt the notion of a trade-off analysis and provide a descriptive summary using a case-by-case approach of the extent and nature of trade-offs between ecosystem goods and services extracted by humans through their land-use practices and any resulting socio-ecological implications. When data permit, we address principally three levels, which are (a) the co-option of natural grasslands, wetlands and original (native) forest or woodland through land-cover conversion, (b) the choice of production system, and (c) the precise way in which production systems are managed at both plot and landscape level (DeFries et al. 2004b; Cassman et al. 2005) – see Box 4.2.

A key issue is how positive or negative impacts of land-use change are distributed among stakeholders. Wherever possible, we mention the social, economic and political concerns of land users and various other groups of stakeholders involved, but do not address their concerns in detail or develop mitigation measures since this is dealt with in more detail in following chapters. Also, inherent to our summarization is an assessment of resistance and resilience of the system (if substantiated by data), because multiple changes and impacts also determine, in part at least, the vulnerability and thus resilience of people and ecosystems to climatic, economic, and/or sociopolitical perturbations (Kasperson et al. 1995; Turner et al. 2003c).

We summarize mostly – but not exclusively – from research carried out by the core projects of the International Geosphere-Biosphere Programme (Lambin et al. 2003; Steffen et al. 2004). We review current state of knowledge on the interlinkages of land-use/cover change with, and consequently the impacts upon, (a) the provision but also lack of food, feed, and fiber; (b) the immediate consequences for human health and the risk of spread of diseases; (c) atmospheric chemistry (methane, NO_x), climate regulation (via albedo, water and carbon cycle) and life support functions; (d) agrodiversity, biodiversity losses and how this relates to human well-being; (e) nutrient cycling and soil conditions (such as degradation and erosion) and how this feeds back upon mainly rural livelihoods; and (f) freshwater hydrology and coastal zones (from where the overwhelming part of humans derive a livelihood).

4.2 Provision and Lack of Food, Feed, Fiber, and Timber

4.2.1 Overview

Since the control over fire and the domestication of plants and animals, human land-use activities spread over about half of the ice-free land surface, mainly by reducing forest cover from 50 to 30% of the Earth's land, but still leaving undisturbed (or wilderness) areas on slightly less than half of the Earth's land surface (Ball 2001; Mittermeier et al. 2003). Agriculture has expanded into forests, but also savannas and steppes, in all parts of the world to meet the demand for food, (animal) feed and fiber. Agricultural expansion has shifted between regions over time, and this followed, and still reflects, the general development of civilizations, economies, and increasing populations (Richards 1990; FAO 2004b). Currently, agricultural land uses occupy about a third of the global land surface, with croplands using about 12% or nearly 18 million km^2 (an area roughly the size of South America) and pastures about 22% or 34 million km^2 (an area roughly the size of Africa) (Foley et al. 2003) (see Chap. 2).

At least over the past half century, the transformation of the land surface into cultivated systems has been highly successful at producing food and fiber, for both human consumption and animal feed (FAO 2004b). This has mainly been due to productivity increases since about 1960 rather than to area expansion of croplands and pasture. Historically, this is unique because humans have increased agricultural output mainly by bringing more land into production. Many of the transformed or modified ecosystems continue to be associated with increases in resource use efficiency, wealth and well-being (Lambin et al. 2003). For example, per capita production of agricultural produce has more than kept pace with population growth globally, and food prices dropped by around 40% in real terms over the last decades (Wood et al. 2000; FAO 2004b; Cassman et al. 2005). However, growth rates in food production are now slowing: it was 3% per year in the 1960s, dropped to 2.3% in the 1970s, and then further decreased to 2% during 1982–1992 (Alexandratos 1995).

At least four aspects need to be taken into consideration. First, there is considerable variation in productivity among various land-use or production systems. Second, local- to regional-scale food shortages continue to persist. Third, the dominant process over the last decades of land-cover modification through agricultural intensification bears immediate and partly detrimental consequences for human health as well as ecosystem conditions. And, fourth, contributions sought from the exploitation of biomass resources to meet human energy demands in the future require more attention.

Box 4.2. Land-use change in Southern Ecuador as a response to population pressure, land-reform programs and increasing demand for food and timber

Veerle Vanacker · Catholic University of Louvain, Louvain-la-Neuve, Belgium

The analysis of aerial photographs over a 33-year period (1962–1995, Fig. 4.1a–c) shows that land use in mountainous catchments in the Southern Ecuadorian Andes is highly dynamic as a response to land-reform programs of the 1960s and 1970s, and a strong population increase. Forest is increasingly replaced by grassland while old grasslands are now used as cropland. Despite the increased pressure on the land, the upward movement of agricultural activity and the concurrent deforestation, the overall forest cover did not decline. Deforestation in the uplands is compensated for by a regeneration of secondary forest on abandoned rangelands and afforestation with Eucalyptus trees on degraded land. Despite the accelerated land-use changes, the area affected by water erosion decreased. Stabilization is mainly due to reforestation of highly degraded areas (Fig. 4.1d), driven by the increasing demand for commercial wood (Vanacker 2003).

Fig. 4.1. Land-use change in the Southern Ecuadorian Andes. **a** 1962; **b** 1989; **c** 1995; **d** view at *point 1* (2000)

4.2.2   Variations in Land Productivity

Over the last half century, productivity increases through land-use intensification has been the primary source of output growth, globally, and in particular in developed countries. In developed countries, low population growth rates go hand in hand with high productivity increases, and the area of cultivated land is stabilizing or even contracting in countries such as Italy, Australia and the United States of America (Cassman et al. 2005). There is evidence that the same process is happening in China, but it has to be noted that national statistics in developing countries often substantially underreport agricultural land area (Young 1999; Ramankutty et al. 2002), by as much as 50% in parts of China, for example (Seto et al. 2000).

The 1.97-fold increase in world food production from 1961 to 1996 was associated with only a 10% increase of land under cultivation, but with a 6.87- and 3.48-fold increase in the global annual rate of nitrogen and phosphorus fertilization (Tilman 1999). Irrigated land had a 1.68-fold increase and showed a rather uneven distribution across the continents. The global cropland area per capita decreased by more than half during the 20th century, a trend which globally has freed more than 200 million ha from agricultural use since 1900. In contrast, area expansion of cultivated land has been the principal source of output growth in many developing countries, mainly located in Sub-Saharan Africa. There, high population growth rates coincide with low productivity (see Chap. 2).

In sum, the mix of cropland expansion and agricultural intensification has varied geographically. Tropical Asia increased its food production mainly by increasing fertilizer use and irrigation, while most of Africa and Latin America increased their food production through both agricultural intensification and extensification. Western Africa is the only part of the world where overall cropland expansion was accompanied by a decrease in fertilizer use (−1.83% per year) and just a slight increase in irrigation (0.31% per year compared to a world average of 1.22% per year). In Western Europe and the northeastern United States, cropland decreased during the last decades, after abandonment of agriculture or, in a few cases, following land degradation on mainly marginal land (FAO 2001c). Despite claims to the contrary, the amount of suitable land remaining for crops is very limited in most developing countries, where most of the growing food demand originates. Where there is a large surplus of cultivable land, land is often under rain forest or in marginal areas (Young 1999; Döös 2002), and both tropical humid and dry forests are lost rapidly. In tropical forest ecosystems, both agricultural expansion through deforestation (for food crops), and land-use intensification (for cash crops) often involves issues of demography, economic livelihood security and differential vulnerability (Lambin and Geist 2003b) – see Box 4.3.

4.2.3   Food Insecurity and Poverty

Land-cover change clearly has implications for land-dependent livelihoods, and a variety of livelihood activities may be influenced by land-cover change. Conversion of forests into agro-industrial agriculture, for example, may lead to fewer opportunities for wild food collection, hunting and forest dependent artisanal activi-

---

Box 4.3. The impact of forest-coffee conversions on Honduran livelihood

The study region comprises an area in western Honduras, including the municipio of La Campa, where both socio-economic and land-cover transformations are occurring, as interactions with world markets intensify (as is the case with much of the country). The population of La Campa nearly doubled between 1961 and 1988, and has since continued to grow. Increasing scarcity of prime land and the introduction of chemical inputs have motivated farmers to abandon their more unproductive and marginal areas of cultivation and establish more permanent agricultural fields. Land-use transformation in the area are both mediated and shaped by relevant policy. Farmers have responded favorably to national initiatives and credit availability for export coffee production. Furthermore, municipal governments have reacted to a national subsidy by making road improvements in coffee producing areas. Between 1987 and 1991, more accessible areas experienced greater deforestation and fragmentation as could have been expected. However, between 1991 and 1996 this trend reversed. Increased deforestation is found at higher elevations reflecting the recent expansion of shade grown coffee for export, while forest regrowth becomes apparent in lower areas that are less suitable for coffee or more intensive agriculture.

Land-cover changes, population growth, and export coffee production have had varying implications for vulnerability. Coffee production has provided many households with income to invest in children's education. As a result, coffee has contributed to increasing levels of education and salaried employment, which may buffer risks related to agriculture. For most households, coffee has added a source of income to a diversified strategy of producing staples for consumption, pottery production, and wage labor (including coffee-picking). For some households, improved transportation and market linkages provided new options to survive the "hungry time" before the harvest. Out-migration may be increasing, thus reducing local population pressure and potentially providing remittances.

However, for some, the expansion of coffee has actually had detrimental effects, increasing vulnerability. This is especially evident for land-poor households. First, it has led to the inequitable accumulation of wealth and land for those able to invest extensively in coffee, and undermined traditional relationships of reciprocity. Second, coffee price volatility has increased vulnerability to market shocks among the few households that have converted all of their land to coffee. If these processes of change continue, the economic advantages that coffee brought for the majority of households may be undermined, resulting in increased vulnerability, particularly, if, as predicted, global environmental changes result in a more variable climate and increased crop failures (Southworth and Tucker 2001; Southworth et al. 2002; Tucker 1999; Nagendra et al. 2003; Munroe et al. 2002).

ties. Declining productive capacity of the land may impoverish land-dependent livelihoods, just as increasing productive capacity may lead to enrichment, with possible interlinkages between the two modes (Barbier 2000a,b; Barrett et al. 2001). Poverty can lead individuals, households, communities and even states to make short-sighted land-management decisions. Degrading land resources, in turn, have an impact on the human livelihoods that depend on them, and these livelihoods are assumed, in many instances, to be those of the poor (Reardon and Vosti 1995). The situation of deprived livelihoods is frequently posited both as a potential cause and/or consequence or impact of land-use/cover change (Gray and Moseley 2005; Moseley 2001, 2006).

As a matter of fact, neither is poverty alleviated nor is food security achieved, with the latter defined as the "access by all people at all times to enough food for an active, healthy life" (World Bank 1986). Local- to regional-scale food shortages, often related to poverty, continue to exist. The number of chronically undernourished people has remained constant at about 20% of the population of less developed countries, which is slightly less than one billion people, and over 200 000 continue to die from lack of food every week. This recent analysis also addresses the potential impact of climate change on food security, highlighting the importance of the current disparities in food production capability and of the differential impacts of climate change. It is expected that 78 individual developing countries that have low per capita incomes and account for 600 of the 800 million undernourished people will be among the losers of climate change (Shah 2002).

Significant, still unsolved trade-offs exist between local livelihood security, national development and global environmental change concerns such as climate change, carbon sequestration and biodiversity. This raises considerable issues of fairness and equity, because developing countries account for about 80% of the global population, but only about 25% of cumulative global CO_2 emissions over the past 50 years or so, and will very likely suffer substantially from changing climate in terms of food production (Steffen et al. 2004).

In this regard, the chronically nutrient-deficient pasture soils in tropical and subtropical Africa, Latin America and Australo-Asia are crucial: *(a)* they concentrate 90% of all global pastures, mainly in developing countries, and *(b)* they have an extraordinarily large number of rural poor associated with livestock systems, especially in drylands (Fisher et al. 1994; Reid et al. 2004).

Deficiencies in data and methodological development, however, impede an exact estimation of current food shortages at a variety of spatial and social scales (Moseley 2006). Historically, many national governments and relief agencies have operationalized food shortages at the national level via an accounting procedure known as the food balance sheet approach (used to establish whether there is adequate food supply to meet demand by calculating national foods needs – population times per capita grain needs, comparing them to the sum of agricultural production, stocks and net imports). However, the unexpected Sahelian famine of the mid-1980s and the nature of exchange entitlements – referring to a person's legitimate claims to available food – demonstrated that the food balance sheet approach was seriously flawed in equating food supply with food access. While food supply or availability might be sufficient at the national level, it may be inaccessible to certain segments of society due to high prices or insufficient income (Sen 1981; Cannon 1991; Watts and Bohle 1993a,b).

Likewise, deficiencies in data and methodological development impede an exact estimation of poverty at a variety of spatial and social scales (Moseley 2006). The way in which poverty is conceptualized has an influence on who is defined as poor and how their interactions with the landscape are perceived. Poverty may be conceived of in relative terms (e.g., comparing households within a community, regions within a state, or states with one another). It may also be measured against a specific benchmark, such as an international poverty line. Finally, poverty may be assessed in a variety of ways, for example, in terms of monetary wealth or income, certain types of assets, or entitlements (Swinton et al. 2003).

Despite these deficiencies, there is general agreement about both a major trend that relates to increasing populations in most of the developing world where a considerable part of people continue to lack means to purchase food for self-sufficiency, and a contrasting trend that relates to rising incomes associated with food consumption and diets richer in meat in most of the developed countries (Naylor 2000). There is growing local case study evidence from developing countries that suggests that very often a circular process of dynamic adaptation is at work (Broad 1994; Scherr 2000; Moseley 2001). In many instances, this process has been depicted as a vicious cycle wherein poor households exert excessive stress on their environment in order to survive, and this degraded environment further impoverishes the household (Kates and Haarman 1992). However, it is also quite possible, and well documented in several instances, that poor households will diversify into other activities (Ruiz-Perez et al. 2004), or simply leave the land (see Chap. 3).

4.2.4 Worsening Conditions for Food and Fiber Production?

Land-cover modifications resulting from agricultural intensification, as the dominant land-change process over the last decades, bear direct and partly detrimental consequences for ecosystem conditions at the local scale, but also for human health (see Sect. 4.3). This may impinge upon worsening conditions for local food production, namely the quality and sufficiency of food. Limiting factors which may potentially trigger reduced production levels, or limits to production growth, include limits to biological productivity, increasing scarcity of water, declining effects of additional fertilizer and pesticide applications, and climate change, i.e., changing atmospheric composition and chemistry (see Sect. 4.4 and 4.7). For example, reduced winter snow cover and summer water supplies in some major agricultural production zones of the temperate and subtropical zone will limit water for agriculture under increasing competition, and increased temperatures may trigger mainly negative impacts on crop yield increases worldwide (Steffen et al. 2004). Likewise, the loss of soil fertility and degradation of agricultural lands has become an important issue in some long-settled agricultural areas as well as in agricultural frontier zones (see Sect. 4.6).

Complementing the direct and more localized effects, there are remote, large-scale regional feedbacks on food production systems stemming from air pollution as a consequence of fossil fuel burning, including land use-related vegetation fires. NO_x, emitted largely though fossil fuel burning, contributes critically to the creation of photochemical smog, including ozone (O_3), which is harmful to crop production (Steffen et al. 2004). Enhanced by summer-time high-pressure systems, repeated exposure to high levels of O_3 potentially reduces crop production by 5 to 10%, with the damage increasing with the magnitude and intensity of exposure (Chameides et al. 1994). Likewise, haze from regional air pollution in China, though largely linked to energy and less to agricultural production, is estimated to reduce about 70% of the crops grown by as much as 5 to 30% below their optimal value (Chameides et al. 1999).

There are potential impacts upon agricultural systems and local to global food production, including the vulnerability and resilience of large sections of human population especially in developing countries. From various modeling approaches, a gain in cereal production is expected in the range of 20 to 230 million t at the global level, if 1995 data are projected into 2080. However, in spite of this positive global outcome, there seems profound concern for many developing countries that will probably lose production due to climate change (Shah 2002) (see Chap. 6). It had been observed that in rice-producing areas the average yield of rice declined by about 10% in the 1979 to 2003 period for each 1 °C increase in growing-season minimum temperature in the dry season, providing a direct evidence of decreased rice yields from increased nighttime temperature associated with global warming (Peng et al. 2004). On a regional scale, several attempts have been made to estimate the impact of climate change especially on the length of the growing period (see Box 4.4).

Box 4.4. Impacts of climate change on the length of the growing season in Africa

Throughout tropical Africa, the length of the growing season will become shorter except for a band extending about 7° north and south of the equator, where the growing period will lengthen. Thus, pastoral lands in the Sahel, in southern Africa, northern central Africa and Ethiopia will become drier. The only rangelands in Africa that will become wetter are in Kenya, northern Tanzania, parts of southern Ethiopia and southwestern Uganda. Despite improvements in areas near the equator, eastern Africa as a region will lose 20% of its land suitable for a variety of crops, with nearly a quadrupling in the area suitable for very short season crops. Southern, western and central Africa will see an overall drying and strong increases in arid land (Jones and Thornton 2002).

The potential future impact of climate change on global food production and food security has been studied using a combination of climate model simulations, crop models, and world food trade system models. Parry et al. (2004) found that climate change would increase yields in mid and high latitudes, and decrease yields in the tropics, and that this effect would worsen with time. They used simulated climate under four different future scenarios of the Intergovernmental Panel on Climate Change (IPCC), and found that, for the most part, the world would continue to feed itself under all climate change scenarios; however, this global outcome arises from increased production in developed countries compensating for the decrease in developing countries. They concluded that while global production may remain stable, regional differences in crop production are likely to exacerbate.

It can be assumed that any of the above factors alone may impede efforts towards increasing production and yield; together, these biophysical factors present problems in increasing sustainable agricultural production. On the other hand, some global and regional environmental changes may benefit agriculture. For example, modeling-driven studies raise expectations that elevated levels of carbon dioxide may trigger increases in crop production and water-use efficiency in some crops, and that in many areas growing seasons will be extended and frost frequencies be reduced, with nearly all of the gains expected to be in the already food-rich countries. However, many of the global change impacts on agricultural land use are highly interactive and the overall consequences on yields and food quality are complex and difficult to predict (Steffen et al. 2004).

Food production may also have to compete for land with other biomass production schemes, above all those aiming at increased availability of biomass for technical energy use. Combustion of biomass – i.e., fuelwood, harvest residues, dried dung, etc. – probably is the oldest source of technical energy used by humans. For a variety of reasons, the use of biomass as source of technical energy as well as raw material for human use has been increasingly promoted: *(a)* biomass is a renewable resource, for which, in contrast to fossil fuels, there should be no problems of resource exhaustion; *(b)* since biomass combustion adds to the atmosphere CO_2 that had previously been absorbed during plant growth, it is often seen as a CO_2-neutral source of energy which, if substituted for fossil fuels, may help to mitigate global warming; *(c)* biomass use is seen as a strategy to diminish the dependency from foreign markets with regard to resource supply and to promote rural economic development, as many countries have productive areas at their disposal on which they can grow biomass for fiber and energy (Sampaio-Nunes 1995).

However, data on the current contribution of biomass to energy supply are considerably uncertain, because biomass used in developing countries, above all in their rural areas, are almost completely unrecorded in statistics. Almost all current reviews assume that biomass currently contributes some 45 EJ yr^{-1} (1 EJ = 10^{18} Joule; uncertainty range: 35–55 EJ yr^{-1}) to humanity's energy supply, that is 9 to 13% of the global supply of technical energy (Hall et al. 1993; Nakicenovic et al. 1998; Turkenburg 2000). Note, however, that all these studies directly or indirectly cite one rather crude estimate (Scurlock and Hall 1990) that was mainly based on an assumption of the per-capita use of biomass for energy provision in rural and urban areas in developing countries (for comparison, 45 EJ yr^{-1} is about 2% of total global terrestrial NPP).

The projected growth of world population – i.e., expected to be around 8 billion in 2030 and between 7 and 11 billion in 2050 (Lutz et al. 2004) – and likely improvements in human diets are strong drivers of further increases in the demand for biomass as food and feed. Many energy scenarios nevertheless also predict strong increases in the amount of biomass used for energy provision. For example, the IPCC-SRES scenarios assume global biomass energy use to rise to 52–193 EJ yr^{-1} in 2050 and 67–376 EJ yr^{-1} in 2100 (Nakicenovic and Swart 2000). Likewise, WEC/IIASA scenarios reach similar values (Nakicenovic et al. 1998). The global potential for biomass energy provision has been estimated to be in the order of magnitude of current global technical energy use, i.e., around 400 EJ yr^{-1} (Fischer and Schrattenholzer 2001), although other studies have even reported a global potential for biomass energy of over 1 000 EJ yr^{-1} (Hoogwijk et al. 2003). Although some modeling studies address potential competition for land resources required to satisfy food and energy demand (Leemans et al. 1996), more in-depth research seems warranted to clarify not only these potential conflicts of interest, but above all the ecological implications which such grand schemes might have (see Chap. 5 and 6).

4.3 Disease Risk and Human Health

4.3.1 Overview

Human health is arguably the most complex of the major types of global change impacts on societies, and understanding how to prepare for the impacts on and improve the resilience of public health systems is surely one of the grand challenges ahead. There is a long history of anthropogenic, mainly land use-driven changes to the environment which were either positive for human health, or posed problems for animal and ecosystem health as well as for human well-being. Health is strongly linked to both ecosystem provisioning and regulating services (such as food production or flood/disease control), but also to cultural services through recreation and spiritual benefits. Included into ecosystem services are those influences, which govern the distribution of disease-transmitting insects and of irritants and pathogens in water and air (Millennium Assessment 2003).

A variety of trade-offs needs to be addressed, with a most important one being between food security and altered habitats promoting deadly diseases (Patz and Norris 2004). For example, wilderness areas in dry tropical zones, but also the abandonment of cultivation and spread of semi-natural vegetation conditions (usually a sign of increasing biodiversity), promote the treat of sleeping sickness. Throughout much of Sub-Saharan Africa, African trypanosomiasis causes disease in people, livestock, and wildlife. Early observations in Nigeria suggested that the distribution of the plant *Mimosa asperata* was almost synonymous with presence of the fly, and, similarly, increased growth of the bush *Lantana camara* in Uganda has been associated with disease outbreak (Berrang Ford 2006a). Changes from cotton and coffee plantations to uncultivated Lantana bush, resulting from social and economic upheaval, thus created ideal tsetse habitat in southeastern Uganda, leading to a substantial sleeping sickness outbreak in the 1980s, for example (Gashumba and Mwambu 1981).

In the past, negative impacts such as those on water and air quality conditions were more or less limited to local to regional scales, affecting millions of people though. The "Black Death", for example, did not occur at the global scale but affected many countries at the same time, notably in 1347 around the Mediterranean. Accumulating evidence, however, suggests that current local land-use actions – such as irrigation farming, urban sprawl, or road construction, in combination with other causative proximate and underlying factors are responsible for altered risk of many diseases such as malaria (Patz et al. 2000), African trypanosomiasis (Berrang Ford 2006b), dengue fever (van Benthem et al. 2005), and onchocerciasis. In addition, there may be synergistic effects of land use and global climate change (Patz et al. 2004).

Whatever biophysically driven impacts of global change on human health can be postulated, it is clear that the differing vulnerabilities of countries or societal sectors will be often the decisive factor in determining whether, for example, a serious infectious disease pandemic breaks out, or not (Turner et al. 2003c). Therefore, indirect changes in the human component of the Earth System may have the most critical effects on health. For example, deterioration of public health systems due to an increased need of society to cope with the more direct impacts of global change could leave populations more vulnerable to disease. Likewise, large population groups in developing countries weakened by poor water quality, malnutrition and hunger will also be more vulnerable to health problems as global change accelerates (Vogel 2006). As in many aspects of global change, one should expect surprises as the Earth System is subjected to a suite of increasing human pressures.

4.3.2   Spread of Vector-Borne Infectious Diseases

In the 1990s, vector-borne infectious diseases killed around 13 million persons annually throughout the world, most of them children of less than 4 years, and they continue to be the major cause of premature death in persons 0 to 44 years old (World Health Organization 1999). These diseases are particularly sensitive to land-use/cover changes because their spatial redistribution is restricted by the geographical range of the vector and by its habitat preferences. Land-use change and vector ecology control the interactions between hosts and vectors, given the use of different land parcels by people, the breeding habitat of specific vectors and their dispersal through the landscape (influenced by landscape pattern and heterogeneity). If the impact of land-use change on vector-borne diseases can be better assessed, then potential disease prevention decisions will (or should likely) affect future land transformation. Effective prevention of vector-borne disease is often at the environmental level through vector control measures, especially in an era of drug and pesticide resistance.

Malaria, for example, is a life-threatening parasitic disease that directly relates to associations with land cover. The Italian term "mal aria" means "bad air" due to an association with typical vector habitats such as fetid swamps and marshes. It is caused by the vector-borne, protozoal parasite *Plasmodium* spp., and transmitted by the Anopheles mosquito. Temperature, rainfall, humidity, sun exposure, soils and hydrology – key to determining surface water availability – combine to affect reproduction and breeding site availability for the mosquito vectors (Patz et al. 1998). *Anopheles gambiae*, for example, which is the principal malaria vector in Africa, breeds well in stagnant water with high sunlight, such as irrigation ditches and pools. Such species respond to rapid

changes in their habitat, and are particularly susceptible to temperature and moisture changes. Thus, land-cover modifications related to measures of agricultural intensification – namely, the spread of irrigation farming resulting in changes to temperature and moisture – can trigger dramatic increases in vector populations and malarial transmission rates. Temperature and moisture also play an important role in transmission through their influence on mosquito survival and parasite development rates, and mosquito feeding behavior. Nonetheless, land-use factors have contributed substantially with both historical and contemporary declines and expansions of malarial distributions (Oaks et al. 1991; McMichael 2001; Patz and Wolfe 2002; Norris 2004; Berrang Ford 2006a).

Today, an estimated 40% of the world's population is at risk of malaria, mostly those living in the world's poorest countries. Malaria is endemic to tropical and subtropical regions where it causes more than 300 million acute illnesses and at least one million deaths annually. Sub-Saharan Africa is the global hot spot of malaria transmissions which causes 90% of global deaths there, mainly among young children. Many who survive an episode of severe malaria usually suffer from learning impairments or brain damage. Pregnant women and their unborn children are also particularly vulnerable to malaria, which is a major cause of perinatal mortality, low birth weight and maternal anaemia (World Health Organization 2004; Vogel 2006).

In some regions of India, for example, malaria is endemic, and the major socio-economic parameters of malarial incidence point to an interaction of new agricultural practices (irrigation farming, in particular), urban settlement extension, and poverty conditions. In particular, rice farming creates large areas of stagnant waters that are suitable breeding grounds for about twenty *Anopheline* species. Forests, where the majority of tribal populations reside, are reservoirs of high levels of malaria. Deforestation, mainly done for development projects out of economic pressures, allows new vectors to invade the forest fringes, producing epidemics, especially in the non-tribal non-immune people who move to these areas for jobs. Therefore, malaria was not only looked upon as a cause of mortality and morbidity in India, but also as a major constraint in ongoing development efforts so that a national malaria eradication programme was started in 1950 (Sharma et al. 1994; Sharma 1996). Taking the case of deforestation in India, this is not to claim that the sign of relationship between forest clearance and disease emergence is necessarily the same everywhere in the tropics – see Box 4.5. In north-eastern Thailand, for example, where the malaria vectors breed in the forest, deforestation reduced malaria risk. This has been the case as long as forest cover was replaced by field crops (such as cassava). However, had field crops been replaced by tree crops (such as rubber plantation), the latter became a suitable habitat for *Anopheles dirus* as a very efficient vector (Rosenberg et al. 1990; Molyneux 1997, 1998). Likewise, taking the case of India, the relationship between malarial incidence and introduction of irrigation schemes is not always straightforward. In most of Sub-Saharan Africa, for example, where malaria is stable, a "paddies paradox" exists, i.e., crop irrigation has little impact on malaria transmission, with higher malaria incidences in surrounding areas than in the irrigation zones (Ijumba and Lindsay 2001). One theory is that rice paddies boost the population density of mosquitoes to a level, for which people begin changing their behavior to avoid being bitten (e.g., increased use of bednets).

Malaria was once more widespread but it was successfully eliminated from many countries with temperate climates during the mid 20th century. However, there is a considerable resurgence of the disease in parts of the Mediterranean and Central America, in northern South America, tropical and subtropical Asia, Russia and newly independent states in Central Asia, i.e., new regions of the world but also areas where malaria had been eliminated (World Health Organization 2004).

Until the 18th and 19th centuries, malaria occurred in the United States, some parts of Canada, and most Northern European countries. A dramatic decline in England and northwestern Europe during the late 19th century was due, in a large part, to land-use changes which included improved drainage and extensive land reclamation, increased cultivation of root crops and increased livestock-keeping. The latter may have diverted mosquitoes from humans to animals, reducing potential transmission of the parasite within the human population (since malaria has no significant animal host). However, climatic conditions may have been responsible, in part, for the relative ease by which malaria was eradicated from temperate climate regions. This relates to the "vectorial capacity" of malaria (that is, the efficiency of spread of disease from the first infected person). In cooler climates, small environmental modifications were enough to lower transmission and see the disease disappear. In the tropics, however, climate conditions are such that major interventions would be required due to the climate and environment highly conducive of transmission. In recent years, malaria has experienced a global resurgence which is primarily associated with local land-use changes related to intensification measures aggregating at the global scale. A significant number of the detailed causative factors of malarial re-emergence are strongly correlated with proximate causes and underlying driving forces of land-use/cover change such as wood extraction and forest removal, agricultural intensification, infrastructure extension, urbanization and human population dynamics. In most of the cases, land-use factors do not only operate in combination with each other, but also operate causative to or aside with other factors such as climate change, increased insecticide resistance, socio-political instability and international travel (Kuhn et al. 2002; Berrang Ford 2006a).

Another example of a vector-borne parasite causing disease in people, livestock, and wildlife, not globally but throughout much of Sub-Saharan Africa, is African trypanosomiasis. It is both affected by land-use/cover patterns, and affects land-use/cover change. The trypanosome parasites (*Trypanosoma* sp.) are predominantly transmitted by the bite of an infected tsetse fly (*Glossina* sp.). Some trypanosome species are infectious to humans, and cause "sleeping sickness," a fatal disease if untreated. Sleeping sickness remains a public health concern in many infected countries, including Uganda, the Democratic Republic of Congo, Angola, and Sudan, where sporadic outbreaks occur. Trypanosome parasites are also infectious to a variety of animal species, including cattle, wild pigs, wild ungulates, buffalo, and some reptiles. The impact of trypanosomiasis on livestock, particularly cattle, has been associated with a significant economic burden in infected regions (Leak 1999; Berrang Ford 2006b).

The disease is affected by land use predominantly due to the role of land cover in driving the habitat of the tsetse fly vector. There are over 25 tsetse species and subspecies, which are generally classified into three main groups, the fusca, palpalis, and morsitans tsetse species. Climate and tropical humid forests control the habitat for the fusca and palpalis groups of tsetse species, and woodland distributions generally control the habitat for the morsitans group. Distributions of specific vegetation covers and species can directly drive the distribution of the tsetse fly, and thus the potential for parasite infestation and disease risk (Leak 1999). Likewise, human population density has also been associated with tsetse distributions, with high densities believed to suppress tsetse presence. This is due to human activity reducing host and reservoir availability, as well as vegetation cover available for habitat (Jordan 1986; Reid et al. 2000; Muriuki et al. 2005).

The close association between land cover and the tsetse vector has resulted in the reciprocal process of African trypanosomiasis affecting land-use/cover patterns in affected regions. For example, it has been early noted that the continental distribution of cattle in Sub-Saharan Africa is inversely correlated with the distribution of the tsetse fly. An estimated 46% of tropical Sub-Saharan Africa is infected with the tsetse fly, which has direct effects on the suitability of land for livestock (Berrang Ford 2006b). Pastoralists in Sub-Saharan Africa usually respond to the risk of disease by avoiding tsetse-infested regions wherever possible (Leak 1999). In many of these areas, livestock are a significant cultural and economic asset, and are central to poor livelihoods. Thus, the presence of the tsetse fly and the trypanosome parasites across large regions of Sub-Saharan Africa is considered to be a significant constraint to livestock and economic development (Perry et al. 2002). This constraint to agricultural development, and its associated influence on land use, has lead to occasional conservationist argument that the tsetse fly, and therein the risk of African trypanosomiasis, has been the savior of African resources, particularly vegetation cover and wildlife populations (Reid et al. 1997).

In association with the impact of trypanosomiasis risk on land use in Africa is the implementation of tsetse control methods which influence land cover. Previous techniques, used particularly during the colonial period, of-

ten focused on removal of vegetation as a habitat for tsetse and elimination of game as a parasite reservoir (Berrang Ford 2006b). These techniques, however, have largely been replaced by the use of insecticides via localized and aerial spraying. During the significant animal epizootics and human epidemics of the colonial period, for example, widespread abandonment and evacuation of land

---

Box 4.5. Land-use and cover change and vector-borne diseases, an example in Thailand, Ban Pa Nai

Sophie O. Vanwambeke · Catholic University of Louvain, Louvain-la-Neuve, Belgium

Vector-borne diseases are linked to the environment by the ecology of the vectors. Landscape features, including land cover, land use, and their pattern, will influence the availability of suitable habitat and hence abundance of the vector. Land use will also determine the location of people in relation to vector habitat and therefore will modify the exposure of people to contacts with the vector.

Important land-cover changes can be observed between the two land-cover maps of Ban Pa Nai and its surroundings shown in Fig. 4.2. In Ban Pa Nai and in the neighboring villages, farmers cultivate irrigated valley fields, and many of them also grow mango or other fruits on the hill slope. In 1995 a new dam was completed (northeast corner of the map), allowing farmers to cultivate their field once more each year (totaling up to three crops

per year, including one or two rice crops). Orchards have also clearly increased, although the aging of orchards make them more visible on a remotely sensed image. Fruit crops are currently very popular in northern Thailand and large areas of field and also of forest are being converted to orchards. Vectors of dengue fever were found in orchards in this area, in which people often go at the end of the day when these vectors are active. However, most dengue control efforts are concentrated on villages, where the largest proportion of dengue vectors is found. Human behavior and the use of preventive measures also play an important role in transmission of vector-borne diseases and could counteract effects of land-cover and land-use changes on transmission (Vanwambeke 2005).

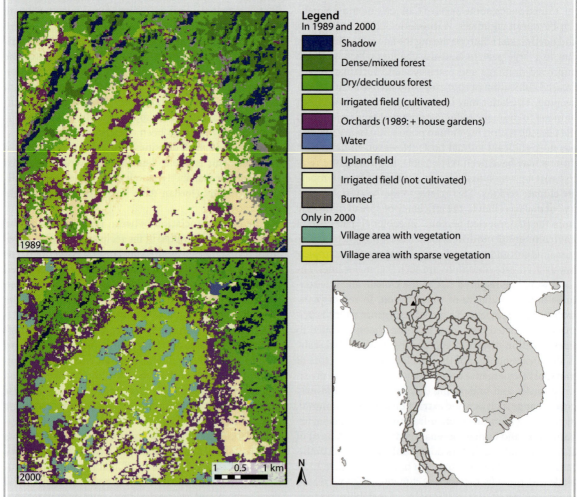

Legend

In 1989 and 2000

- Shadow
- Dense/mixed forest
- Dry/deciduous forest
- Irrigated field (cultivated)
- Orchards (1989: + house gardens)
- Water
- Upland field
- Irrigated field (not cultivated)
- Burned

Only in 2000

- Village area with vegetation
- Village area with sparse vegetation

Fig. 4.2. Land-use and cover change in Ban Pa Nai, Chiang Mai province, Thailand. Source: 1989: Landsat 4TM; 2000 7ETM+

near rivers and shores occurred around Lake Victoria in Uganda (Jordan 1979). This abandonment of land cultivation and occupation resulted in increased bush growth, causing further expansion of tsetse-infested lands. Proceeding re-occupation of these lands may have contributed to more recent epidemics of disease. These examples illustrate the reciprocal relationship that exists between African trypanosomiasis and land use/cover (Berrang Ford 2006b), and the inherent potential of land-use conflicts (Muriuki et al. 2005).

4.3.3 Biocide Usage and Land-Use Intensification

The widespread application of biocides (pesticides, fungicides, insecticides, and larvazides) triggered large-scale land-use intensification and agriculture output growth worldwide, thus contributing to food security in most regions of the world over the past 50 years. It also implied negative health consequences which are not (or no longer) limited to highly intensive production zones of the developed world.

Taking the case of food cropping in Africa, trade-offs exist between food provision and disease risk, namely malaria. The decline of malaria following the Second World War was predominantly the result of wide-scale spraying of dichloro-diphenyl-trichloro-ethane (DDT), a pesticide lauded for its efficacy at drastically reducing mosquito densities. However, when widespread application of DDT became used in agricultural as well, the emergence of DDT resistance in mosquitoes (in conjunction with detection of persistent environmental impacts) reversed the progress to date. As per today, about fifty mosquito species have now been found to be resistant to one or more insecticides worldwide, including DDT. In addition to DDT, other chemicals or biocides are used to reduce mosquito populations and malarial rates as well as for increased crop production, but associated drug resistances and environmental impacts have likewise been noted (Oaks et al. 1991). By contrast, 19th century disease decline in Europe has been linked to agricultural change as the result of improved livestock husbandry, therefore diverting mosquito feeding from humans to animals (Berrang Ford 2006a). Today, according to the World Health Organization, DDT spraying limited to interior walls of huts is still recommended for malaria endemic and epidemic areas.

Health problems in desertified areas comprise the spread of infectious and chronic diseases which are exacerbated by the impacts of biocide usage. The collection of agricultural chemicals in irrigation canals and drinking water, in the Aral Sea Basin provide a striking example (Geist 2005). Various trade-offs exist between food security and clothing (cotton growing) on the one hand, and the degradation of highly fragile dryland ecosystems and health impacts on the other hand.

4.3.4 Health Concerns from Indirect Land-Use Effects

Many concerns about the impacts of climate change and biodiversity losses (including stratospheric ozone depletion) are adding to the problems posed by the more local and/or near-term effects of land-use/cover change on health. They touch fundamental life-support functions and can be exemplified as follows. While it is evident, for example, that biodiversity losses may impede the future development of medicines based on wild-living species, many other concerns are beginning to be elucidated (McMichael and Martens 1995; Martens and Moser 2001). For example, recent research has suggested that forest fragmentation, urban sprawl, and biodiversity loss could be linked to increased Lyme disease risk in the northeastern United States (Schmidt and Ostfeld 2001).

Emissions from biomass burning contribute significantly to the injection of pollutants into the atmosphere, with greenhouse gases and carbonaceous aerosols not only exerting climate impacts but also impacting on air quality and thus human health (likewise on acidification of precipitation). The 1997–1998 fire events in Indonesia caused hundreds of death via respiratory problems of haze and smoke. The haze extended across Southeast Asia, and cost more than $4.5 billion in lost tourism and business. The burning peat resulted in the largest annual increase up to 40% in levels of carbon dioxide in the atmosphere since records began in the 1950s (Crutzen and Andreae 1990; Eva and Lambin 1998a,b, 2000; Stolle et al. 2003).

There are concerns originating from anticipated changes in physical and biological systems which are likely to interact strongly with socio-economic factors such as human welfare and economic development. For example, urban sprawl, in particular, poses health challenges stemming from heat waves which are exacerbated by the "urban heat island" effect, as well as from water contamination due to expanses of impervious road and concrete surfaces (Patz and Norris 2004). Urban "heat islands" result from lowered evaporative cooling, increased heat storage and sensible heat flux caused by the lowered vegetation cover, increased impervious cover and complex surfaces of the cityscape. Dark surfaces such as asphalt roads or rooftops can reach temperatures 30–40 °C higher than surrounding air. At a scale of an entire country, for example the United States, land-cover changes (from both agriculture and urban areas) caused a surface warming of ~0.27 °C (0.49 F), which is a substantial portion of the total warming seen in the U.S. to date (Kalnay and Cai 2003). And in southeast China, where significant urbanization has occurred, land-use change effects on surface temperatures and estimate warming of mean surface temperatures attributable to urbanization has been 0.05 °C per decade since 1978 (Zhou et al. 2004).

Another concern which originates from anticipated changes in physical and biological systems are the indirect effects of ecosystem change on health, for example, from plant and animal diseases related to highly industrialized agricultural production methods. Antibiotics are routinely used for prophylaxis and growth promotion in high-production livestock agriculture, rather than being used sparingly for medical purposes. Such subtherapeutic levels exert selective pressure on the emergence of resistant bacteria (Patz et al. 2005). For example, *Campylobacter* bacteria and *E. coli* strains cultured from piggeries show widespread resistance to multiple antibiotics. Livestock have also been shown to be reservoirs of drug-resistant *Salmonella* bacteria and other *E. coli* that are resistant even to newer generation antibiotics, like cephalosporins. In short, concentrated and intensive animal production (either in agriculture or aquaculture) carry both ecosystem and human health risks.

Likewise, "mercury pollution from deforestation" (Veiga et al. 1994) is another example of indirect land-use effects, relevant for human health and linking to the hydrological cycle (see Sect. 4.7). In Amazonia, the conversion of tropical forests to cropland and pasture plays a major role in the mobilization of mercury through the ecosystem with important implications for human health. Mercury in the Amazon Basin originates from at least three sources – i.e., gold mining, biomass burning and soil erosion – and is transported via the atmosphere to the forest soil and vegetation which serve as important sinks for this heavy metal. Remobilization of mercury occurs when the rain forest is burned, releasing Hg from the vegetation and evaporating it from the soil. The soil is exposed to accelerated erosion and leaching, which brings mercury into the waterways, lakes and floodplains. In the aquatic environment, microorganisms in anoxic waters and sediments transform mercury into methlymercury (MeHg) which is more bioavailable than mercury in elemental form. This organic mercury bioaccumulates up through the aquatic food chain to reside primarily in the muscle tissue and gills of fish, particularly piscivorous species, which are consumed as a vital part of the regional diet. Elevated concentrations of mercury have been found in several riverine populations of the basin, and concern runs high as to the health effects of this neurotoxin which has been mobilized by land-use change along the colonization fronts of the Amazon Basin (Roulet et al. 1999; Lacerda et al. 2004; Wasserman et al. 2003).

The human immunodeficiency virus (HIV), and its associated disease, acquired immunodeficiency syndrome (AIDS), rank also high among health concerns related to land-use/cover change. On the one hand, HIV/AIDS is known to drive the depletion of household resources and assets, mainly inducing loss of labor force. This, in consequence, easily triggers a decline in agricultural yields, a shift to less labor-intensive crops, increases in the area under fallow, or even land abandonment (Yamano and Jayne 2004). On the other hand, HIV/AIDS relates to land-use/cover change in terms of the differential vulnerability of people using the land (Turner et al. 2003c). This means that in the very same area, some people's coping capacity may be markedly different to those in the same or similar setting or environment (Berrang Ford 2006c). In southern Africa, for example, those at risk to climate impacts are also often the resource poor who lack access to markets and information, and they are also vulnerable to a variety of health risks such as HIV/AIDS and malaria (Vogel 2006).

4.4 Atmospheric Chemistry, Climate Regulation, and Life Support Functions

4.4.1 Overview

Terrestrial biota had a major influence on the development of contemporary atmospheric conditions, and they continue to do so (Scholes et al. 2003). However, land-use/cover change, in conjunction with fossil fuel burning, has resulted in major and globally significant alterations of the naturally evolved synergies over the last 1 000 years (Mann et al. 1999), but particularly in the last 200 to 250 years (Crowley 2000). The ensuing increases in atmospheric greenhouse gases and aerosol load can, at least partially, be attributed as a consequence of these human activities (Penner et al. 2001; Prather et al. 2001; Foley et al. 2003). For example, over the past 100 years, the global average temperature has increased by approximately 0.6 °C and is projected to rise at a rapid rate (Houghton et al. 2001; Steffen et al. 2004).

The multiple interactions, and related impacts, between land cover, atmosphere and other components of the climate system occur at various spatial scales and time frames: climate near the ground (micro and local), regional climate (meso), and global climate (macro) (Kabat et al. 2004), and, in terms of time, short-term interactions (minutes to a few weeks), long-term interactions (months to 100 years), and very long-term interactions (greater than 100 years) (Pielke et al. 1998). At all of these spatial and temporal scales, land cover holds biophysical control on the physical properties of the land surface, determined by the physiology and structure of vegetation present within the land cover (Pielke and Avissar 1990; Sellers 1992). A deeper understanding gained so far of these interactions stems mainly from atmospheric modeling, partly coupled with remote sensing analysis. Though the historical dominance of the magnitude of direct and remote land-use effects on climate was recognized (Stohlgren et al. 1998; Chase et al. 2001), knowledge about the effects on weather and climate at the local to regional scale remains still very limited.

4.4.2 Micro- and Meso-Level Impacts

At the local to subnational, national and regional scales, the impacts of land-cover changes on surface radiation budgets, surface hydrology, surface energy balance and surface friction are not straightforward but rather complex.

In the Three-Lakes Region of Switzerland, for example, most of the landscape transformations occurred from 1850 onwards, with the leveeing and draining of formerly marshy and often inundated plains. Investigating the changes in local and possibly regional climate due to documented historical land-use/cover change, the magnitude was estimated to an average warming of more than 1.0 °C in areas where afforestation took place, and in a cooling of up to 2.0 °C in areas of deforestation on a typical July day (Schneider et al. 2004). The effects, however, are difficult to generalize as they depend on season, climate, and soil conditions.

Through an examination of the impact of drainages of marshes and water meadows on the local atmosphere, it was found that surface temperatures over marshland were up to 2 °C higher than over grassland (Mölders 1999). When further investigating effects from a broad set of land-use change classes – such as deforestation, urbanization, afforestation, drainage and recultivation of open-pit mines – results indicated that areas dominated by grassland and forest are much more sensitive to concurrent land-use changes than are agricultural areas (Mölders 2000).

Micro- to mesolevel impacts of land-use/cover change upon climate include remote impacts upon local circulation regimes. They can be labeled land-use driven "biological teleconnections" where changed ecosystem characteristics affect, for example, local weather or livelihood conditions such that effects were communicated to regions distant from actual changes in surface characteristics (Eastman et al. 2001).

For example, the influence of irrigation on precipitation in the Texas High Plains of the United States is complex as irrigation enhanced summer precipitation by 6% to 18% (for areas up to 90 km downwind), with storms of greater duration, length, and total accumulation. However, cool, and wet surface also increases low-level instability and triggers storms (Moore and Rojstaczer 2002).

Likewise, weather influences in the high Rocky Mountains were found to be related to the presence of irrigated farmland in the plains below, with irrigated regions affecting the daily summer mountain-plains breeze by altering temperature patterns and thereby allowing communication between the two regions (Chase et al. 1999).

Other examples relate to lowland deforestation in Costa Rica which has an impact on cloud immersion of adjacent tropical montane cloud forests during the dry season, with serious impacts on the moisture dependent ecosystem (Lawton et al. 2001).

At the regional scale, Bonan (1999) showed that conversion of forest to cropland in the central and eastern United States may have led to cooling. Using climate model simulations, it has been shown that mean annual temperature decreased by 0.6 to 1.0 °C, east of 100 W, with the coolest temperatures found in the Midwest in summer and autumn. Furthermore, daily maximum temperature decreased more than daily minimum temperature, leading to a decrease in diurnal temperature range. Bonan (2001) found further observational evidence for the decrease in daily maximum temperature.

4.4.3 Macro- or Global-Scale Impacts

At the global scale, changes in land-surface properties associated with changes in vegetation can have impacts on continental and global atmospheric circulation, with possible large impacts on regional and continental climate. Extensive reviews are provided on the complex relationships that exist between vegetation and other components of the climate system at the local to global scales, detailing the differences in magnitude and sign that similar vegetation change investigations have identified in different geographic localities over the Earth (Betts et al. 1996; Pielke et al. 1998; Kabat et al. 2004) (see Box 4.6, for an ancient/historical example).

There is accumulating evidence that large-scale land-cover changes, particularly in the tropics, generate re-

---

Box 4.6. The impact of historical land-cover change upon climate in the Mediterranean

Historical land-cover and climate changes in northern Africa were investigated from 2000 years ago to see whether the changes in land cover could be responsible for changes in Mediterranean climate. Climate proxies from the region suggested a widespread drying trend across the Mediterranean since the Roman Classical Period (RCP). The study also showed that the desert areas of northern Africa corresponding with modern Egypt, Tunisia and Algeria were significantly moister during the RCP, with wealthy agricultural economies making the area the most productive in the Roman World. To investigate if land-cover changes associated with deforestation and cultivation across the Mediterranean may have contributed to the widespread drying, the study modeled the climate of the region with a fine resolution general circulation model using vegetation that existed in the RCP compared with the climate modeled with the vegetation representing modern day cover. The modeling experiments showed significant changes in the climate under the two vegetation scenarios, with a northward shift in the Inter Tropical Convergence Zone, and the creation of a sea-land circulation over northwestern Africa under the RCP vegetation. The changed atmospheric circulation resulted in substantially moister conditions in northern Africa, with the speculation that the vegetation of the RCP would be sustained under the wetter conditions in areas that under current day conditions are too dry. The conclusion from the research was that clearing of the Mediterranean by human activity since the RCP may have triggered a positive climate feedback with a drift towards the dryer conditions of modern day (Reale and Dirmeyer 1998).

mote climatic effects of global extent far from where the surface has been directly affected by land-cover changes (Franchito and Rao 1992; McGuffie et al. 1995; Chase et al. 1996; Zhang et al. 1995; Sud et al. 1996; Snyder et al. 2006). Through an examination of global circulations model (GCM) simulations of the effect of observed levels of land-cover change globally, strong evidence was found of changes in global scale circulations (Chase et al. 2001).

Again, it was demonstrated that remote effects of land-cover change were prevalent in a variety of models under a range of configurations and model assumptions, and that remote temperature anomalies resulting from land-cover change could be similar in magnitude as effects of the historical increase of the radiative effect of increased CO_2 (Pitman and Zhao 2000; Zhao et al. 2001; Bounoua et al. 2002).

Also using a GCM, the effects of a wholesale removal of the Amazonian rainforest on remote climates was examined and significant evidence found for a reduction in large scale circulations generated by tropical convection and for propagating atmospheric waves which affected rainfall in Northern Hemisphere winter (Gedney and Valdes 2000).

Likewise, statistically significant remote effects due to deforestation in Amazonia, Central Africa and Southeast Asia were found (Werth and Avissar 2002, 2004), and, through an examination of the potential impacts of future land-use changes, regional temperature anomalies were found of up to 1.5 °C in regions not directly affected by land-cover changes (DeFries et al. 2002a).

Other impacts of land use and related large-scale vegetation-cover changes upon regional land-atmosphere-ocean systems such as the Asian monsoon (Yasunari 2002) relate to the question whether human-induced land-cover changes will modify a system of humid climate and dense "green" vegetation in the eastern half of Eurasia (Fu 2002). However, such globally averaged changes remain small (Chase et al. 2000).

4.4.4   Land Use and Greenhouse Gas Forcing

Agricultural land uses are estimated to contribute to changes in atmospheric concentrations of three greenhouse gases (GHG) – methane (CH_4), nitrous oxide (N_2O), and carbon dioxide (CO_2) – in total accounting for about 20% of current annual GHG forcing potential (Houghton et al. 2001). The expansion of crop and pastures to the detriment of forests results in an increase in atmospheric CO_2. This decreases the sink capacity of the global terrestrial biosphere, and thereby may amplify the atmospheric CO_2 rise due to fossil and land-use carbon release. Grassland conversion into croplands and ecosystem degradation is widespread due to high growth rate of human population and political reforms of pas-

toral systems. These dramatic changes in land use with widespread reduction of forest and grasslands have increased carbon emission in arid and semi-arid lands of east and central Asia (Chuluun and Ojima 2002).

The implications of biomass burning associated with agricultural land use are not fully understood, but indications point to high relevance. Fire is used in agricultural practices (such as in southern Russia), land-management practices (such as in African national parks) and for forest clearance (such as in Amazonia). Emissions from biomass burning contribute significantly to the injection of pollutants into the atmosphere, with greenhouse gases and carbonaceous aerosols impacting, among others, on the radiation balance at the surface (but also on the acidification of precipitation and air quality) (Crutzen and Andreae 1990; Eva and Lambin 1998a,b, 2000; Stolle et al. 2003). Biomass burning is thought to contribute up to 40%, 16% and 43% of the total emissions of anthropogenic origin for carbon dioxide, methane and carbon monoxide, respectively (Tansey et al. 2004).

As can be seen, for example, from patterns of regional vulnerabilities to fire, large differences need to be considered between temperate and tropical agricultural land uses.

Methane (CH_4) is one of the most potent contributors to the atmospheric greenhouse effect and plays an important role in tropospheric chemistry. Depending on the time scale, it is 24.5 times more powerful a greenhouse gas than carbon dioxide.

Microbial processes in natural wetlands have always been a major source of atmospheric methane, and are considered even today to represent a chief part of total emissions. Due to its strong greenhouse radiative potential, methane emission will result in wetlands such as rice paddies having a positive radiative forcing, but also a tremendous mitigation potential. In China, for example, midseason paddy drainage, which reduces growing season CH_4 fluxes, was first implemented in the early 1980s, and has gradually replaced continuous flooding in much of the paddy area. As a consequence, decreased methane emissions from paddy rice may have contributed to the decline in the rate of increase of global atmospheric methane (CH_4) concentration over the last 20 years (Li et al. 2002).

Due to its capability to act as sink for OH-radicals, methane will indirectly participate on atmospheric chemistry and aerosol dynamics. Its global emissions are therefore not only of importance for the radiative forcing in the Earth's energy budget but also of significance for the oxidative capacity of the atmosphere (Christensen et al. 1996).

Agriculture, clearly, is the largest source of anthropogenic methane. Besides land use and management in wetlands, most of the global rangelands support ruminant, grazing herbivores (cattle, sheep and goats, mainly) that emit methane (CH_4), either directly or through the management of livestock manure. Emissions from manure management are important in more intensive livestock systems such as intensive dairy cattle and pig systems in Europe and North America, while emissions from manure deposited on extensively used rangelands is likely to be a relatively small source only, if compared to enteric sources (Reid et al. 2004). Worldwide, these emissions are responsible for 23% and 7%, respectively, of all anthropogenic sources of CH_4 gas emissions (Houghton et al. 2001). They contribute 30% of global warming potential of all agricultural emissions and about 5% of the global warming potential from all anthropogenic sources (U.S. EPA 1999). Cattle alone is the largest contributor (73% of global CH_4 emissions), with more than half of the global cattle population located in the tropics. Thus, pastoral lands in tropical, mainly developing countries are a significant global source.

Agricultural land use is also a significant contributor to increases in atmospheric nitrous oxide (N_2O) concentration. The greenhouse gas has 320 times the warming strength of carbon dioxide. In pastoral lands, sources include land conversion, manure, fertilizer, and changes in temperature. With the exception of some areas in South America, the amount of manure and use of fertilizer in tropical grazing lands are low, but land conversion remains important. All in all, and given the dispute about rates of dryland transformation (see Chap. 2), N_2O is esteemed to be of minor importance in extensive grazing systems (Reid et al. 2004).

The carbon cycle is the best-studied trace gas exchanged between land surfaces and the atmosphere (Scholes 2002). Over the last 150 years, carbon atmospheric concentration has increased by >30% due to fossil fuel burning and following land-use/cover change (Prentice et al. 2001; Field and Raupach 2004). In contrast to other GHG emissions, constant agricultural land use plays a relatively small role in total carbon dioxide emissions. However, the conversion of (semi)natural ecosystems, especially forests, to agriculture is responsible for greater emissions of carbon than any other land-use change. Carbon is lost from soils in the first years of cultivation, as little as 3 years in the tropics, and as much as 40 years in temperate zone ecosystems. Many cultivated systems, in addition, have the potential to sequester carbon or, in other words, to capture and secure the storage of carbon in soil organic matter (SOM) that would otherwise be emitted to or remain in the atmosphere, with improved crop and soil management practices. Agricultural land uses emit CO_2 through the decomposition of SOM and crop residues directly. However, vegetation fires act as a direct disturbance to terrestrial ecosystems which play a role as sources or sinks of carbon at local, regional and global scales (Stolle and Lambin 2003; Stolle et al. 2003). And, in a wider sense, agricultural systems emit further CO_2 through the (direct) use of fossil fuels in

(non) food production and the (indirect) use of embodied energy in inputs that require the combustion of fossil fuel in their production. Nitrogen (ammonium) fertilizer is by far the most important. Some of the climatic impacts of land-use systems are mediated through erosion – see Sect. 4.6. Including the conversion of forests and grasslands to agricultural land, the direct effects of land-use/cover change are estimated to have led to a net emission of 1.7 GT C yr^{-1} in the 1980s and 1.6 GT C yr^{-1} in the 1990s (Watson et al. 1996).

In the early 1980s, terrestrial ecosystems were highlighted as sources and sinks of carbon, underscoring the impact of land-use/cover change on the global climate via the carbon cycle, with most insights stemming from tropical deforestation (Woodwell et al. 1983; Houghton et al. 1985) – see Box 4.7. Quantification of global carbon pools and fluxes remains mainly based on land-cover mapping and measurements of cover conversions worldwide (Dixon et al. 1994; Houghton et al. 1999; McGuire et al. 2001), and decreasing the uncertainty of terrestrial sources and sinks of carbon remains a serious challenge today, mainly because the translation of vegetation changes into net CO_2 fluxes to or from the atmosphere is non-trivial (Steffen et al. 2004). Where uncertainty can be limited such as in the United States (Pacala et al. 2001), projections of the future of sources and sinks are possible. For example, ecosystem recovery processes are primarily held responsible for the contemporary U.S. carbon sink resulting from land-use changes and fire suppression since 1700 onwards. They are predicted to slow down over the next century resulting in a significant reduction of the sink.

Despite uncertainties as outlined above, the mass balance of carbon in the field remains crucial in determining whether there is a net loss or gain in SOM. Currently, the global average soil organic carbon density is estimated at 100 to 135 metric t of carbon per hectare of land,

---

Box 4.7. Recent estimates of net carbon emissions from land-cover change in the tropics for the 1990s

Three recent estimates of the net flux of carbon from land-cover change in the tropics show the uncertainties in both rates of tropical deforestation and associated carbon emissions. DeFries et al. (2002a) used coarse resolution AVHRR to determine an average rate of deforestation of 5.56 million ha yr^{-1} during the 1990s and calculated an average annual emission of 0.91 Pg C yr^{-1}. Houghton (2003) calculated a net flux of 2.1 Pg C yr^{-1} based on estimates of deforestation from the FAO (12 million ha yr^{-1}). And Achard et al. (2004) reported net emissions of 1.1 ±0.3 Gt C yr^{-1} from a net deforestation rate of 9.7 million ha yr^{-1}, determined by sampling with high resolution Landsat data. All of these estimates include the emissions of carbon from cleared vegetation and soil and from forest degradation, as well as the uptake of carbon in regrowing forests. The estimate by Achard et al. (2004) also includes emissions from the exception 1997-1998 Indonesian fires (see Chap. 2 for a detailed discussion on the rates of deforstation).

and the total global store of SOM at 1 567 to 2 011 Gt C (Prentice et al. 2001). During much of the past century, most cropping systems worldwide assumedly have undergone a steady net loss of SOM, except for few land-use systems only in which net carbon sequestration occurred. Factors which exert the greatest impact on the carbon balance relate to crop yield levels, removal of crop residues for fuel and livestock forage, crop rotations that include a pasture phase or perennial forage legume, and tillage.

When putting carbon sequestration into the context of land-use/cover change, two major patterns bear different policy implications. In developing countries of the tropical zone, most of the carbon flux components stem from forests or the forestry sector, while in developed countries outside the tropics they stem from other sectors, mainly (Fearnside 2000; Gower 2003; Cramer et al. 2004). The Indo-Gangetic Plains Region in northern India and the United States of America provide paramount examples of the various impacts of land-use/cover change, and how differently they need to be addressed.

In the U.S., where the annual net flux of carbon is small, forests may account for half or less of the total carbon sink. Carbon accumulation in forests has been attributed to historical changes in land use and the enhancement of tree growth by CO fertilization, N deposition, and climate change. The accumulation of carbon in agricultural soils, harvested wood products, and aquatic sediments, and through the expansion of woody plants into herbaceous lands are significant, although the latter flux (woody encroachment) is highly uncertain (Houghton et al. 1999; Pacala et al. 2001) – see Table 4.1.

As for India, a full carbon accounting has not been achieved yet, so that it remains difficult to estimate sources and sinks of carbon from different land-use change processes and land covers in a detailed manner. Forests cover about 67.55 Mha or 20.5% of the country's geographic area. Nonetheless, the current estimates of Indian forest phytomass carbon pool are in the range of 2.0 to 4.4 Pg C. For example, it was estimated in the range of 3.8 to 4.3 Pg C, based on a growing stock volume approach (Chhabra et al. 2002), and recent forest soil organic carbon has been estimated as 6.8 Pg C (Chhabra

et al. 2003). Using a simple book-keeping model approach, the cumulative net carbon emission from Indian forests due to land-use/cover changes such deforestation and phytomass degradation in the period 1880 to 1996 was estimated to be 5.4 Pg C, while the net carbon release to atmosphere from forests has dominated the terrestrial carbon emissions during the 20th century (Chhabra and Dadhwal 2004). The broad picture can be detailed by a study of long-term historical land-use/cover changes and its impacts on the agro-ecosystem carbon cycle in states of the Indo-Gangetic Plains (Dadhwal and Chhabra 2002). For this region, an increase of 435.6 Mt in crop biomass was estimated for the period 1901 to 1991. High cycling of the produced biomass through livestock as fodder and use as domestic fuel sustains the high population density in the area. The intensification of agriculture with modern technology based on mechanization, high fertilizer and energy inputs have also led to increased agricultural contribution to carbon emissions. Using IPCC methodology, the total CO_2 emissions in 1990 from energy, industry, agriculture, forestry, waste and land-use change in the Indo-Gangetic Plains was estimated to be 585 Tg (Asian Development Bank et al. 1999).

4.4.5   Feedbacks, Surprises and Unresolved Issues

Is seems important to note that relatively small alterations in the magnitude of greenhouse gases and aerosols in response to climate forcing will influence climate in turn through a number of important biogeochemical feedbacks associated with land-use/cover change. This would have immediate consequences for impacting upon life-support functions in terms of sudden, unexpected and very likely cascading effects (Steffen et al. 2004).

It was recognized, for example, since the mid 1970s (Otterman 1974; Charney and Stone 1975; Sagan et al. 1979; Lofgren 1995) that land-cover change modifies surface albedo and thus surface-atmosphere energy exchanges directly, with an impact on regional climate. Global drylands show feedbacks such as droughts in the African Sahel and their effects on vegetation, which are reinforced

Table 4.1. Sources (+) and sinks (–) of carbon (GtC per year) from different processes and land covers in the United States

	Pacala et al. (2001)		Houghton et al. (1999)	Houghton (2003)	Goodale et al. (2002)
	Low	High			
Forest trees	0.11	0.15	0.072	0.046	0.11
Forest organic matter	0.03	0.15	–0.010	–0.010	0.11
Cropland soils	0.00	0.04	0.138	0.00	–
Woody encroachment	0.12	0.13	0.122	0.061	–
Wood products	0.03	0.07	0.027	0.027	0.06
Sediments	0.01	0.04	–	–	–
Total sink	0.30	0.58	0.35	0.12	0.28

Box 4.8. Global emissions from biomass burning: vegetation fires in the tropics

Hugh Eva · Joint Research Center, Ispra, Italy

Biomass burning (Figs. 4.3, 4.4) has been shown to be a major source of greenhouse gases at a global level, contributing possibly as much as 40% to the global budget of major gases such as carbon dioxide and is almost the sole source in tropical countries (Hao and Liu 1994).

Vegetation fires in the tropics contribute around 60% of the global total of biomass burning sources. It is recognized, however, that there are major uncertainties in the magnitude of emissions from certain sources both in the calculation of the areas involved and in the emissions factors to be applied. Whilst uncertainties in industrial processes are thought to be around 10%,

Watson et al. (1996) have estimated that the overall uncertainty for carbon dioxide emissions due to land-use change and forestry is 60%, and 100% for methane and nitrous oxide from biomass burning. By combining field data with satellite imagery research teams can reduce two key elements in this uncertainty: *(a)* the actual areas burnt in particular vegetation types and *(b)* the burning efficiency.

Field validation of algorithms to detect and map the extent of the burnt area is of utmost importance. Burning efficiency is a new area of research, and field data is required to characterize the intensity of burn (Eva et al. 2004).

Fig. 4.3.
ASTER data over the Park W in West Africa, showing large burned (*blue*) areas. At the top of the image (*boxed and inset*) is a fire in progress. Gallery forests (*dark green*) and unburned savannahs (*green*) dominate the region. The extract is some 30×30 km

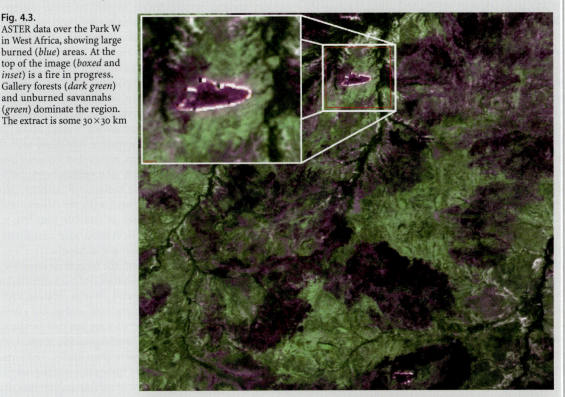

Fig. 4.4.
The characteristic advance of the fire front is shown below from the ground. The provision of these data from satellites to park managers in real time can give a valuable input for ensuring an effective burning campaign and for helping in the battle against illegal activities such as poaching in protected areas

at the decadal timescale through a mechanism that involved surface changes caused by the initial decrease in rainfall (Zeng et al. 1999). Grazing and conversion of semiarid grasslands to row-crop agriculture are the source of another positive desertification feedback by increasing heterogeneity of soil resources in space and time (Schlesinger et al. 1990; Seixas 2000). And, the reduction of precipitation from clouds affected by desert dust can cause drier soil, which in turn raises more dust, thus providing a feedback loop to further decrease precipitation, with land-use change exposing the topsoil and initiating such a desertification feedback. The latter means an example of an indirect effect through the radiative forcing potential of emitted trace gases, their tropospheric chemistry and cloud-formation properties (Andreae and Crutzen 1997; Monson and Holland 2001).

Likewise, and adding to the overall observation of multiple and multi-directional impacts, are the relatively slow changes in vegetation cover but also the sudden, extreme and/or episodic events like fire (see Box 4.8 and 2.7 in Chap. 2). They set additional constraints in the surface-atmosphere system, affecting vegetation patterns and causing dynamic changes in ecosystem structure and species composition. For example, atmosphere-vegetation interactions and greenhouse-gas induced climate changes were seen to be a function of land cover in North Africa (Claussen et al. 2002). It was further found that the expansion of woody shrubs in the western United States grasslands, following fire suppression and overgrazing, may have contributed to a large carbon sink (Houghton et al. 1999; Pacala et al. 2001). Time series of remote sensing data revealed that there are short-term land-cover changes, often caused by the interaction of climatic and land-use factors, which show periods of rapid and abrupt change followed either by quick recovery of ecosystems or by a non-equilibrium trajectory (Taylor et al. 2002b; Stolle and Lambin 2003). The interaction of land use and climate in the West African Sahel, including the social and biological responses, are a paramount example of this (Xue et al. 2003; Henry et al. 2003, 2004).

Comparably, and as a result of interactive effects – i.e., between deforestation, abandonment of agricultural land reverting to forests, fires, and interannual climatic variability – the role of the Amazonian forest as a carbon sink and source varies from year to year. Likewise, periodic El Niño-driven droughts trigger an increase in the susceptibility of forests to fires (such as in Indonesia). Thus, accidental fires are more likely and lead to the devastation of large tracts of forests and to the release of huge amounts of carbon from peatland fires (landholders also use drought conditions to burn large tracts of forest to convert them to plantations). As a matter of fact, forests affected by fragmentation, selective logging, or a first fire subsequently become even more vulnerable to fires as these factors interact synergistically with drought (see Sect. 2.4.2.2).

Despite of the detailed understanding of some complex impacts, a range of key processes in the land-atmosphere system still needs to be better understood (Steffen et al. 2004).

Increasing evidence, for example, suggests a tight coupling of inert and reactive trace gas exchange, which means the emission and deposition of gaseous substances and particles that is directly or indirectly controlled by feedback loops involving water and thus the surface energy balance. For example, the important contribution of local evapotranspiration to the water cycle (precipitation recycling) as a function of land cover highlighted yet another considerable impact of land-use/cover change on climate, at a local to regional scale in this case (Eltahir and Bras 1996), with a strong role of surface vegetation in maintaining the regional West African monsoon circulation (Eltahir 1996). Also, modeling investigations into the climate impacts of tropical deforestation in South America, Africa, and Southeast Asia identified the main impact of removing tropical forests on the hydrological cycle through reduced transpiration. The reduction in atmospheric moisture combined with the increase in sensible heat flux, resulted in reduced cloud formation and precipitation. The reduced cloud cover significantly increased solar radiation flux which offset the increased albedo associated with the deforestation, to result in further increases in sensible heat flux and further reduction in precipitation and cloud cover (Polcher and Laval 1994; Zhang et al. 1995).

Second, aerosols represent the largest uncertainty in climate forcing. Their so-called indirect effects, i.e., changes induced in clouds and on precipitation, represent a chief uncertainty in regional and global climate issues. Organic aerosol particles are emitted by the terrestrial biota as well as by anthropogenic sources, and they are effective cloud condensation nuclei. The formation of aerosol particles and their subsequent growth to CCN size has been observed at remote continental sites, and their connection to biospheric processes has been noted. The exact aerosol formation processes assumedly depend on the presence of organic vapors, and a feedback loop linking the type of vegetation, its photosynthetic activity and the capacity to emit volatile organic substances, which contribute to aerosol growth, can be proposed (Andreae and Crutzen 1997; O'Dowd et al. 2002).

To separate "natural noise" in the climate system from human-driven changes remains a crucial issue. Remote sensing data from wide-field-of-view satellite sensors, for example, reveal patterns of seasonal and interannual variations in land-surface attributes that are driven not by land-use change but rather by climatic variability. These variations include the impact on vegetation and surface moisture of the El Niño Southern Oscillation (ENSO) phenomena (Eastman and Fulk 1993; Behrenfeld et al. 2001; Lambin and Ehrlich 1997a,b), natural disasters such as floods and droughts (Lupo et al. 2001), changes in the length of the growing season in boreal regions (Myneni et al. 1997), and changes in vegetation productivity due to erratic rainfall fluctuations in the African Sahel which lead to an expansion and contraction of the Sahara (Tucker et al. 1991; Tucker and Nicholson 1999). Despite of these uncertainties, natural environmental change and variability mostly interact with human causes of land-use change. Highly variable ecosystem conditions driven by climatic variations tend to amplify the pressures upon terrestrial ecosystem condition and services which is especially true for global dryland resources (Geist 2005).

Finally, some of the climatic impacts of land use are mediated through, or actually caused by land-use driven soil erosion. An ecological off-site impact of accelerated erosion is the emission of erosion-induced GHGs into the atmosphere. While some of the organic carbon transported to depositional sites and aquatic ecosystems is buried and sequestered, a large fraction as high as 1 Pg C yr^{-1} is assumed to be emitted into the atmosphere (Lal 2002a,b).

4.5 Agrodiversity and Biodiversity Loss

4.5.1 Overview

Agrodiversity, or the many ways in which farmers use the natural diversity of the environment for production, including crop choice (or the management of species in land use), soil and water-management practices, and marketing arrangements, can definitely have aggregated impacts from the level of fields, farms, communities or landscapes up to the global scale. These must not be negative impacts, necessarily, but the loss rather than the provision of species habitat certainly are. In fact, the loss of biological diversity is one of the only truly irreversible global environmental change at present (Dirzo and Raven 2003).

Biodiversity is conventionally interpreted as diversity in genetics, population, species and the ecosystem. Biodiversity, in fact a property of the natural ecosystem, is a product of complex historical interactions among physical, biological and social systems over time (Pei and Sajise 1993; McNeely 1994). The functional roles of biodiversity have economic, cultural and ecological aspects. Therefore it is also linked to local knowledge system and cultural diversity. Increasingly, both scientists and development planners are recognizing the importance of social factors in biological resource management and biodiversity maintenance, particularly in agro-ecosystems. These social factors include institutional arrangements, policy, knowledge and technology and marketing. Fire, agriculture, technology and trade have been particularly powerful human influence on biodiversity and land cover (McNeely 1994). For instance, although swidden cultiva-

tion is often cited as a cause of deforestation, it is a century-old agricultural practice in the tropics and may actually increase, rather than reduce, the diversity of habitats and species in the landscape. It does this by creating a swidden-fallow succession, transplanting both domesticated and wild plant species, and attracting related insect and animal species. This implies that biodiversity conservation efforts may need to give greater attention to ecosystem processes than to ecosystem products (McNeely 1994). People have played important role in maintaining biodiversity.

In managed ecosystems, biodiversity is one facet of agrodiversity, which also includes biophysical diversity (i.e., features of soils, hydrology and micro-climate which control the intrinsic quality of the natural resource base to be exploited by agricultural systems, including ecosystem resilience), management diversity (i.e., all biological, chemical and physical methods of managing land, water and biota for crop and livestock production and the maintenance of soil fertility and structure, based on the local knowledge of farmers), and, underpinning management diversity, organizational diversity (i.e., diversity in the manner in which farms are owned and operated, and in the use of resource endowments, such as labor, gender, household features, and off-farm employment) (Brookfield and Stocking 1999; Brookfield 2001). These categories are closely interrelated, they are shaped by mediating factors, and play out differently at various time and spatial scales. For example, crop choice and type of conservation practices often differ between poor and rich farmers, thus affecting the pattern of management diversity, and feeding back to enlarge differences in natural land quality. There are several manifestations of biological diversity, in particular, – such as the latitudinal gradient, megadiversity countries, centers of endemism and hot spots – which point to the fact that global biodiversity is highly concentrated in a few patches of the Earth. In particular, tropical humid forest ecosystems stand out as highly significant reservoirs of global biodiversity.

The issue of biological diversity is probably best characterized by the key elements given in the preamble of the Convention on Biological Diversity (CBD) of the United Nations Environmental Programme (UNEP 1992, 2004b). Thus, biodiversity is hypothesized to influence in a positive way human well-being (i.e., welfare and development), with both direct and indirect links to be considered. At first instance, people derive material or spiritual enjoyment from having, or bequeathing, a biologically diverse planet. In the case of the indirect link, it is argued that material well-being depends on the provision of ecological services such as climate regulation, nutrient cycling, maintenance of hydrological cycles, and so forth. The ability of natural ecosystems to provide these services, in turn, is assumed to have a positive relationship with biodiversity. Economic activities fall into consumption and production activities that generally increase aspects of human well-being such as welfare and national development. However, economic activities are also believed to have a negative influence on biodiversity via a range of proximate causes, intermediate factors and fundamental or underlying driving forces (Cervigni 2001).

Among the various levels of organization surrounding biodiversity, ecosystem diversity is considered here in more detail, but less so genetic and organism diversity. This is due to biological communities which assure the functioning of the ecosystems and provide a number of ecosystem services, thus presenting the joint answer of species to various impacts. Herewith, ecosystem functioning represents the assemblage of processes such as primary production, decomposition, nutrient cycling, and their interactions, and ecosystem services are vital life support functions such as flood control, soil protection, water quality, and pollination. Species perform diverse ecological functions differently. A species may regulate biogeochemical cycles, modify disturbance regimes or change the physical environment. Other species regulate ecological processes indirectly, through trophic interactions such as predation or parasitism, or functional interactions such as seed dispersal and pollination. To better assess the biodiversity-related impacts of land use, cultivated systems can be classified according to their diversity of species and the biophysical complexity of the production system (Swift et al. 1996; Freitas 2006).

4.5.2   Conservation and Loss of Biodiversity

Despite of claims to the contrary, biodiversity is more rich and varied now than ever before. As a result of more than 3.5 billion years of organic evolution, biodiversity has clearly reached an unprecedented magnitude of diversity. Over the last 600 million years of life on Earth, the diversity of families of multi-cellular marine organisms increased steadily since the Permian (with the trend in species number being even more extreme), and, likewise, terrestrial organisms increased rapidly in diversity until present for each group, be they plants, fungi, vertebrate animals or arthropods (Groombridge and Jenkins 2002). Fossil records of both marine and terrestrial organisms indicate maximum diversity at the present time. The current level of terrestrial diversity is estimated to be about twice its historical average since organisms first invaded the land surface about 440 million years ago. Despite five major extinctions in palaeo-times, the reasons of which had very likely been modifications of the physical environment after impacts such as from meteorites or volcanic activity, the trend of biodiversity increases has been continually upward, with an overall estimated half of the global species richness resting in just 6 to 7% of the land in the humid tropics (Dirzo and Raven 2003).

However, neither the total number of organisms nor the total number of extinct species can be known, and only very few groups of organisms are well enough known to be assessed for extinction (Dirzo and Raven 2003). Complete catalogues of described, valid species exist for only a few groups of organisms, so that the total number of species of organisms can only be estimated, based upon expert opinions, extrapolations from an initial estimated number, or a combination of these methods (Hammond 1995; Pimm et al. 1995; May 2000). The most recent, comprehensive and careful effort reveals that the best estimate for the total number of (eukayotic) organisms "possibly" lies in the 5 to 15 million range, with a best guess of around 7 million species (May 2000). However, it is still conceivable that this figure would be greatly increased by new discoveries (Heywood 1995; Pimm et al. 1995), even by as much as an order of magnitude (Dirzo and Raven 2003). There seems agreement, though, that the average species has a life span of about 5 to 10 million years (May et al. 1995), and that the rate of extinction is 1 to 0.1 species per million species per year (Pimm et al. 1995). These species lifetimes may be overestimated, while the million-year life span of species seems supported as a conservative estimate. It should further be noted that for tropical moist forests some 19 of each 20 species would be unknown to science at present so that, for example, the continued effects of forest burning mean a considerable though difficult-to-measure impact on biodiversity (Dirzo and Raven 2003).

The processes described above, which maintain ecosystem functioning, are adversely affected by the loss of diversity, mainly as an irreversible result of land-use/ cover change activities which bear a truly global character (Wilson 1992; Magurran and May 1999; Cassman et al. 2005).

In tropical (moist) forest ecosystems, evidence indicates that the proximate causes of biodiversity losses are identical with those of deforestation, linked to underlying causal synergies among demographic, economic, technological, institutional and socio-political or cultural factors (Cervigni 2001; Dirzo 2001; Xu and Wilkes 2004; Van Laake and Sánchez-Azofeifa 2004). In many more cases, human population dynamics in combination with economic growth are mentioned as the major underlying drivers of the conversion and modification of vast areas to settlement, agriculture, and forestry. This leads to the assumption that mainly human population dynamics over the near future is the crucial factor in determining the loss of biodiversity, pushing animal and plant populations past critical thresholds of tolerance and renewal (Freitas 2006). Biodiversity hot spots clearly show the trend. They are priority areas in conservation efforts as they have high biological diversity, high levels of endemism, and are currently threatened by anthropogenic impacts such as habitat destruction and species extinction.

Box 4.9. Balancing biodiversity, carbon, beef and soybean in the Brazilan Cerrado?

- Together with the Venezuelan and Colombian Llanos, the Brazilian Cerrado forms one of the principal pastoral lands of Latin America which have the worldwide highest concentration of cattle in the tropics (sometimes >20 km²) and thus highest methane emissions per person (4–8 tropical livestock units per person) (Reid et al. 2004).
- The Cerrado is a tropical savanna covering 22% of Brazil's territory, with only around 5% of the area protected by conservation. The mosaic landscape of grassland, scrub and woodland environments create complex habitats for fauna, and the region ranks twelfth as global biodiversity hot spot due to high levels of plant endemism. The Cerrado claims an estimated 160 000 species of plants, fungi and fauna. While large areas are insufficiently inventoried or underrepresented in data sets, the best scientific measurements estimate that 44% of vascular plants, 9.3% of mammals, and 3.5% of avifauna are endemic species (Silva and Bates 2002; Jepson 2006).
- Available biome-scale land-cover loss estimates vary between 40% and 80% of the original cover, but important trends exist also in vegetation regeneration and agricultural intensification. Soybeans, cattle ranching and mechanized commercial agriculture, including cotton, rice and maize, have expanded rapidly during the past three decades. Especially soybean-related land-use technology, state development policy

and agricultural colonization projects (e.g., POLOCENTRO, PRODECER) contribute to exceptionally high conversion and modification since about 1970 in Rio Grande do Sul and Paraná, and since 2003 in Mato Grosso. This stimulated the further expansion beyond designated project areas of diverse agricultural frontiers in the states of Minas Gerais, Goiás, Bahia, Tocantins and Maranhão, with the consequence that the Cerrado soybean boom pushed Brazil into its current position as the world's second leading producer. Likewise, livestock remains an important economic activity as it supports a large proportion of the national herd that sustains Brazil's beef exports. Over the past thirty years, Brazilian scientists developed a suite of soil conservation and agronomic techniques which partly address climate or biodiversity concerns, but are designed for further stimulation of agricultural output growth. Measures include, for example, crop-pasture rotation, no-till agriculture, and the heavy application of lime and fertilizers to encourage increased and intensified production on the chronically nutrient deficient soils. Strategies are discussed to integrate annual cultivation with livestock operations through pasture-crop rotation schemes, and maize and soybeans varieties have been selected which tolerate the high aluminum toxicity and low pH of Cerrado soils (Jepson 2006).

On the other hand, there is evidence from an array of land-change studies that it is not the sheer size and distribution of population numbers but institutional and policy factors which appear to be crucial in determining changing landscapes and the loss of biodiversity (Tri Academy Panel 2001; Homewood et al. 2001). Rules used for making policies are important to ensure that local users are able to influence resource-management institutions, and very often a mismatch needs to be addressed between environmental signals reaching local populations and the macrolevel institutions (Lambin et al. 2003; Poteete and Ostrom 2004). Thus, trade-offs between local livelihood security, biodiversity concerns, carbon sequestration and (inter)national development or economic growth render solutions difficult (see Box 4.9), which is especially true for pastures in global drylands.

4.5.3 Habitat Destruction, Degradation and Fragmentation

The combination of proximate causes and underlying drivers triggers processes of habitat destruction (conversion), degradation (modification) and fragmentation which are the most important chains of events leading to worldwide species decline and extinction (Heywood 1995; Magurran and May 1999; Van Laake and Sánchez-Azofeifa 2004). For the past 300 years, in which most rapid land-cover changes have been estimated to occur with consequences for most of today's landscape configurations (see Chap. 2), recorded extinctions for a few groups of organisms reveal rates at least several hundred times the rate expected on the basis of geological record (Pimm and Brooks 2000). Likewise, projected extinction

rates for animals, plants and birds over the next 50 years are very likely some 1 000 times higher than the background rate of 1 species per million species per year, or less. Human activities have greatly accelerated the historical and recent rate of species extinction far beyond the natural rate, and the background rate of 0.1 to 1.0 can provide a yardstick for quantifying the current and projected impact of land-use/cover change in terms of expected rates of extinction (Dirzo and Raven 2003).

Extinction of species has occurred since the beginning of life on Earth. In fact, most of the organisms that have ever existed are now extinct, and the species currently living amount to 2 to 3% of those that have ever lived, with all others have become extinct, typically within about 10 million years of their first appearance (Dirzo and Raven 2003). Five significant extinction events occurred during palaeo times, but these events collectively seem to have ended no more than 5 to 10% of the species that ever lived. Most strikingly, however, the species at risk now represent an unusually high proportion of all those that ever lived (May et al. 1995). It appears that a recent pulse of extinction started during the late Quaternary period. Evidence indicates that a massive (sixth) extinction event has been underway for some 40 000 years, driven by human activities, and not ended yet. Human activities have greatly accelerated the rate of species extinction far beyond the natural rate, and the occurrence of species extinction is not random. Most have taken place, and still takes place, on islands and in the tropics (Hilton-Taylor 2000), but the percentage of threat in continental areas may be underestimated (Manne et al. 1999). A large majority of the threatened mammal and bird species, for example, occurs in tropical countries such as Indonesia, Brazil, China, Mexico, and the Philippines, and –

Box 4.10. Tropical deforestation and land-use conversions in the occidental Brazilian Amazon: the Pedro Peixoto colonization project

Rodrigo Lorena · Catholic University of Louvain, Louvain-la-Neuve, Belgium

Tropical deforestation represents one of the biggest environmental threats of our time. This process will affect global climate change; increase habitat degradation and fragmentation, and cause a series of unprecedented species extinction. Massive deforestation in the Brazilian Amazon, the largest continuous region of tropical forest in the world, seriously influences the distribution of plant and animal species, soil and water resources, and regional and global climate patterns.

There is no single reason for deforestation (see Chap. 3); it is driven by a combination of factors like population growth, inappropriate colonization policies, absence of assistance and infrastructure for the small producer. The Brazilian Amazon has received over one million migrant farm households from other regions of the country in the last 30 years; many of them are attracted by the government-sponsored frontier settlement programs that offer free tropical forestland. As a result, pressures on tropical forest have intensified along several settlement corridors throughout the region.

The colonization project "Pedro Peixoto" (Fig. 4.5) in the extreme east of the state of the Acre, Brazilian Amazonian,

resulted in a strong reduction of the area of primary forest during the 1990s. The deforestation rate increased from 86 km^{2}yr^{-1} until 1997 to 165 km^{2}yr^{-1} in the remaining period. The development of the colonization project led to a fast increase of cultivated areas, which cover about 300 km^2 today. The pastures represent currently about 1 000 km^2 in the region of study, inserted in all parcels. Similarly, the area of "secondary succession" also increased in the last three years. This was either caused by a temporary abandonment of the cultivated parcel, due to a lack of financial returns to pursue investments or because of the low productivity rates as the soil is often poor in nutrients. This situation of abandonment also reveals the absence of assistance and attention of the state in relation to the small family producers in the new agricultural border of the Brazilian Amazonian.

Fig. 4.5.
Land-use and cover evolution in the occidental Brazilian Amazonian

though flawed with some statistical deficiencies – the tropical-insular predominance is also evident from the numbers of threatened plants in leading countries such as Malaysia, Indonesia, Brazil and Sri Lanka (Hilton-Taylor 2000).

To explain the end-Pleistocene megafaunal extinction, overhunting – rather than low temperatures or pathogenic diseases – has been the most consistent explanation for human impact over the past 40 000 years, until historical times (Martin and Klein 1994; Alroy 2001). In addition, so-called first contact extinctions of species are numerous over the last 1 000 years on islands such as Madagascar and New Zealand, i.e., extinction correlated with the arrival of humans including the introduction of alien invasive or exotic species. Differently, historical and contemporary causes of threat to and, finally, extinction of species are chiefly habitat loss and degradation related to land-use/cover change, followed by direct exploitation of ecosystems and the introduction of exotic species. This was found in the 2000 report of the International Union for Conservation of Nature and Natural Resources (IUCN), addressing the causes of threat to 720 mammalian threatened species, 1 173 threatened birds, and 2 274 plants (Hilton-Taylor 2000). The comprehensive list of threatened species includes 11 167 species facing a high risk of extinction in at least the near future from mainly land-use activities. It might appear as a small number relative to the total number of species (i.e., less than 1%), but it includes 24% of all mammals and 12% of all birds, respectively. For plants, the IUCN number represents only 2 to 3% of the known species, but this is probably due to a serious underestimation (Dirzo and Raven 2003). In contrast, local-to national scale studies demonstrate the wider significance of the issue. For example, it is estimated that in the United States about 33% of the plant species native to the country is threatened with extinction, including 24% of the conifers (Master et al. 2000) (see Box 4.11 on Austria).

Habitat loss and degradation is the most important threat to the three groups, affecting, respectively, 89%, 83%, and 91% of the sampled threatened mammals, birds, and plants. Proximate causes of land-use/cover change are the leading causes of habitat losses, such as agricultural activities (plantations, crop and livestock farming), extraction activities (logging, harvesting, mining, fishing), and the development of infrastructure (human settlements, industry, roads, dams, power plants and lines) (Hilton-Taylor 2000) (see Chap. 3).

Of these specific causes, agricultural activities affect 70% of the threatened species of birds, 49% of the plant species, and less so of the threatened species of mammals (13%) (Hilton-Taylor 2000). Impact figures with view upon birds and plants are considerable, while the low impact numbers for mammals in the 2000 IUCN report assumedly is due to the lack of information on mammals (Dirzo and Raven 2003). In contrast, local-scale studies indicate the relevance of land-use/cover change for mammal extinction. In the tropical forest zone of southeast Mexico, for example, forest fragmentation was found to be the leading cause for the local extinction of several mammal species with medium or large body size. Fragmentation hereby means the reduction of the area of the original habitat available to a particular species, thus decreasing its potential for dispersal and colonization. An empirical study of these predictions in fragments in the forest zone of western Kenya revealed that the fragments will lose half of all species in fifty years, and about three quarters in a century (Brooks et al. 1999). Only recently, fine resolution, spatially explicit data on landscape fragmentation helped to better understand the impact of land-use/cover change on habitat fragmentation as one of the greatest threats to biodiversity (Marguels and Pressey 2000; Liu et al. 2001; Van Laake and Sánchez-Azofeifa 2004). In drylands, the coupled effects of rainfall variability, fire and land-use activities such as over-

---

Box 4.11. Land-use driven endangerment of species and habitats in Austria

Changes in land use and land cover have resulted in the endangerment of many valuable biotopes in Austria, where the Federal Environment Agency is currently working on red lists of biotopes. About 47% of Austria's total area is covered by forests, so first results which are only available for forest ecosystems, are significant. According to this recently published assessment (Federal Environment Agency 2004), there exist 93 different forest biotop types in Austria of which 57% are endangered. Only 24% biotope types were not classified as endangered, while 19% were found to be different types of managed forests with little or no value for nature conservation.

Biotope loss resulting from changes in land use and land cover is also thought to be one of the most important drivers of species loss and endangerment. Figures given in Austria's red list (see Federal Environment Agency 2004) are alarming. For example, 44.6% of all mammals and 57.4% of all bird species in Austria are endangered. 4.0% of all mammals are already extinct, and another 4.0% are critically endangered. The situation is worse with birds, of which 8.6% are extinct and 13.6% critically endangered.

Recent Austrian studies suggest that changes in the availability of trophic energy in ecosystems caused by land use may be a critical driver of species loss. Such changes can be assessed by calculating the "human appropriation of net primary production" (HANPP) which is defined as the difference between the NPP of potential vegetation and the proportion of the actually prevailing vegetation's NPP remaining in ecosystems after harvest (NPP_t). A recent study on an East-Austrian transect consisting of 38 plots sized 600×600 m (Haberl et al. 2004a) found that the species diversity of seven groups – vascular plants, bryophytes, orthopterans, gastropods, spiders, ants, and ground beetles – was linearly and highly significantly related to NPP_t. HANPP, which reduces NPP_t, was found to be strongly and negatively correlated with species diversity of these seven groups. Another study on bird species richness which covered Austria's total area on four spatial scales from 250×250 m to 16×16 km confirmed that NPP_t is a good predictor of bird species diversity, also implying that HANPP should result in species loss (Haberl et al. 2005).

grazing and the encroachment of grain cropping onto extensive rangelands or wilderness zones – rather than density dependence or competition between natural dryland ecosystems – are driving forces of habitat degradation and destruction, with immediate consequences for threats to species (Serneels and Lambin 2001a; Homewood et al. 2001; Geist 2005) – see Box 4.12. In coastal zones, coral reefs are degraded through eutrophication as a consequence of intensive agricultural uses and through sedi-

mentation, in addition to global warming (coral bleaching), with declines in coral abundance leading to corresponding declines in the abundance of coral-dwelling fishes, etc. (Munday 2004) (see also Sect. 4.7.3).

Extraction acitivities such as logging, harvesting, mining, and fishing have greatest impact on plants, affecting 34% of the species, but 53% of the bird species were also affected (Hilton-Taylor 2000). In a study exploring causal chains leading to biodiversity losses in northwest Yun-

---

Box 4.12. Impacts of land-cover change on East African wildebeest populations

Suzanne Serneels · Catholic University of Louvain, Louvain-la-Neuve, Belgium

Trends in biodiversity of large mammals and land cover in the Serengeti-Mara ecosystem (SME) in East Africa over the period 1975–1995 show the relative importance of coupled biophysical, socio-economic, demographic and policy factors in driving those trends. The area comprises some 25 000 km^2 of rangelands, encompassing a network of conservation areas with extensive adjacent buffer zones inhabited by (agro-)pastoralists. It constitutes a natural experiment in which matched and contrasting policy zones are replicated across an area where ecological, ethnic and micro-economic continuities make rigorous control of confounding factors possible. The system is divided by the Kenya/Tanzania border which demarcates the contrasting macro-political and -economic systems of the two countries. It is roughly defined by the movements of the migratory wildebeest (*Connochaetes taurinus mearnsi* Burchell). Migratory wildlife species such as wildebeest, zebra and Thomson's gazelle show similar seasonal movements between habitats, using the short grasslands in the south of the Serengeti National Park in Tanzania during the wet season (January to June), and using the tall grasslands in the north of the Serengeti and in the Masai Mara National Reserve (Kenya) during the dry season (August to November). Another, smaller, wildebeest population is covering a smaller migration range in the Kenyan part of the ecosystem. The Loita Plains in the north make up the wet season range and the main calving area. When the short grasslands are depleted, the herds migrate to the Masai Mara National Reserve (MMNR), where they meet with the Tanzanian wildebeest population.

Whereas the wildebeest migration in the Tanzanian part of the ecosystem is almost entirely confined to protected areas, the wildebeest population in Kenya resides in unprotected land for most of the year. The wildebeest population in the Kenyan part of SME declined drastically over the past twenty years to about 31 000 animals, 25% of the population size at the end of the 1970s. The Serengeti wildebeest population fluctuated around a mean of 1.2 million animals since the late 1970s. The population is regulated by green biomass availability in the dry season (Mduma et al. 1999). There has been little evidence of changes in resident wildlife densities over the past 20 years in Serengeti, except for rhino, roan antelope and buffalo, whose numbers declined, mainly due to poaching (Campbell and Borner 1995). Meanwhile, a decline of 58% for all non-migratory species, with the exception of elephant, ostrich and impala, was observed in the MMNR and adjacent rangelands. There was no significant difference in the rate of decline between the protected area and the unprotected rangelands (Ottichilo et al. 2000).

The temporal changes in the wildebeest populations in the Kenyan and Tanzanian parts of the ecosystem and their relationship with possible driving forces of change were analyzed, such as rainfall, normalized difference vegetation index (NDVI) (as a proxy for green biomass), livestock numbers, human population growth rates and land-use changes. Result show that land-use changes were most important in the unprotected buffer zones in the Kenyan part of the SME (Serneels and Lambin 2001a). By 1995, more than 50 000 ha of land in the Loita Plains has been converted to commercial cereal cultivation. Hence, about 20% of the

wet season range for wildebeest was lost (Serneels et al. 2001). Land-use change in the Tanzanian buffer zones consisted of conversion of rangeland to small patches of subsistence agriculture, scattered in the landscape. Livestock populations in both Kenyan and Tanzanian parts fluctuated but did not show any trend between 1975 and 1995. Human population growth rates as estimated from demographic survey and uptake of subsistence cultivation (measured by household survey) did not differ significantly between Kenyan and Tanzanian buffer zones. Despite high interannual variability, there are no significant differences in rainfall or NDVI time series for both parts of the ecosystem, nor was any trend found in the data.

Among the possible driving forces behind the downward trend in wildebeest numbers in the Kenyan part of the ecosystem, only land-use change showed a clear and concomitant trend over time. The first decline in wildebeest numbers in the Kenyan part of the SME that occurred between 1980 and 1982 was most probably caused by high wildebeest mortality due to prevailing drought conditions. Subsequent declines in the Kenyan wildebeest population are clearly attributable to changes in land-use, as the decrease in wildebeest densities is limited to those parts of their wet season range that were converted to mechanized agriculture. The expansion of wheat farms forced wildebeest to either use the dryer rangelands or to move to wetter areas where competition with livestock and other wildlife was higher. It suggests that, from the 1980s onward, competition for food in the Kenyan rangelands has put a stress on the wildebeest population throughout the year. Dry season food availability is probably responsible for the smaller inter-annual fluctuations in wildebeest numbers, while the reduction of the wet season range in the Loita Plains has caused wildebeest numbers to drop considerably since the early 1980s. The shrinking wet season range amplifies the impact of low rainfall and prevents the population from recovering from drought impacts, due to limited per capita availability of food.

Detailed cost benefit analyses showed that conversion to commercial agriculture is being driven by relative returns to cultivation *versus* wildlife tourism *versus* livestock rearing, and more importantly, by the selective capture of the returns to wildlife enterprises by local and national elites, making wildlife based activities an unattractive option for most of the rural population (Thompson and Homewood 2002).

The Serengeti-Mara case provides compelling evidence that external processes such as the expansion of mechanized agriculture in response to market opportunities and/or policies (Homewood et al. 2001; Homewood 2004) may have a major impact on the dynamics of ecosystems within protected areas. Over the last decades, the decline in the Kenyan wildebeest population did not seem to affect the much larger Serengeti wildebeest population. However, ongoing land privatization in the Kenyan Group Ranches adjacent to MMNR will open the way for individual land owners to make land-use decisions over cultivation, livestock and/or wildlife-based activities, land lease or sale on the basis of relative returns to them individually. This might lead to further habitat loss and therefore in turn to wildlife loss across the SME.

nan of China, it was found that logging followed by monocultural forest plantation, cash crop plantation and livestock grazing have contributed significantly to past species losses, while at present, triggered by national policy changes, the market-driven demand for non-timber forest products (NTFPs) such as wild fungi and medicinal plants poses the largest threat to species (Xu and Wilkes 2004).

Activities related to the expansion of infrastructure such as human settlements, industry, roads, dams, power plants and lines affect 34% of the threatened plant and 32% of the bird species, but only 8% of the threatened mammals.

Given a high concentration of threatened species in tropical ecosystems and an unbroken trend of deforestation, degradation and fragmentation in this zone (see Sect. 2.3.1), a tremendously high rate of extinction can be expected in the humid tropics. Studies based on modeling habitat loss as major process associated with species extinction, for example, reveal nil extinctions of birds in areas that have long been partly deforested such as North America, or on the brink of extinction in the region (Pimm and Askins 1995), but medium-term extinctions in areas of recent deforestation such as in insular Southeast Asia (Brooks et al. 1997) and Brazil's Atlantic forest (Brooks and Balmford 1996). At the global scale, a reasonable interim estimate would be that, at present, a third of the plant species of the world are threatened. As for birds and mammals, a global picture of extinction is as follows: at least 500 (but probably closer to 600) out of 1192 threatened bird species and some 565 of the 1137 threatened species of mammals will go extinct in the next fifty years, due to habitat loss and fragmentation in the tropical forest zone mainly (Dirzo and Raven 2003). At any event, more than a third of the existing species on Earth could disappear with the destruction of tropical forests, only, and it is reasonable to envision the loss of two thirds of the species on Earth by the end of the 21st century (Dirzo and Raven 2003).

Direct exploitation is second in importance as contemporary driver of threats to species, and possibly extinction. The threats to 37%, 34%, and 8% of the sampled bird, mammal, and plant species arise from hunting, trading, and collecting (Hilton-Taylor 2000). At a regional scale, hunting is particularly critical for mammals, especially in the tropics (Dirzo and Raven 2003). It was estimated, for example, that subsistence hunting alone may be responsible for the killing of about 14 million animals per year in the Brazilian Amazon (Redford 1992). And, (subsistence) hunting is a driver of extinction in Africa and Asia as well.

Third in importance in the IUCN study is the introduction of alien invasiv species, which affects 30% of all threatened birds and 15% of the plants, but only 10% of the mammals. In particular the extinction of birds on islands since about 1800 can be attributed to the activities of introduced exotic species. On Hawaii, virtually all of the threatened species are in danger of extinction because of exotic species, and, for the mainland United States, between 25 and 40% of the threats to extinction for native plants stem from the activities of introduced plants and animals (Hilton-Taylor 2000).

4.5.4 Pollination Losses and Other Impacts

In addition to habitat conversion (destruction) and modification (degradation, fragmentation), pollinator loss is further mentioned as a pressure directly related to land-use/cover change – apart from pollinator loss due to other pressures such as climate change and global warming (Root et al. 2003). For example, cotton growing is known to bear many impacts associated with pesticide use, triggering losses of plant pollinators, among others. Heavy pesticide use has a negative impact on the majority of plant pollinators, especially bees, which not only pollinate cotton but a number of important food crops in addition (Bingen 2004).

Finally, evidence is accumulating that the coupled impacts of (anthropogenic) climate change and land-use/ cover change will lead to the differential loss of populations especially at the warmer margins of species' ranges. The problem is compounded by the effects of habitat fragmentation. Whereas in the past, individuals dispersing from a marginal habitat in a warming (or cooling) world may have found more suitable conditions close by, in the 21st century the likelihood is that they will be separated from the nearest patch of a more appropriate habitat by a considerable expansion of agricultural or other unsuitable habitats (Freitas 2006). Fragmentation thus greatly reduces the probability of successful dispersal and establishment. Artificial and natural corridors between remaining habitat patches may become increasingly important in favoring range shifts through highly fragmented landscapes, although it is suspected that large protected areas, having a series of climatically discrete habitats, may be of even greater value. In facing major environmental changes, species that fail either to relocate their ranges or to adapt accordingly elsewhere simply go extinct. Species restricted to isolated habitat fragments and reserves must rely either on their limited physiological tolerances, or on evolutionary adaptation in situ, to survive quick global warming (Pimm et al. 1995; Magurran and May 1999; Sala et al. 2000).

Some species and ecosystems might be more sensitive to land-use change and land-management practices. Both island and mountain flora is more vulnerable to invasion, due to a high percentage of endemic species, coupled with extreme vulnerability to habitat destruction. Many endemic species often depends on indigenous pollinators. However both island and mountain pollinators are disappearing due to heavy use of pesticides, habitat destruction, land-use change and introduction of exotic species (Cox and Elmqvist 2000; Joshi et al. 2004) – see Box 4.13.

> **Box 4.13. Declining cliff bee (*Apis laboriosa*)
> in the Himalayan mountain ecosystem**
>
> *Apis laboriosa*, the largest honey bee species of the world, lives
> in the Himalayas on inaccessible cliff faces. It lives at high al-
> titudes ranging from 1200 m to 3600 m, forages at up to
> 4100 m, makes a seasonal migration depending upon the avail-
> ability of the bee forage. It provides pollination for hundreds
> of plant species along different altitudes in mountain ecosys-
> tems. Traditional honey hunting is an important livelihood
> for indigenous people such as Gurungs and Magars in Nepal.
> However, the indigenous bee population of *A. laboriosa* has
> been declining rapidly during recent decades (number of cliffs
> with bee colonies and number of nests per cliff, as well as nest
> size). These include destructive honey hunting, loss of forage
> and loss of nesting sites as a result of land-use change and
> landslides, livestock overgrazing, destruction of forests, intro-
> duction of modern technology, particularly improved crop va-
> rieties with the attendant application of pesticides, as well as
> diseases introduced by exotic bee species (Joshi et al. 2004).

A recent international assessment found many diverse
"ecoagriculture" systems around the world whereby local people modified land-use patterns and resource management systems to raise both agricultural productivity
and biodiversity and ecosystem services (McNeely and
Scherr 2003). Shifting cultivation is often held to be the
principal driving force for deforestation in tropical world.
To view swiddens as just temporary fields surrounded
by abandoned land under wild regrowth, however, is
wrong. More than four decades ago, Harold Conklin
(1957) pointed out that "shifting cultivation may refer to
any one of an undetermined number of agricultural systems." Spencer (1966) described 18 distinct types of shifting agriculture within Southeast Asia alone. Brookfield and
Padoch (1994) argue that swidden agriculture is not one
but many hundreds or thousands of systems. Alcorn (1990)
calls swidden farming "managed deforestation," a system
built around patchy, pulsed removal of trees but not of the
forest. Indigenous farmers work to manage deforestation
in sequential agroforestry systems that integrate secondary successional vegetation – everything from grass and
bushes, to young open-canopy tree communities, to mature closed-canopy tree communities. Studies in Southeast Asia (Fox et al. 1995; Xu et al. 1999) suggest that land-
use change has begun to occur in the region as farmers
switch from swidden cultivation to cash crops including
both paddy rice and plantation tree crops. These results
suggest that most upland areas of Asia will eventually see a
major change in land-use with the conversion from swidden agriculture to commercial crops and a change in land
cover from secondary vegetation to permanent mono-cul-
tural agriculture albeit tree crops in many cases (see
Chap. 3). Permanent agriculture could result in a tree-
dominated land cover (e.g., rubber, palm oil, cardamom,
or tea), or it could result in a land cover composed of an-
nuals (e.g., maize, cassava, and upland rice). In either case,
biodiversity, as measured by the number of species found
on the landscape, would probably decline (Nagata 1996).

4.6 Soil Quality and Land-Use/Cover Change

4.6.1 Overview

Soil quality is the ability of soils to function within natural and managed ecosystems (Karlen et al. 1997), and
depends on physical (e.g., bulk density, depth, texture),
chemical (e.g., organic C, extractable N, extractable P),
and biological (e.g., soil respiration, soil enzyme activities, microbial biomass) properties of the soil. Soil quality influences five functions of the soil, namely the ability to *(a)* accept, hold and release nutrients; *(b)* accept,
hold and release water both for plants, and for surface
and groundwater recharge; *(c)* promote and sustain root
growth; *(d)* maintain suitable biotic habitat; and *(e)* respond to management and resist degradation (Larson
and Pierce 1991; Brejda et al. 2000). The notion of soil
quality is much broader, but includes soil fertility, which
is related to the nutrient supplying capacity of the soil.
The concept of soil quality was introduced more than
30 years ago, but has undergone its most rapid adoption in the 1990s as a result of the effects of land-use
practices on soil quality (Karlen 2004).

Soils have both inherent and dynamic qualities. Inherent soil quality is a soil's natural ability to function,
and is determined by climate, parent material, topography, time and vegetation under which it has formed.
Dynamic soil quality on the other hand is a measure of
how soils change in response to use or management.
Soil quality assessment focuses on dynamic qualities
to evaluate the sustainability of soil management practices. Most of these qualities cannot be measured directly, but are typically inferred from soil properties
that serve as indicators. Many of these indicators exhibit high correlation, function together, and are mutually influenced by land use at various levels of management such as the cropping system, the farming system,
and the catchment (Dumanski et al. 1998; Tan et al. 2003;
Braimoh et al. 2005).

Many authors – such as Larson and Pierce (1991) and
Doran and Parkin (1996) – have proposed several minimum data sets for use as soil quality indicators. However, to date, there is no universally accepted standard
data set, nor are their universal critical values of soil quality parameters. This is because the magnitude and direction of change in soil quality and the equilibrium contents of parameters are dependent on climate, mineralogy, soil conditions and land-use practices which vary
from region to region (Sanchez-Maranon et al. 2002;
Sparling et al. 2003). Nonetheless, the conservation of
organic matter in soil is supported by nearly all soil scientists because it is a source of mineralizable nutrients,
acts as substrates for soil microbes, influences soil structure and aggregate stability, increases cation exchange
capacity and improves moisture retention (Craswell and

Lefroy 2001). It has been indicated, however, that low soil C contents could sometimes be beneficial by reducing pesticide application rates as a result of lower absorption (Sojka and Upchurch 1999). The decline of soil organic matter in the soil usually occurs in a curvilinear way, and related soil properties usually change with it along a continuum of use and management (Sparling et al. 2000).

Soil resilience is an important component of soil quality. It is the capacity of the soil to resist change or recover its functional and structural integrity after a disturbance (Lal 1997), with disturbance or perturbation being any event (stress) that leads to a significant change from the normal functioning of the soil ecosystem (Forman and Godron 1986). Human activities such as logging, urban and industrial development, and agricultural practices such as tillage can be classified as disturbance (Seybold et al. 1999). A soil's capacity to recover is measured by the rate of recovery and the degree of recovery. The recovery rate (that is, elasticity) measures the amount of time it takes the soil to stabilize, whereas the degree of recovery is the magnitude of restoration to some stabilized potential relative to the pre-disturbance (antecedent) state (Seybold et al. 1999). If a disturbance is too drastic (e.g., subsidence or terrain deformation through gully erosion or mass movement), or if the soil is inherently fragile (e.g., shallow soils on steep slopes), the soil can undergo profound degradation, leading to a long time for its capacity to function to be restored or a high amount of monetary investment to ensure restoration. The mechanisms that affect the ability to recover and rate of recovery of soils include the rate of new soil formation, aggregation, organic matter accumulation, nutrient recycling, leaching of excess salts, and increases in biodiversity, including species' succession (Lal 1997).

The extent of soil-related inherent constraints to agricultural production across world regions is presented in Table 4.2. Erosion hazard implies susceptibility to erosion caused by very steep slopes (>30%) or moderately high slopes (8–30%) accompanied by a sharp textural contrast within the soil profile. It is the major inherent soil constraint for the world, occurring in 16% of its total land area. Strong acidity (that is, aluminum toxicity) occurring in 15% of the soils is the next dominant inherent soil constraint on a global basis, followed by shallowness; the occurrence of rocks close to the soil surface (14%), and poor soil drainage (13%). Erosion hazard is a major challenge in all the regions, with frequency ranging from 10% for soils in North Africa and Near East to 20% for soils of Europe. Poor soil drainage is also a widespread inherent constraint (16 to 27%) in North America, Europe and North Asia east of Urals. The occurrence of rocks in a sub-surface horizon is highest for North Africa and Near East (23%) and lowest for South and Central America (11%). Soil acidity is the principal inherent constraint for soils of Sub Saharan Africa (18%) and South and Central America (39%). Low inherent fertility is the next principal constraint for Sub Saharan Africa (16%), whereas high P fixation resulting from a preponderance of ferric oxides in the clay fraction constitutes the third dominant constraint (15%) for soils of South and Central America. The presence of free soluble salts leading to salinity and sodicity is highest for Asia and Pacific (11%) and North Asia east of Urals (10%).

Table 4.2. Global distribution of the major inherent soil-related constraints to crop production[a]

Constraint	Sub-Saharan Africa (24×10^6 km²)	North Africa and Near East (12×10^6 km²)	Asia and Pacific (29×10^6 km²)	North Asia, east of Urals (21×10^6 km²)	South and Central America (20×10^6 km²)	North America (21×10^6 km²)	Europe (7×10^6 km²)	World (134×10^6 km²)
Proportion (%)								
Erosion hazard	15	10	16	16	19	18	20	16
Aluminium toxicity	18	0	14	4	39	10	8	15
Shallowness	13	23	17	13	11	12	12	14
Poor soil drainage	8	1	11	27	10	16	17	13
High salt content	4	6	11	10	5	1	3	6
Low inherent fertility (cation exchange capacity)	16	2	4	0	5	0	1	5
High phosphorus fixation	4	0	5	0	15	0	0	4
Soils difficult to till due to high proportion of swelling and shrinking clays	5	1	5	0	2	1	1	2

[a] The estimated land area for each region is indicated in parenthesis. Constraints are not mutually exclusive; for instance P fixation is common in soils with Al toxicity. *Source:* Modified from FAO (2000).

4.6.2   Extent of Human-Induced Degradation

Soil degradation implies a loss in soil quality, that is, the inability of the soil to perform any of the five functions specified above. More specifically, soil degradation is defined as an anthropogenic process that reduces the capability of soils to support life on Earth (Oldeman et al. 1991). It is a biophysical process driven by socioeconomic (land tenure, marketing, institutional support) and political (incentives, government stability) factors. A high population density may not necessarily result in soil degradation given proper resource management technologies, the right social and economic milieu and supportive environmental policies (Tiffen et al. 1994a). Soil degradation is an additional challenge to the inherent constraints to agricultural productivity. Of the estimated total Earth's land area of 134 million km², over 14% are estimated to have been degraded by anthropogenic activities (UNEP 2002). Aggregated globally, the five ma-

jor human causative factors of soil degradation in order of magnitude are estimated to be: overgrazing, deforestation, agricultural mismanagement, fuelwood consumption, and urbanization – see Table 4.3 and Chap. 3.

Four major types of soil degradation can be distinguished, namely water erosion, wind erosion, chemical degradation and physical degradation. The severity of these types of degradation varies across world regions, with about 50% of the world soil resources being in the moderate to extremely degraded class – see Table 4.4 and Chap. 2.

Water erosion is the most frequent type at the proximate level across world regions with deforestation as the primary causal mechanism of soil degradation in almost all the regions – see Table 4.5.

In Central and Eastern Europe, soil compaction which occurs in about 11% of total land area in association with crusting (about 5%) is the predominant type of soil degradation, followed by erosion of top soil by water in 8% of the land area – see Table 4.6. Compaction and crusting are most likely due to the effects of me-

Table 4.3. Five major causes of human-induced soil degradation at the global level

Cause	Extent	
	× 1000 km²	Proportion (%)
Overgrazing. The stripping of soils of vegetation by animals in pastures and rangelands exposes the soil to wind and water erosion	6800	34.6
Deforestation. The clearing of large forest reserves for farmland and urban use and large scale logging	5800	29.4
Agricultural mismanagement. Poor land management strips the soil of vegetation cover leading to erosion; salinization and waterlogging are caused by poor drainage of irrigated land, whereas soil nutrient loss occurs through bush burning and continuous cropping with little or no fertilizers to replenish the soil	5500	28.0
Fuelwood consumption is high in areas where firewood and charcoal are the primary sources of energy	1370	7.0
Urbanization. Expansion of built up areas, road building, mining and other industrial activities lead to the loss of valuable agricultural land in Asia and the USA	195	1.0
Total	19665	100

Source: Modified from FAO (1996).

Table 4.4. Severity of anthropogenic degradation across world regions

Country	Degradation severity classes				
	None	Light	Moderate	Severe	Very severe
	% of total land area				
Sub-Saharan Africa	34	24	18	15	10
North Africa and Near East	30	17	19	27	7
Asia and Pacific	28	12	32	22	7
North Asia, east of Urals	53	14	12	17	4
South and Central America	23	27	23	22	5
North America	51	16	16	16	0
Europe	9	21	22	36	12
World	34	18	20	20	6

Source: Modified from FAO (2000).

Table 4.5. Severity, causes and types of soil degradation for selected countries[a]

Country	Total land area (×1 000 km^2)	Degree of degradation					Cause	Type
		None	Light	Moderate	Severe	Very severe		
		% of land area						
Sub-Saharan Africa								
Cameroon	465	40	5	18	14	23	A, O, D	W
Chad	1 259	40	30	7	23	1	O	N, W, P
Ethiopia	1 104	4	10	57	8	20	O	W
Nigeria	1 267	3	38	4	27	28	D, O	W
Uganda	200	4	1	43	41	12	O, D, A	W
North Africa and Near East								
Iran	1 643	8	6	28	41	17	V, O, D	W, C, N
Oman	271	16	28	17	39	0	O, D	W, N
Saudi Arabia	2 396	21	31	15	28	6	O	N
Yemen	480	4	18	33	45	0	D, O	W, N
Asia, East of Urals								
China	9 550	28	8	30	25	10	D, A	W, N
Indonesia	1 916	1	36	26	32	6	D, A	W, N
India	3 157	37	1	4	43	16	D, A	W, C
Malaysia	333	0	0	17	83	0	D, A	W, C
Uzbekistan	446	75	2	9	13	0	A	C, P
Tajikistan	143	83	0	10	7	0	A	C, W, N
Latin America								
Argentina	2 772	8	29	51	11	0	A, O, D	W, N, C
Ecuador	283	5	66	23	2	4	D	W, C
Colombia	1 136	10	53	19	18	0	D, O	W, C
Costa Rica	51	0	0	0	32	68	D, O	W, P, C
Europe								
Turkey	779	1	0	0	69	30	O, D, A	W, N, P
Poland	313	16	0	1	49	34	A	W, P
Hungary	93	0	14	21	65	0	D, A	W, C, P
Romania	238	0	0	0	89	11	D, A	W
Slovenia	505	2	20	39	35	3	D, A, O	W, P
Iceland	103	17	0	31	32	20	D, O	W, N

[a] Cause A: agriculture; O: overgrazing; D: deforestation; V: overexploitation of vegetation; Type W: water erosion, N: wind erosion, C: chemical deterioration; P: physical deterioration. *Source:* Modified from FAO (2000).

chanical land clearing and mechanized cultivation with heavy agricultural machinery (van Lynden 2000). Wind erosion is common in the southeast portion of the region which is notably drier. Most of the water and wind erosion is caused by poor agricultural land practices. Fertility decline is the most predominant form of chemical degradation occurring in about 6% of the land area. Heavy metal and radioactive pollution both occur in about 3% of the land area as a result of industrial activities, whereas acidification and salinization are attributed to agriculture.

Human-induced changes in nutrient cycling in terrestrial ecosystems significantly affect the sustainability for food production, the state of the natural resource base, and the health of the environment (Craswell et al. 2004). It has been estimated that 230 Tg (1 Tg = 1 million t) of plant nutrients are removed yearly from agricultural soils, whereas global fertilizer consumption of N, P_2O_5 and K_2O is 130 Tg (Vlek et al. 1997). The principal mechanisms contributing to nutrient depletion include runoff and erosion, leaching, crop residue removal and harvested products (Smaling et al. 1997).

Table 4.6.
Types and severity of degradation in Central and Eastern Europe

Degradation type	Land area affected (×1 000 km²)	Light	Moderate	Strong–extreme
		Proportion (%)		
Physical				
Compaction	625	40	59	1
Crusting	273	36	62	2
Aridification	267	16	51	33
Waterlogging	85	40	44	17
Chemical				
Fertility decline	313	26	68	6
Acidification	250	18	76	6
Pesticide pollution	108	40	59	1
Heavy metal	80	18	79	3
Radioactive contamination	62	45	54	1
Salinization	51	31	55	14
Water erosion				
Top soil erosion	455	19	39	42
Terrain deformation	51	2	34	64
Wind erosion				
Top soil erosion	199	28	33	39

Source: van Lynden (2000).

Nutrient imbalance and depletion is a major cause of declining crop yields in developing countries of Africa, Latin America and the Caribbean. In Latin America where erosion hazard resulting from steep slope is already a major problem for sustaining agricultural production, estimates of nutrient balance in arable soils show negative balances of N, P_2O_5 and K_2O ranging from –156 kg ha^{-1} for Guyana to –15 kg ha^{-1} for Mexico in the 1996 to 1999 period – see Table 4.7.

Erosion resulting from continuous cropping and crop residual removal constitute about 70% of all N losses, nearly 80% of all K losses and 95% of all P losses (Craswell et al. 2004). Gains in nutrients through mineral fertilizer application, nutrient deposition and nitrogen fixation, however, occurred for Chile, Costa Rica, Uruguay and Venezuela. In Sub-Saharan Africa with highly eroded, intensely leached soils with low inherent fertility, the average application rate for mineral fertilizer is below 10 kg ha^{-1}, and few countries – Nigeria, Zimbabwe, Kenya, Sudan and Ethiopia – account for about 75% of total fertilizer consumption (Craswell et al. 2004). Fertilizers are mostly applied on export crops such as cocoa, cotton, coffee, groundnuts and oil palm.

The negative nutrient balances due to insufficient external inputs, and the inequitable distribution of nutrients between and within countries are further exacerbated by the transport of nutrients in traded agricultural products. Population (growth), rapid urbanization and change in consumption patterns are driving the global structure of food demand (see Chap. 3), and hence the flow of nutrients across countries (Rosegrant et al. 2001).

China uses large amount of fertilizer to produce food for its teeming population – see Table 4.8. The NPK in its net export in 1997 is only 2% of the domestic fertilizer it consumed in the same year. Japan's net import of NPK in 1997 was 87% of its domestic fertilizer consumption. The proportion of imported NPK is expected to increase to 101% of its fertilizer consumption by 2020. This reflects the overdependence on agricultural imports to feed the growing population on the one hand, and the expansion of urban areas at the expense of cropland, on the other. Settled areas increased from 1.1% in 1900 to 5.5% of Japan's land area in 1998, and are estimated to increase to 7.8% by 2020 (Himiyama 1998). The U.S. export of nutrients is expected to increase from 18% of its domestic fertilizer consumption in 1997 to 28% in 2020. This largely reflects government subsidy policies that encourage agricultural exports by American farmers. Whereas Sub-Saharan Africa is a net importer of nutrients in food, wastages resulting from food consumption are hardly used to ameliorate deficiencies in rural soils. Rather they constitute waste disposal problems in cities that consume a large proportion of the food imports. This is a significantly high potential for nutrient cycling in urban and peri-urban agriculture that is yet to be explored in Sub-Saharan Africa (Craswell et al. 2004).

Recent studies indicate that soil erosion and organic matter mineralization are the two major processes of soil degradation in grasslands in Northern China (Wu and Tiessen 2002). The intensification of grazing in grasslands led to topsoil loss of 0.3 kg m^{-2} yr^{-1} on lands with 80% vegetation cover and 1.3 kg m^{-2} yr^{-1} on those with

Table 4.7. Nutrient balance of soils in selected countries of Latin America, the Caribbean and Africa

	Annual balance (1996–1999) of N + P$_2$O$_5$ + K$_2$O (kg ka^{-1})
Latin America and Caribbean	
Belize	−152
Guyana	−156
Nicaragua	−50
Mexico	−15
Brazil	−60
Bolivia	−110
Argentina	−130
Paraguay	−152
Venezuela	25
Chile	185
Costa Rica	182
Uruguay	28
Trinidad and Tobago	−70
Panama	−5
Jamaica	−70
Africa	
Zimbabwe	−42
Rwanda	−105
Malawi	−70
Cameroon	−41
Kenya	−64
Sudan	−51
Botswana	8
Cote d'Ivoire	−58
Swaziland	−94
Lesotho	−58
Nigeria	−48
Ghana	−52
Gambia	−54
Guinea Bissau	−82
Libya	43

Source: Craswell et al. (2004).

less than 60% cover. Soil losses on croplands were considerably higher: 9.5 kg m^{-2} yr^{-1} on soils cultivated for 8 years, 6.5 kg m^{-2} yr^{-1} on those cultivated for 16 years, and 3.1 kg m^{-2} yr^{-1} for those cultivated for 41 years. Furthermore, the degradation of pasture due to intense grazing led to decline in organic C (by 33%), total N (by 28%) and cation exchange capacity (by 18%). Similarly, the conversion of native pasture to cropland led to organic C decline by 22%, 37% and 55% after 8, 16 and 41 years of cultivation, respectively, as a result of erosion and mineralization. The effective cation exchange capacity (ECEC) and total N also decreased significantly on steep slopes as pastures were converted to cropland, whereas about half of the organic C and organic P losses in a site cultivated for 16 years were due to erosional topsoil loss and the remaining due to mineralization and reduced organic matter inputs under cultivation.

The interactive effects of tillage and soil quality are a concern for sustainable land management owing to the effects of tillage on soil stability, soil resilience, and soil quality. Global analysis of organic C loss following conversion of forests or grasslands to agriculture indicated a 30% loss in organic C from the entire soil profile within 20 years following cultivation, with a higher proportion of the losses occurring within the first 5 years. The effects of tillage on soil properties manifest by affecting processes like infiltration, runoff and sediment loss (Moorman et al. 2004).

In soils of the Northern Great Plains of North Dakota noted for their high fertility, it has been observed that conventional tillage involving the use of disk and chisel plough for seed bed preparation led to reduction in soil organic C (by 7.28 Mg ha^{-1}), particulate organic matter C (by 4.98 Mg ha^{-1}), potentially mineralizable N (by 32.4 kg ha^{-1}), microbial biomass (by 586 kg ha^{-1}), aggregate stability (by 33.4%) and infiltration rates (by 55.6 cm h^{-1}) over a 20-year period compared to no-tillage plots in which soil surfaces were not disturbed except at planting (Liebig et al. 2004). Also, it has been observed that continuous cropping (that is, consecutive

Table 4.8. NPK flows in net trade for selected countries/regions

Country/region	1997 total NPK consumption (Tg)	1997 total NPK in net trade (Tg)	1997 total NPK in net trade as a proportion of 1997 NPK (%)	Projected 2020 total NPK in net trade (Tg)	2020 total NPK in net trade as a proportion of 1997 NPK consumption (%)
USA	17	−3.06	−18	−4.76	−28
Japan	1.11	0.97	87	1.12	101
Australia	1.53	−0.61	−40	−0.86	−56
Eastern Europe	3.24	0.15	5	−0.24	−7
Sub-Saharan Africa	1.01	0.27	27	0.63	62
China	29.87	0.60	2	2.15	7
Latin America	8.95	−0.65	−7	−1.95	−22
South East Asia	6.46	0.33	5	0.77	12

Negative value indicates that a country or region is a net exporter of nutrients. *Source:* Craswell et al. (2004).

maize monoculture over an 8-year period) decreased soil chemical quality, and the rate of decrease was more with plough-based than no-till methods (Lal 1997). Further, significantly lower contents ($p < 0.05$) of clay (by 31%), silt (by 15%), total N (by 33%), available P (by 28%), ECEC (by 32%) and organic C (by 21%) were found in soils continuously cultivated for 15 years compared to those under natural vegetation in Ghana (Braimoh et al. 2004). In a study to examine the effects of tillage on soil quality and cereal yields in Switzerland, it has been noted that reduced soil tillage increased earthworm populations, reduced *Pseudocercosporella herpotrichoides* infection in wheat and increased colonization by symbiotic arbuscular mycorrhizal fungi. It had been observed that except for direct drilling of maize, where maize yields decreased by more than 10% over a 14-year period, maize yields obtained from no tillage and other ploughless cultivation techniques were not significantly different. The no-tillage technique did not differ from conventional tillage methods in terms of bulk density, leading to the conclusion that reducing soil tillage intensity generally leads to an improvement in soil quality without substantial reductions in yield (Anken et al. 2004).

4.6.3   Impacts of Soil Degradation

Soil quality is not often considered a policy objective by policy makers unless soil degradation threatens other objectives. Two reasons possibly account for this. Reason one is the absence of comprehensive data linking soil quality to agricultural productivity (Lal 2000), with the implication that past rates of soil degradation are merely inferred from historical yield trends. The second reason is that soil degradation processes often occur so creepingly to the extent that land managers hardly contemplate initiating timely ameliorative or counterbalance measures (Glantz 1998). However, the decline in long-term productivity potential currently constitutes a threat to food security in many developing countries (Vlek et al. 1997), necessitating more than ever the development of indicators for soil quality management on the one hand, and the development of intervention programs to specific soil degradation issues, on the other (see Chap. 7).

There are at least three circumstances in which the impacts of soil degradation should be of interest to policy makers. The first is when lands with degrading soils are a critical source of food security for subsistence households with very few alternative livelihood options. The least cost response to declining soil fertility from the farmers' perspective is agricultural extensification, especially when the soil quality is suitable and the land accessible. However, the cultivation of marginal lands is increasingly inevitable in areas where population density is high. Soil degradation has been observed to have the highest impact on poor areas in Asia and Africa. For instance, Barbier (1996) indicate that the cost of soil degradation to farmers in Burkina

Faso in the Sahelian zone of West Africa by 2020 will be about 20% of village income as a result of declining soil productivity. Apart from extensification, rural household options are the diversification of livelihood activities to widen income earning portfolio or seasonal or permanent migration to seek economic opportunities elsewhere.

The second situation is when degrading soils are a significant source of supply for national consumers or export markets, and alternative sources of supply are either not viable or totally unavailable. A study to project food demands in the 21st century (Crosson 1995) estimated an aggregate global loss of 12% to 13% of agricultural supply, assuming a 15%, 35% and 75% yield decline, respectively, for light, moderate and strongly degraded cropland soils, and 5%, 18% and 50% decline in yield for pasture soils. A global reduction in yield has been estimated as a result of erosion at 10% in cereals, 5% in soybean and pulses and 12% in root and tubers (Lal 1997). Regional estimates indicate that since World War II, soil degradation has led to a loss of productivity in Asia by 13% in cropland, and 4% in pastures; in Africa by 25% in cropland and 7% in pasture; in South America by 14% in cropland and over 2% in pasture; in Europe by 8% in cropland and 6% in pastures; and in Central America 37% for cropland and over 3% for pasture (Oldeman 1998). Considerable decline in productivity under agricultural intensification are being experienced in several parts of Africa. As formerly fertile lands are continuously cultivated without sufficient fertilizers, yields of cereals have declined from 2 to 4 t ha^{-1} to under 1 t ha^{-1} (Sanchez et al. 1997). A recent study to establish a relationship between maize yield and soil quality index, comprising organic C, ECEC, soil drainage, pH and sand and clay contents in Ghana showed that maize yield declined at the rate of 32 kg ha^{-1} for every 1% decrease in soil quality index (Braimoh et al. 2004) – see Fig. 4.6. Organic C, clay and ECEC were identified to be the most limiting soil properties to maize yield.

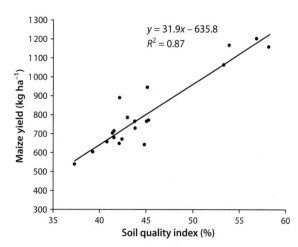

Fig. 4.6. Relationship between soil quality index and maize yield in Northern Ghana. *Source:* Redrawn from Braimoh et al. (2004)

The last situation is when soil degradation significantly reduces agricultural income as a result of lower agricultural production or higher costs, and alternative sources of income are expensive to develop. The estimated annual cost of different types of degradation in South Asia were water erosion $5.4 billion, wind erosion $1.8 billion, soil fertility decline $0.6–1.2 billion, water logging $0.5 billion and salinization $1.5 billion. the total annual cost $9 billion to $11 billion is equivalent to 7% of agricultural GDP (Young 1994). In South Asia, the annual nutrient loss due to erosion was estimated at $600 million, whereas the loss due to soil fertility depletion was $1 200 (UNEP 1994).

4.6.4 Preventing Soil Degradation, Improving Soil Quality

Soil degradation is not a sudden event; it is a gradual process. The cost of preventing degradation is not high if the degradation process is recognized early and appropriate actions promptly taken. Once degradation reaches a point where reclamation is economically prohibitive, the land will be abandoned (Vlek 2005). As late diagnosis adds to the cost of reclamation, it is important to develop quantitative indicators with threshold limits to predict the onset of soil degradation. Such indicators should be sensitive to give early warning signals of change, be able to assess present status and trend, be able to distinguish changes due to natural cycles as opposed to anthropogenic perturbations, and be relevant to ecologically significant phenomena (Rubio and Bochet 1998). Indicators for monitoring soil degradation should also arouse measures to amend the on-going unsustainable land-use practices to prevent further soil degradation (see Box 4.14).

The identification of soil-specific properties that affect resilience is important in improving soil quality. Appropriate soil management will facilitate recovery but may not totally alleviate all constraints to productivity. Changes in agricultural practices intended to increase soil organic C must either decrease the mineralization of organic matter or increase organic matter inputs to the soil, or achieve both. Conservation tillage and changing from monocropping to crop rotation are a promising approach to improve soil quality. No-tillage has been shown to also enhance water-use efficiency in dryland cropping systems (Farahani et al. 1998).

Box 4.14. Impact of land consolidation on erosion and generation of muddy floods in the Belgian loam belt

Olivier Evrard · Catholic University of Louvain, Louvain-la-Neuve, Belgium

Numerous villages of the European loam belt are confronted with muddy floods originating directly from cultivated areas. Central Belgium is particularly confronted with this phenomenon. 80% of the municipalities experienced at least one muddy flood during the last decade.

Land cover in 1957 (Fig. 4.7a) has been mapped from aerial photographs and compared with field observations in 2003 (Fig. 4.7b). After the 1957 consolidation, the mean size of the fields in the study area increased about four-fold from 1.02 ha in 1957 to 4.34 ha in 2003. The construction of a new road in the thalweg leads to runoff concentration in case of heavy rainfall. A simulation with a hydrological model shows that runoff volume and peak discharge at the outlet increase by 20% and 33%, respectively, following the land consolidation operation (Evrard et al. 2006). Runoff concentration on the road leads to the sudden arrival of silt-laden water in the village located downstream, causing damage to public infrastructure and housing property.

A grassed waterway has been installed in the catchment thalweg to mitigate the floods (Fig. 4.7b). This measure leads to a slowing down of runoff. The spread of runoff over a longer time period decreases the flood risk for the village located 500 m downstream. However, implementation of additional conservation measures is needed to limit runoff generation within the catchment.

Fig. 4.7.
Impact of land consolidation on land cover in a catchment of the Belgian loess belt

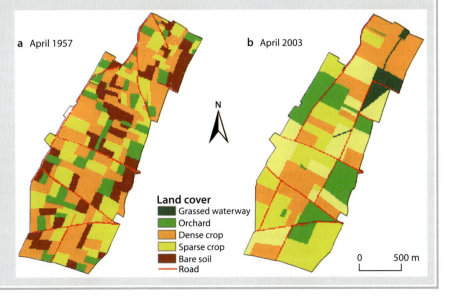

a April 1957

b April 2003

N

Land cover
- Grassed waterway
- Orchard
- Dense crop
- Sparse crop
- Bare soil
- Road

0   500 m

Using a global database of 67 long-term experiments, is has been demonstrated that no-tillage, because of its ability to sequester C, can reverse the loss of organic C that may have occurred during intensive cultivation. On average, a change from conventional tillage to no-tillage can sequester 57 ± 14 g C m^{-2} yr^{-1}. Increasing crop rotation complexity (that is, changing from monoculture to continuous cropping, changing from crop-fallow to continuous monoculture or rotation cropping, or increasing the number of crops in a crop rotation system) can sequester less, an average 20 ± 12 g C m^{-2} yr^{-1} (West and Post 2002). With a change from conventional tillage to no-tillage, carbon sequestration rates can be expected to peak in 5 to 10 years, with soil organic C reaching a new equilibrium in 15 to 20 years. However, increasing the complexity of crop rotation results in a longer time (approximately 40–60 years) for C to reach a new equilibrium. A major inference from this study (West and Post 2002) is that if a decrease in tillage and enhancement in crop rotation complexity are simultaneously implemented, the short term (approximately 15 to 20 years) increase in organic C will be primarily caused by change in tillage, whereas the long term (approximately 40 to 60 years) increase in organic C will be due to rotation enhancement.

4.7 Freshwater Hydrology, Agricultural Water Use, and Coastal Zones

4.7.1 Overview

On its transit overland, water is exposed to the properties of the terrestrial surface which, in conjunction with land-use/cover change, are important determinants for both water quantity (i.e., sufficient supply of freshwater to support human and natural systems) and water quality (i.e., suitability of supply for an intended use). It is well recognized that direct land changes such as forestation, cropland change, mineral extraction and urbanization – but also indirect effects – hold considerable potential to significantly modify or even disrupt hydrological cycles (DeFries and Eshleman 2004; Eshleman 2004; Mustard and Fisher 2004).

The recent intensification of agriculture, in particular, has largest impacts on both freshwater and marine ecosystems, which are and continue to be greatly eutrophied by high rates of nitrogen and phosphorus release from agricultural fields. Aquatic nutrient eutrophication triggers the growth of blue-green algae that renders water unpalatable and increases the growth of weed. It increases the turbidity of water and links to several other impact types (Steffen et al. 2004): it can lead to shifts in the structure of food chains, including fish killings (see Sect. 4.2), outbreaks of nuisance species (see Sect. 4.3), and loss of biodiversity (see Sect. 4.5). Water shortages already exist in many regions of the world (with more

than one billion people without adequate drinking water), and 90% of the infectious diseases in developing countries are transmitted from polluted water (Tilman 1999; Pimentel et al. 2004).

It has early been estimated that withdrawals (i.e., water removed from a source and use for human needs) from streams, rivers, and aquifers, combined with instream flow requirements (all together totaling 6 780 km^2 yr^{-1}), already account for more than 50% of total accessible runoff (Postel et al. 1996). Claims have arisen that the human impact on the terrestrial water cycle during the last 50 years (actually dating back to about 4 000 years with water engineering in association with cropland expansion; see Chap. 2) has likely exceeded natural forcings of continental aquatic systems in many parts of the world (Meybeck and Vörösmarty 2004).

Several interconnected properties intervene into the relationships shaping the impact of land-use/cover change upon freshwater hydrology. First, the life history of water on a landscape encompasses its appearance already in precipitation until its exit to the ocean. Indeed, any land-use decision very often turns out to be a water decision (Falkenmark 1999), be it the conversion of coastal marshes into agriculture, settlements or industrial zones, or the modification of farmland through the development of freshwater resources for irrigation. In a broader Earth System perspective, responses to these influences are already discernible to reverberate through the hydrological cycle which go well beyond the direct human appropriation of freshwater and coastal zones for drinking, agriculture, and industry (Kabat et al. 2004; Steffen et al. 2004).

Various trade-offs need to be addressed between the potential benefits of land-use/cover change and potentially negative consequences upon the hydrological cycle (DeFries et al. 2004a). Clearly, agricultural ecosystems have become incredibly good at producing food, and it has been the irrigation of croplands which contributed enormously to food security, with irrigated lands being three times as productive as non-irrigated cropland, also providing for greater economic value than non-irrigated cropland or rangeland (Mustard and Fisher 2004). Currently, around 40% of all agricultural production comes from irrigated areas, and global food production foreseeably becomes largely dependent on artificial irrigation systems (Gleick 2003). On the other hand, current agricultural practices involve deliberately maintaining ecosystems in a highly simplified, disturbed and nutrient-rich state – with threats to biodiversity and the supply of water for food production (Matson et al. 1997; Tilman 1999; Pimentel et al. 2004). Especially in irrigated dryland zones of the world, the negative hydrological consequences of land-use/cover change are most pressing. They illustrate modern or industrial society's capacity to transform large coupled human-environment systems rapidly such as in the case of the Aral Sea Basin

(Turner and McCandless, forthcoming), or the collapse even of ancient societies (Diamond 2005). In any case, and not only in drylands, the increased yields of food production have environmental costs that cannot be ignored, especially if the rates of nitrogen and phosphorus triple, and the amount of irrigated land doubles in the coming decades (Tilman 1999). Not included in trade-off considerations so far are the large social costs associated with social disruptions due to dam construction (and population displacement) and other large-scale water infrastructure constructions particularly along transnational rivers in dryland zones of the world (Gleick 2003; Pimentel et al. 2004).

4.7.2 Hydrological Consequences of Land-Use/Cover Change

Many insights into the hydrological consequences of land-use/cover change stem from the experimental manipulation of land cover at rather small spatial, observable scales such as research plots, hill slopes, and small catchment areas (e.g., 100 to 1 000 ha) – see Table 4.9. These manipulations prove that human activities can modify or disrupt interception losses by different plant species, soil infiltration, storm runoff, water yields, flood peaks, evapotranspiration rates, concentration of water quality constituents, snow accumulation and snow melting. However, extrapolating from such studies to larger systems such as river basins is confounded by the diversity of land-use as well as hydrological systems (Newson and Calder 1989; Eshleman 2004; Mustard and Fisher 2004).

There is an unequal distribution of water resources or hydrological systems, for which the demand likewise varies greatly. Regions in which water quantity is particularly stressed by human demands are western North America, areas bordering the Sahara, the Arabian Peninsula, and several densely populated zones in Asia, namely India, Pakistan, and northeastern China (Vörösmarty and Sahagian 2000).

The dominance of different processes changes at different scales. For example, land-use/cover changes in the upstream of a catchment may have a different impact on hydrology than changes downstream, and processes of interception, infiltration and storage dominate at the plot scale, while channel processes assume a greater role with increasing catchment size (Archer 2003). Only few examples exist for controlled long-term studies of the impacts of permanent land conversions at multiple scales such as forest to agriculture or agriculture to urban cover (DeFries and Eshleman 2004).

The impact of land-use/cover change also varies in terms of time scale. On the scale of catchments, changes usually occur at irregular time intervals, while crop planting, drainage of afforestation and other changes at the plot scale occur fairly regularly (Archer 2003). Furthermore, river and lake quality can be restored in quite a short time, while destroyed biodiversity can take several thousands of years to recover to the original condition. And, especially in the case of sediments, even after a "beneficial" land-use change there will be still enough sediment in the system from prior human-induced or natural erosion that would lead to increased sediment loads in the rivers. In most of the cases, the time taken to contaminate a system is only a fraction of the time that is required to later clean up the same system. Remedies against high loadings of pathogens as a result of large population centers, for example, can be effective within less than one year, while eutrophication and micro-pollutants may contaminate the system for up to 100 years. Agrochemicals, in particular, may have a large impact over a long period of time. Also mining and other sources of suspended load may have an impact on the ecosystem over many years (Peters and Meybeck 2000) – see Fig. 4.8.

Table 4.9.
Spatial dimensions of land-use effects on the hydrological cycle

Impact	Basin size (km²)						
	0.1	1	10	100	1 000	10 000	100 000
Average flow	×	×	×	×	–	–	–
Peak flow	×	×	×	×	–	–	–
Base flow	×	×	×	×	–	–	–
Groundwater recharge	×	×	×	×	–	–	–
Sediment load	×	×	×	×	–	–	–
Nutrients	×	×	×	×	×	–	–
Organic matter	×	×	×	×	–	–	–
Pathogens	×	×	×	–	–	–	–
Salinity	×	×	×	×	×	×	×
Pesticides	×	×	×	×	×	×	×
Heavy metals	×	×	×	×	×	×	×
Thermal regime	×	×	–	–	–	–	–

Legend: ×: Observable impact; –: no observable impact. *Source:* Adapted from Kiersch (2001).

Fig. 4.8.
Logarithmic diagram showing
the time scale of impacts
on hydrology after land-use
changes. *Source:* Batchelor
and Sundblad (1999)

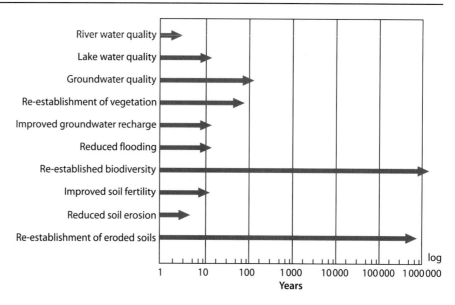

Identifying a linkage between land-use and hydrological change often implies to isolate the impact of biophysical forces and climatic variability, in particular. There is some reason to assume that climate change and long-term climatic fluctuations are particularly inherent to extreme weather events such as flooding and droughts (Clark et al. 2001). Likewise, human interventions at the micro-scale appear easily possible and are well documented, but the change in flood peaks, sediment load and base flow at the large scale becomes much stronger dominated by natural processes (Ives 1989). Also, distinguishing the impact of land-use/cover change on hydrology from the impact of climatic variability is more difficult at the catchment scale than at the plot level (Archer 2003).

In forest zones, land clearance through deforestation – with the subsequent loss of the surface organic layer and decline in soil organic matter – increases overland stream flow through decreased evapotranspiration as well as raindrop detachment of soil particles, sheet erosion, rill erosion, gullying, and downstream sedimentation, though the latter processes are often episodic (Bosch and Hewlett 1982; Eshleman 2004; FAO 2004b; Mustard and Fisher 2004). The hydrological consequences following deforestation, however, are highly variable, and depend upon a wide array of host factors. At the watershed scale, forest clearing generally results in a significant increase in annual water yield, but generalization across different streamflow response measures remains difficult. On deforested slopes in particularly steep terrain, rates of erosion are maximized if the terrain is subsequently subjected to intensive cropping, fire, or both. Barren wastelands, unable to support any vegetation, can be the end result of this process, and the foothills north of Mexico City are an example of this. In forest zones where either revegetation occurs rapidly or secondary regrowth is part of the land-use cycle, the effects of forest clearing on hydrological cycles are transient and less

extreme (Eshleman 2004). It has been noted though that geological conditions can often override the effects of land-use/cover change.

Reforestation (or afforestation) is seen to contribute to a reversal of the hydrological responses to deforestation. Complete reversal of trends, however, depends upon the restoration of both vegetation and soil properties that had been characteristic of native forests within a particular climate on a particular parent material. In temperate climates, for example, hydrological processes are restored quite gradually, given the slow speed of soil development there (Eshleman 2004). It is estimated that old-growth forests help to completely recover water flows to original levels only after about 150 years past disturbance (Falkenmark 1999; Giambelluca 2002). There appears to be a clear link between forests and the quality of water, a much more sporadic link between forests and the availability of water quantity, and a variable link only between forests and the constancy of flow at the catchment level (Chomitz and Kumari 1996; Dudley and Stolton 2003; Mustard and Fisher 2004).

At the scale of river basins, modeling studies for the Amazon Basin suggest that complete conversion of the rainforest to degraded pasture would cause annually a decrease of 26% in mean precipitation, 30% in evapotranspiration, and 18% in runoff (Shukla et al. 1990), while other studies have produced similar, if less dramatic responses to deforestation (Werth and Avissar 2002). It has been shown that changes in riparian forest and vegetation have a major impact on the instream biota as well as on the pollutants entering the stream or river stretch (Sweeney et al. 2004).

There is little scientific evidence for the largest, most damaging flood events being caused by deforestation at the global scale (Eshleman 2004). Likewise, studies in South America, South Africa and in the Asian Himalayas indicate that the increase in infiltration capacity of forested lands over non-forested lands is insufficient to influence major downstream flooding events (Hewlett and Helvey 1970; Hewlett and Bosch 1984; Gilmour et al. 1987; Hamilton 1987). Rather, the intensity, amount and spatial distribution of rainfall appear to be the key elements determining the extent and magnitude of damage caused by such disasters, with local geology, land use, and topography being important, concomitant factors (Xu and Rana 2005).

Agriculture is the largest consumer of water by humans worldwide (nearly 85% of total human consumptive use), and land-use practices associated with agricultural intensification have been identified to exert an impact on hydrological cycles, especially where improper cultivation techniques were applied in environments with high natural variability (Tilman 1999; Gleick 2003; Pimentel et al. 2004) – see Box 4.16. In the United States – where per capita withdrawals were about 1 700 m^2 in 1995 – crop irrigation accounts for more than 40% of the withdrawn freshwater on average, but the figure is as high as 85% in California where agriculture accounts for only 3% of the state's economic production (Myers and Kent 2001; Gleick 2003). In Africa – where reported water uses range from approximately 600 to 800 m^3 pers^{-1} yr^{-1} in Egypt, Libya, the Sudan and some other countries to under 20 m^3 pers^{-1} yr^{-1} in the poorest countries of the continent – as much as 90% (or more) of reported water uses go to agricultural uses. As a matter of fact, large countries which produce grain in monoculture – such as Canada, United States, Argentina, and Australia – all have significantly higher per capita water uses than average (Gleick 2003), and in another group of large countries with extensive irrigation – such as India and China – agricultural water use numbers approximately 90% (Steffen et al. 2004).

The central argument here are as follows: first, those agricultural practices that can retard soil erosion (see Sect. 4.6.4) are practices that will also increase infiltration, thus reducing and delaying surface runoff (eroded soils absorb 87% less water through infiltration than uneroded soils), and most problems such as salinization are no problems at all with rainfed crops, because the soils are naturally flashed away, for instance (Eshleman 2004; Pimentel et al. 2004). In the U.S. Southern Great Plains, for example, the paucity of farm-level conservation strategies – combined with a period of extreme drought and the Great Depression – caused the Dust Bowl of the 1930s, during which 2 to 12 inches of topsoil were removed by wind and water erosion and hundreds of thousands of farming households were deprived of their economic livelihood (Worster 1979; Puigdefábregas 1995). Most of the region eventually recovered from the Dust Bowl episode. For example, changes in streamflows and sediment yields were observed and simulated for several watersheds in Texas, and both observational data and modeling results indicated a significant decline in erosion and reservoir sedimentation during the period 1910 to 1984 – attribut-

able to the combined effects of conversion from rural to urban land, changes in agricultural crops (i.e., replacement of cotton by wheat and sorghum), and the implementation of soil and water-conservation measures beginning in the 1940s (i.e., terracing, contour plowing, strip cropping, and no-till cropping) (Arnold et al. 1995).

In the high-intensity agricultural production zones across the world, the use of commercial fertilizers – containing primarily nitrogen (N), phosphorus (P) and potassium (K) – bear several effects on water quality, extending into stream chemistry across both watershed zones and coastal systems (Vashishta et al. 2001; Merz et al. 2004; Mustard and Fisher 2004). Indeed, there is a direct and quantitative link between the amounts of nitrogen in the major rivers of the world and the magnitude of agricultural nitrogen inputs to their watersheds (Tilman 1999). In particular, small but severely disturbed agricultural areas such as animal feedlots export ca. 100 times more N and P than other types of agricultural land use. One of the major (unintended) consequences of land-use intensification has been the contamination of shallow groundwater with nitrate, especially due to an explosive growth in fertilizer application following the end of World War II. Nitrate concentrations of 10 to 20 mg NO$_3$-N l^{-1} are frequently observed in most shallow aquifers of agricultural areas in North America, for example. Due to the tendency of nitrate concentrations >10 mg to cause methemoglobinemia in infants and to form carcinogenic nitrosamines in the human intestine, these waters are undrinkable (Merz et al. 2004; Mustard and Fisher 2004). In the United States, about 40% of water is deemed unfit for drinking or recreational use due to contamination by microorganisms, pesticides and fertilizers, and more than 76 million Americans are infected (and 5 000 die per year) as a result of pathogenic *Escherichia coli* and related foodborn pathogens (Pimentel et al. 2004). In recent decades, more U.S. livestock production systems have moved closer to urban areas, contaminating water and food with manure there (Board on Agriculture and Natural Resources 2003).

It has been noted that the effect of decreasing forest cover and increasing agriculture is not linear. As agriculture expands to >70% of land cover, the transformation of last remaining normal landscape traps for NO$_3$ to agriculture – such as wetlands or riparian forests – makes NO$_3$ concentrations in streams rise exponentially (Mustard and Fisher 2004). For example, in an estimation of the biogeochemical effects of land-use change in the Choptank Basin of Chesapeake Bay area in the eastern U.S. over the last 150 years, it was found that conversion of forest to agriculture in the 18th and 19th centuries increased N and P by a factor of 2, but application of fertilizers in the 20th century resulted in a factor of ca. 5 increase in N (Benitez and Fisher 2004). Modeling results suggest that not all NO$_3$ in the groundwater may appear in the base flow, with hydric watershed soils driving denitrification by as much as 80% below the expected concentrations based on land use (Mustard and Fisher 2004).

Irrigation farming, or the consumptive and nonrecoverable use of water by irrigated crops, is a major component of the water balance at many scales, and the availability of freshwater is a key factor for intensification and expansion of agriculture (Falkenmark 1999; Mustard and Fisher 2004). The hydraulic control on previously natural river systems – with damming, water extraction, and redirection of flows as most important engineering works – alters the behavior of rivers immediately and exerts an array of impacts reverberating through the coupled land-water system (Steffen et al. 2004).

Soil salinization is one of the problems associated with irrigation farming. It has been estimated that about half of all existing irrigated soils worldwide are adversely affected by salinization and that the amount of world agricultural land destroyed by salinized soils is 10 million ha (Pimentel et al. 2004). Another problem associated with crop irrigation is waterlogging, which means that – in the absence of adequate drainage – water levels rise in the upper soil levels, including the plant root zone, and crop growth is impaired. Such irrigated fields are sometimes referred to as "wet deserts" because they are rendered unproductive. In India, for example, waterlogging adversely affects 8.5 million ha of cropland and results in the loss of as much as 2 million to grain every year (Pimentel et al. 2004).

In most developed countries, the total area of irrigated land has been relatively stable over the last decade, but irrigation has remained a major off-stream use of both surface and groundwater resources. In the western U.S., for example, groundwater withdrawals for irrigation have been among the most significant impacts of agricultural activities, contributing to dramatic increases in evapotranspiration, excessive declines in water tables, surface subsidence, and soil salinization (Eshleman 2004). Following recoverage from the Dust Bowl episode, changing regional and national economies have promoted substantial pumping of groundwater for crop irrigation. Groundwater aquifers provide worldwide an estimated 23% of water per year available for sustainable use – with approximately 60% of the water intended for crop irrigation never reaching the crop due to water losses during pumping and transport (Wallace 2000) –, while the United States relies disproportionately on water pumped from aquifers, i.e., 65% (Pimentel et al. 2004). The capacity of the large Ogalla aquifer, for example, which underlies parts of Nebraska, South Dakota, Colorado, Kansas, Oklahoma, New Mexico, and Texas, has decreased 33% since about 1950, and water withdrawal is three times faster than its recharge rate. This continues to trigger decreasing water levels and subsequent abandonment of agricultural land, thus, raising new concerns about the long-term sustainablity of cultivation in the wider region (Brooks and Emel 1995; Pimentel et al. 2004).

The abandonment of irrigated agricultural land – due to soil and water degradation (but also as a consequence of shifting priorities in water allocation) – has been identified as a common feature of land use irrigation zones of arid and semiarid regions across the world (Steffen et al. 2004; Geist 2005). In the United States, about 150 000 ha of agricultural land have already been abandoned – because of high pumping costs only (Pimentel et al. 2004). Despite of land stability in irrigation farming, depleted freshwater resources and land-use legacies have been playing together in the western United States at some locations in producing an abundance of opportunistic shrubs and non-native annual plants as well as lower species diversity than lands that were never cultivated (Okin et al. 2001; Eshleman 2004).

In contrast, developing countries which hold large dryland zones – such as Central Asia or Middle Eastern countries along the Euphrates River – have been experiencing an explosive growth in irrigation over the last decade, often driven by huge government-funded water engineering projects (see Sect. 3.3.2). As in developed countries, groundwater withdrawal, declining water tables and subsequent land abandonment are common. In the agriculturally productive Chenaran Plain in northeastern Iran, for example, the water table has been declining by 2.8 m annually since the late 1990s, likewise in Guanajuato in Mexico by as much as $3.3 \, \mathrm{m \, yr^{-1}}$ (Pimentel et al. 2004). In countries like Turkey and Syria, a pattern of reallocation of land and water use has been typical, i.e., irrigated lands along river bottoms and floodplains got abandoned for upland sites, due to a myriad of factors including ground water depletion, reallocation for surface water, salinization and waterlogging (Mustard and Fisher 2004). In Central Asia, the impacts of land-use/cover change upon the hydrological cycle in low-lying ecosystems over the last 300 years, or so, are linked to a typical pathway of land- and water-use transition. The transition more or less paraphrases those impacts which are typical for the transformation from a predominantly rural mode to a largely industrialized mode of farming and society (Mustard et al. 2004) – see Sect. 3.5.3 and Box 4.15.

Large hydraulic structures for river regulation are often designed to support the extension of irrigation farming – besides their many other purposes, e.g., for electricity generation, domestic water supply, and flood control. Examples are dams, reservoirs, diversions, levees, artificial channels, and detention ponds. Only 23% of the flow in 139 of the largest rivers in the Northern Hemisphere remains unaffected by reservoirs, and the equivalent of 40% of the total global runoff to the oceans is intercepted by large dams (Steffen et al. 2004). There is growing evidence that land-use changes associated with the introduction of hydraulic control of especially large river systems in drylands contribute to rapid water degradation, disruption of hydrological cycles, and the partly irreversible collapse even of regional human-environ-

Box 4.15. Land- and water-use transition at lowland sites in Central Asia

In dry and hot lowland plains, depressions or basins of Central Asia, which carry river, delta or lake ecosystems under (sub)desert conditions, long-settled traditional land uses based on irrigation were supported for centuries, if not millennia due to rich groundwater resources and constant river flows. These dryland sites entered a pathway of contemporary desertification during the 20th century, partly starting in late 19th century, the features of which are water degradation such as salinization leading to vegetation degradation and sandification, at the utmost. A key factor is the transition from small-scale irrigation farming to large-scale irrigation schemes. The latter expand even onto hitherto marginal or completely unsuitable sites for irrigation farming. Examples are widespread across the Central Asian desert and steppe region, i.e., northern China, Turkmenia Plain, Caspian and Aral Sea Basin regions.

Advances in water technology, mainly large-scale hydro-technical installations, and strong directional policies, motivated out of economic and demographic reasons, led to the expansion and simultaneous intensification of irrigated farming land. Intensification meant also changes in the composition of crops, in particular, a shift towards water-demanding crops which did not appear in drylands prior to the availability of inexpensive energy and irrigation water: cotton monocultures, high intensity grain and rice productions, occasionally aside with vegetables, fruits, and grapevines. Pressures on water resources were amplified by the influx of booming industries (such as oil/gas, but also mining) and related infrastructures (such as power plants and factories) as well as by expanding settlements. Thus, original sites of traditional oasis agriculture, where formerly productive land had mainly been used for small-scale food production, often became the primary sites of contemporary desertification in river and delta ecosystems. These sites got desertified due to the decay or destruction of traditional irrigation systems, due to soil salinization and the advancement of surrounding desert sands (Geist 2005).

ment systems such as the Aral Sea Basin (Niasse 2002; Eshleman 2004; Mustard and Fisher 2004; Geist 2005; Turner and McCandless, forthcoming). There is further evidence that some of the responses reverberate through the coupled land-water cycle, strongly impacting on hydrological conditions through changes in the partitioning of incoming solar radiation between evapotranspiration and sensible heat, which in turn affect the amount of water that runs off into riverine systems or infiltrates into soil (Kabat et al. 2004).

In the Lake Chad Basin of northern Africa, for example, long-term decreases in lake area, lake level, and river discharges were primarily attributed to climatic variations (rainfall has been declining since the 1960s), but increases in water losses from rapidly growing irrigation explained a large proportion of the variation (Coe and Foley 2001). Likewise, a decline in potential evapotranspiration in a regional-scale water development project of arid, southeastern Turkey could be attributed to increasing irrigated land area, playing together with decreasing wind speed and increasing atmospheric humidity (Ozdogan and Salvucci 2004). For both regions, palaeo records prove that pumping for crop irrigation would mean the mining of fossil water, because groundwater reservoirs have not been fully recharged for thousands of years (Steffen et al. 2004). Worldwide, the Nile and Syr-Darya Rivers are among the most heavily regulated rivers, partly demonstrating drastic changes in coupled land-water cycles. The Nile, after the erection of the Aswan high dam in 1968, shows reduced overall discharge, truncated peak flows, higher lower flows, and a seasonal shift in the timing of the natural hydrograph. In the case of the Syr-Darya River, the progressive losses of discharge are associated with expanded water use for irrigation and the contraction of the Aral Sea (Meybeck and Vörösmarty 2004). In the Aral Sea Basin, water from the actually two rivers entering the inland sea was diverted from the 1960s onwards, and led to an enormous decrease in the area and volume of the sea over the next

35 years, i.e., from 6.7×10^4 km^2 to 3.2×10^4 km^2 and 1 064 km^2 to 310 km^2, respectively. Most studies anticipate its complete disappearance within the next 25 years. Associated with the hydrological consequences of land-use/cover change in the sea basin have been an array of ecological influences reverberating in the coupled land-water-atmosphere complex (e.g., surface temperature and local climate changes) as well as an array of social influences on economic livelihoods (e.g., decline in agricultural productivity due to salinization and water logging) and human health (Glazovsky 1995; Saiko and Zonn 2000; Mustard and Fisher 2004; Geist 2005).

Owing to impoundments, still waters (i.e., the standing stock of river channel water) have worldwide increased by more than 700% relative to the natural state (Dynesius and Nilsson 1994; Vörösmarty and Sahagian 2000). The consequences of such water aging for associated material transport are to trap a substantial proportion of the incoming suspended sediments and to modify the concentration of dissolved components of N, P, and silicon. It has been estimated that the current registered 45 000 largest reservoirs (i.e., >0.5 km^3 storage capacity) trap nearly 30% of global sediment flux destined for the ocean, and that this estimate of sediment retention rises further with inclusion of about 800 000 smaller impoundments – and, of course, with continued dam construction (McCully 1996; Meybeck and Vörösmarty 2004).

Unlike agriculture, mineral extraction accounts for less than 1% of the terrestrial surface, although the degree of land and related water transformation is often extreme. It is an inherently dynamic form of land use, the "boom and bust" histories of which translate into relatively localized, yet intense forms of land-use change. Minerals wrested from the Earth – such as aluminum, nickel, oil, and natural gas – had no significant commercial applications 150 years ago, but account for a significant portion of the flow of minerals into the economy today. Despite an increase in recycling, resource extraction remains the primary means for meeting new metal

Box 4.16. Agricultural intensification in Almería province, Spain

David Dabin · Catholic University of Louvain, Louvain-la-Neuve, Belgium

Having an insulation between 2 800 and 3 000 h yr^{-1}, the stony semi-desert of the Campo de Dalías (Fig. 4.9) has become one of the most important horticultural centers of Europe generating more than U.S.$1.5 billion per year after important aquifers were discovered in the 1950s (Pulido-Bosch et al. 2000b).

This development has been achieved at high environmental costs. Each year 130 × 10^6 m³ are pumped in the aquifers to supply greenhouses with water (Orgaz et al. 2005), whereas the total recharge is estimated at only 65 × 10^6 m³ yr^{-1} (Dominguez and González-Ascensio 1995). This has led to an overexploitation of the water resources. By places, the piezometric level has decreased up to 20 m under sea level (Pulido-Bosch et al. 2000a). Associated with this fall of the piezometric level, marine intrusion and salinization have been noticed, making some wells unusable (Vallejos et al. 2003; Pulido-Bosch et al. 2000b). Pollution of the aquifers by fertilizers, pesticides and nitrate is also detected where agriculture is most intensive. All those elements lead to water scarcity in the Campo de Dalías. A well-thought water management is thus urgently needed.

Fig. 4.9. Evolution of greenhouses in the Campo de Dalías (Almería province, Spain) between 1987 and 2000, colored compositions (band 1, 2 and 3) of Landsat 5 TM and Landsat 7 ETM+ scenes. The progression and densification of greenhouses in the endoreic basin (central part of the Campo) is clearly visible between the two dates

demand, the vast majority of construction minerals, and the sole means of supplying demand for fuels and industrial minerals. Access to low-entropy forms of energy at low cost has underpinned the scaling up of "earth moving" capacities during the 20th century such as the introduction of open pit mining for metals, or the shift from underground methods to "mountain-top removal" in coal mining. Surface mining for coals and other minerals (and the subsequent reclamation of the altered land surface) represent significant land-use/cover changes with the potential to alter hydrological processes in watersheds (acid mine drainage is a common problem associated mostly with underground mining). Major activities are the excavation of previously consolidated geological strata which is followed by replacement of unconsolidated fill materials and approximate restoration of original slopes. Recent shifts in the location of mining investment in developing countries have intensified long-standing concerns about the impact of mining on critical ecosystem services such as water provision (as well as on global biodiversity). It has been estimated that 75% of active mines and exploration areas overlap with areas of high conservation value and areas of watershed stress, and that nearly one third of all active mines and exploration sites are located within intact ecosystems of high conservation value (Eshleman 2004; Bridge 2006).

Soil compaction by heavy machinery during the reclamation process has been shown to reduce soil bulk density, porosity, and infiltration capacities (Chong et al. 1986), but measures exist to promote infiltration technically. Usually, infiltration-excess overland flow is considered to be the dominant flow pathway in mined and/or reclaimed watersheds. At three watersheds in eastern Ohio, for example, peak flow rates were shown to have increased during the coal mining and reclamation phases relative to the pre-mining period (Bonta et al. 1997). In situations of infiltration-excess overland flow the restoration of normal hydrological functioning takes considerably longer than the normal five-year period associated with active reclamation and land management (Eshleman 2004).

The ramifications of land transformation towards urban cover have been qualitatively described for various stages of urbanization. They include decreases in transpiration from loss of vegetation, decreases in infiltration due to decreased perviousness associated with urban development (streets, roofs, sidewalks, parking lots, etc.), increases in stream runoff volumes, increases in flood peaks, declines in water quality from discharges of sanitary wastes to local streams and rivers, and reductions in baseflow (Eshleman 2004).

Both empirical analyses and modeling results suggest that urbanization and suburbanization increase stream flow through increased runoff (and thus flood potential), but also decrease water quality when the amount of impervious surface in a watershed exceeds 10 to 15% of the total land cover (Schueler 1994; Falkenmark 1999; Mus-

tard and Fisher 2004). In catchments, the degree of impervious areas in catchments is often directly related to the size of floods (Wissmar et al. 2004). Several studies point to the fact that increased nitrate-nitrogen exports across river basins (such as that of the Mississippi River) can be associated with the percentage of developed land there. Together with agricultural intensification, the disposal of human waste in septic systems are clearly the principal causes of elevated NO_3 in groundwater, and contribute to the eutrophication of aquatic ecosystems (Eshleman 2004; Mustard and Fisher 2004).

4.7.3   Land-Ocean Interactions in the Coastal Zone

The interactions between natural processes and human activity are most active in the coastal zone which is the transitional area where land and ocean meet, stretching from the coastal plains to the outer edges of the continental shelves. More than 50% of the word population lives within 100 km of a coast, and eight of the top-ten largest cities in the world are located by the coast, with coastal cities having the highest rates of growth than any other areas (Steffen et al. 2004; Hwang 2006).

At the underlying level, there are coupled effects of considerable human population concentrations and multiple economic activities, including increasing coastal tourism which is one of the largest and fastest growing sectors of the global economy, all together adding to the pressure on coastal ecosystems (Hwang 2006; Crossland et al. 2005a). There are trends counterbalancing processes of conversion and modification in the coastal zone, but these are limited to cases from developed countries only. For example, some river and harbor areas of major western coastal cities have been significantly cleaned of pollutants (e.g., the Thames in London and many cities along the Rhine). Likewise, the transport of excess nutrients and other pollutants through the coastal zone has been diminished in many cases (Lomborg 2001). In a broader Earth System perspective, however, virtually no large stretches of coastal areas – outside of Greenland, northern Canada and Siberia, and remote areas of South America and Australia – are now without significant human influence (Goldberg 1994; Steffen et al. 2004; Burbridge et al. 2005).

At the proximate level, the geomorphology of the coastal zone is altered – e.g., through the construction of shoreline engineering works, port and harbor development, and extension of urban, industrial and infrastructure covers –, and coastal wetlands are drained, reclaimed, and converted to agricultural and other uses (Walker 1990; Goldberg 1994; Vernberg and Vernberg 2001; Solecki and Walker 2001; Hwang 2006). It is estimated that globally about one third of the coastal land – excluding Antarctica – has been altered or semi-altered in some way (Steffen et al. 2004; Burbridge et al. 2005).

Drainage of wetlands has occurred extensively in coastal areas worldwide since ancient times. It has been mainly done for the provision of aerated agricultural soils, most efficiently through the use of clay tiles or plastic drainage pipes that promote saturated groundwater flows through the peat to open ditches. For example, in the United States of America more than 400 000 km² of the national territory had been subjected to drainage development by 1950, which is an area roughly the size of the state of California (Eshleman 2004). It appears as if the key hydrological problems associated with wetland drainage – especially the problems associated with peat shrinkage – had not changed since they were encountered in drainage of the Fens in England which started under the Roman rule and peaked in the middle of the 17th century under the Stuart kings (Purseglove 1988): dewatering of peatlands promotes peat aeration and increased rates of peat decomposition, with the consequence that excessive peat decomposition in coastal systems can cause a lowering of the land surface, relative to river and sea levels which in turn increases the susceptibility of the land to both inland flooding and coastal inundation (Eshleman 2004). In the case of the Fens, reclamation of the marshes caused a gradual reduction in the river gradient, leading to a reduction in scouring, siltation of the river outfall to the estuary, and an increase in inland flooding. In addition to land level declines relative to sea levels, coastal inundation of the Fens in 1673 and again in 1713 provided some extreme examples of the consequences of wetland drainage (Purseglove 1988).

Another paramount example is the conversion of mangrove forests to prawn farms. Mangroves cover about one-quarter of tropical coastlines, and some 112 countries and territories have mangroves. It has been estimated that anywhere from 5 to 85% of original mangrove area in various countries have been lost, with extensive losses occurring in the past 50 years. Many coastal countries in Southeast Asia have lost half or even more of their mangrove forest since the mid-1960s because of industrial timber logging, fuelwood harvesting, conversion to rice fields, urban encroachment and other uses such as fishponds and prawn farms (Walker 1990; Naylor et al. 1998; Vernberg and Vernberg 2001). Globally, approximately 50% of mangrove systems have been converted to other uses since 1900 (Steffen et al. 2004; Hwang 2006).

As for the encroachment of urban-industrial uses, the coastal lowlands of western Korea, southern Japan, and southern China are prime examples of high-density populations coupled with rapid economic development, resulting in huge demands of land for urban-industrial uses; many of the coastal mudflats have been reclaimed for industrial, infrastructural and urban development, and agriculture alike; and even large-scale modern airports (in Korea and Japan) have become located on reclaimed wetlands, with sea walls drying out the tidal flats (Vernberg and Vernberg 2001; Seto et al. 2004; Hwang 2006; Tan et al. 2006).

Adding to these pressures are the amount, quality and timing of the through-flow of water and suspended and dissolved materials from upstream areas through the coastal zone to the continental shelves which has been significantly affected (Salomons et al. 2005). As could be seen from the amount of sediments delivered to the coastal zone, impacts can work in two opposing directions: there have been regional increases in some areas in the delivery of sediments through increased soil erosion upstream, while in other areas the delivery decreased through sediment trapping within reservoirs and other impoundments upstream (Steffen et al. 2004).

Coastal ecosystems provide the important service of maintaining water quality by filtering or degrading toxic pollutants, and absorbing nutrient inputs. This capacity is easily exceeded by direct human injection with the coastal zone itself, i.e., chemical pollutant discharges from agriculture and industry, oil spills in the ocean, oil discharges from land-based sources, excessive nutrient inputs from urban runoff, and sewage effluent, with additional nutrient loadings, for example, coming from upstream agriculture (Smith et al. 2005). It has been estimated that the increase of nitrogen delivery entering the North Atlantic, for example, has increased by a factor of between 3 and 20 (Steffen et al. 2004). In addition to nutrients, contaminants such as heavy metals, persistent organic pollutants, various other synthetic chemicals, radioactive materials, bacteria, and slowly degrading solid waste like plastics are transported from land to the coastal regions (Hwang 2006; Salomons et al. 2005). Coastal ecosystems are losing much of their capacity to produce fish because of overfishing in the shelf zone and destruction of nursery habitats there. Likewise, as the extent of mangroves, coastal wetlands and sea grasses declines, coastal habitats are losing their pollutant-filtering capacity (Burbridge et al. 2005).

With a coast line as long as about 7 500 km, India constitutes a paramount example of land-ocean interactions worldwide, where both human induced disturbances – such as pollution, sand mining, tourism, and shipping – and natural disturbance – such as storm, sedimentation and tsunamis – interact and play out differently in terms of extension, severity, and frequency. There are sixty coastal districts and richly diverse coastal zone areas in India which have highly productive ecosystems as diverse as coral reefs, mangroves, mud flats, lagoons, estuaries, beaches, and dunes (island territories such as Lakshdweep, Andaman and Nicobars constitute 22% of the coastline). Most of India's coastline is densely populated with the consequence that natural ecosystems have been largely converted to various types of land uses: in particular, wetlands are reclaimed for agriculture, settlements, and aquaculture, and, as a consequence, natural lagoon ecosystems have shrunk. Among today's most pressing issues along the Indian coastal zone are rapidly growing populations and economic activities, deteriorating environmental quality, loss of critical habitats, diminishing levels of fish and shellfish populations, reduced biodiversity and increased risk from natural hazards. Significant changes in habitat and diversity due to intense anthropogenic pressures were observed, for example, in Mumbai and the Gulf of Kachchh areas. The latter area holds India's most degraded coral reefs due to destructive fishing, mining, sedimentation, and invasion by alien species. Almost the entire Mumbai Coast, previously rich in biodiversity and some rare species, is characterized by the current distribution of mono-species mangrove (*Avicenia*). About 75% of mangrove ecosystems in India are in good condition, in total occupying an area of about 4 460 km^2. In the Godavari-Krishna deltaic regions (Tamil Nadu), they have been degraded and destroyed due to their use as fuel, fodder and conversion of these areas for agriculture, aquaculture and industrial purposes. Likewise, mangrove habitat located in the inter-tidal zone along the Mahul Creek in the Mumbai region coast has been severely degraded, mainly as a result of land reclamation in the 1996–2000 period. The mangrove ecosystem of the Sunderbans Delta, actually the largest single block of mangroves in the world, is rapidly depleted. Along the Gujarat Coast, both degradation and loss of ecosystems continued up to 1985 due to the mining of coralline sand and use of mangroves as fuel and fodder, until the Jamnagar area was declared a marine park in 1983. Currently, about 0.5 million ha have been declared marine and coastal protected areas, i.e., 3 national parks and 13 wildlife sanctuaries (Nayak et al. 1997; Rajasuriya et al. 2000; Nayak and Bahuguna 2001; Selvam 2003). Land-use driven changes of inland freshwater cycles do also easily extend into the coastal zone. An example is the deforestation of the Andaman and Nicobars Islands in India which has resulted in an increased flow of freshwater and sediment that have affected the corals along the coast (Nayak et al. 1997; Nayak and Bahuguna 2001).

4.8 Conclusions

Estimates or generalizations exist about some (isolated) impacts, human "imprints" or ecological "footprints", but no globally valid statement is possible yet about an aggregated, overall or generalized impact of land-use/cover change upon ecosystems and people in a coupled manner. This is due to the multiplicity of impacts in terms of various types, time scales, hierarchical scales, feedbacks or repercussions as well as actors and causes involved. Nonetheless, it can be stated in normative and descriptive terms that human well-being – as a context- and situation-dependent state, comprising basic material for a good life, freedom and choice, health, good social relations, and security (Millennium Assessment 2003) – depends on a sufficient and safe supply of food, but also on the provision of timber, clean air and water, stable climate, freedom from floods, landslides and droughts, rec-

reational opportunities, and a full range of cultural and aesthetic pleasures. These are just some of the services provided by ecosystems, and most often, but not necessarily, their supply is more abundant in the absence of land cultivation (DeFries et al. 2004b; Cassman et al. 2005).

From a wide array of case studies it can be put forward – as a hypothesis, at least – that with the major forms of current land use (i.e., agriculture, forestry, mineral extraction, pastoral systems, and urbanization), "there is often a decreasing capacity of ecosystems to provide 'native' resources" (Asner et al. 2004). However, we are only now beginning to gain a rudimentary understanding of ecosystem interactions with land use and the implications for society, not to mention the absence of a full understanding of land-use interactions with societies and ecosystems in a coupled manner for large civilizations – maybe, apart from some isolated cases of small societies in the ancient past (Redman 1999; Diamond 2005; Turner and McCandless, forthcoming).

As for ecosystem types and impacts, there are few if any universal patterns among the myriad combinations of land use and ecosystems, but there are some repeating patterns regarding trade-offs in ecosystem goods and services following land-use/cover change. Land-use decisions often involve trade-offs between intentional appropriation of ecosystem goods for human use or consumption such as food, feed, fiber and timer, and unintended ecosystem responses such as flooding, habitat loss, and nutrient runoff (DeFries et al. 2004b). Balancing the trade-offs, if known (or not), clearly depends on societal values, and any quantification of the biological responses to land-use/cover change would help (DeFries et al. 2005). Often, there are *obvious* trade-offs between the types of services that an ecosystem can provide to humans, but just as often there are *hidden* (or indirect) trade-offs that go unaccounted for by society (Asner et al. 2004). As demonstrated in the case of global forest transitions, it is most revealing to relate classes or types of impacts to the various stages of land-use transition worldwide (see Chap. 3).

One such pattern relates to the mobilization and alteration of fluxes of materials, energy and species in ecosystems, always taking place initially after land-use change, and often evolving and persisting over some extended period of time (Asner et al. 2004). The pattern mainly relates to the human appropriation of net primary productivity (Haberl et al. 2001, 2004a). These are direct effects squarely linked to the intended responses of ecosystems to land-use/cover change, while the unintended and indirect effects on the mobilization of ecosystem resources are often substantial and of a magnitude that exceeds the direct impacts. It has to be noted though that the cascading indirect effects – such as nutrient and soil carbon losses – always remain related to the direct impacts on the mobility or flux of key ecosystem resources at multiple temporal and spatial scales (Asner et al. 2004).

A second emergent pattern is that there are clear bioclimatic and edaphic controls over the vulnerability of ecosystems to degradation during land use. It has been suggested that there is a limited geographic area of the Earth that can endure land-use extensification and intensification without enormous negative ecosystem responses, i.e., the mesic regions of the world – often referred to as "temperate zones" and "breadbaskets" – that have the soil substrates and climatic conditions conducive to major agricultural, pastoral, and timber harvesting activities, with examples stemming from India, East Asia, central Europe, and the central United States of America (Foley et al. 2004; Klein Goldewijk 2004; Ojima et al. 2004). The social controls over the vulnerability of coupled socio-ecological systems seem less clear. Mainstream thinking points to unchecked human population dynamics, especially fertility transition. Others relate magnitude and severity of the land-change impact over the past 300 years, or so, to the unchecked spread of European colonization (and related land-management techniques) across the ecosystems and cultural systems of the world (Richards 1990; Klein Goldewijk 2004). Further, we are about to gain a basic understanding of the general conditions that have controlled a transition towards sustainable land use, again with most evidence originating from the global forest transition (Rudel et al. 2005).

A third repeating pattern is that indirect, unintended responses are ubiquitous and represent additional trade-offs in ecosystem services that go well beyond the direct impacts of land-use/cover change (Asner et al. 2004). A paramount example is the agricultural use of freshwater – currently determined by complex interactions among market forces, cultural preferences, institutional dynamics, and regional and international politics – when put into the context of coupled atmosphere-land-water interlinkages, with systemic changes to the environment occurring along the catchment-coastal zone continuum (Salomons et al. 2005). Many of these changes in the hydrological cycle, reverberating to the larger Earth System, are already discernible now (Steffen et al. 2004):

- Precipitation appears to be increasing over land in most of the mid- to high-latitudes in the Northern Hemisphere (0.5 to 1.0% per decade) over the 20th century (Folland et al. 2001); the trends are less pronounced for other parts of the Earth, but there appears to be decreasing precipitation in the northern sub-tropics and small increases over tropical lands and tropical oceans (McCarthy et al. 2001).
- In those regions where total precipitation has increased, extreme precipitation events are increasing, perhaps by 2 to 4% over the last half of the 20th century (Folland et al. 2001); by the same token, regions experiencing diminishing total precipitation appear to be experiencing more severe and extended droughts.
- Increased aerosol particle loading in the atmosphere is likely affecting the water cycle through changes in

precipitation caused *(a)* by the number and size of cloud condensation nuclei in the atmosphere and hence the efficiency of rain droplet formation and *(b)* diminished evapotranspiration, and hence ultimately diminished precipitation, through decreases in incident solar radiation at the Earth's surface (Folland et al. 2001).
- Land-cover change is strongly impacting on the water cycle through changes in the partitioning of incoming solar radiation between evapotranspiration and sensible heat, which in turn affect the amount of water that runs off into riverine systems or infiltrates into soils (Kabat et al. 2004).
- Subtle indirect effects are occurring through the effects of increasing atmospheric CO_2 concentration on the water-use efficiency of terrestrial ecosystems, which ultimately influences the balance between evapotranspiration and soil moistures (Mooney et al. 1999).

The case of freshwater demonstrates that any future research of the coupled atmosphere-land-water system very likely needs to address the potential effects of climate change (e.g., on water quantity and water quality) – see Box 4.17. Further, a coupled socio-ecological vul-

---

Box 4.17. Indirect and unintended effects of agricultural freshwater use

As for water quantity, there is some evidence that human driven climate change in terms of extreme weather events has already changed the number of severe floods around the world. For example, 16 out of 21 extreme floods were found to have occurred after 1953, and the sharp increase in floods went consistent with the projections of climate models suggesting that the increase in frequency of extreme floods will continue into the future (Milly et al. 2002). These changes will have serious implications for water resource infrastructure (such as dams and reservoirs) which controls much of river regulation and thus the provision of water for irrigation, but also for human domestic use and industry (Steffen et al. 2004).

As for water quality, human impacts in terms of river engineering and waste disposal have by now exceeded the influence of natural variability in water quality in many parts of the world (Vörösmarty and Meybeck 2004). Agricultural runoff remains to be the major issue, and high fluxes of nitrate as well as contamination by pesticides remain important water-quality issues in industrialized countries, while urban and industrial pollution loading are likely to increase in developing countries (Steffen et al. 2004). The impacts on the current level of water quality – and that projected for the near future – are clearly dominated by land-use/cover change and increasingly intersect with systemic changes in the hydrological cycle due to climate change. For example, any increases in extreme rainfall events and hence soil erosion will be especially important for loadings of phosphorus and pesticides (Steffen et al. 2004). Indeed, marked improvements have occurred such as reductions in the contamination of major European rivers, reductions in effluent discharges from the surrounding countries of the Baltic Sea, afforestation and reduced usage of pesticides and fertilizers, maintenance and restoration of wetlands (Lomborg 2001; Steffen et al. 2004). As for most parts of the developing world, needs to mitigate or reverse negative impacts of land-use/cover change upon the hydrological cycle are still pressing, though. Adding to this pressure is the notion that for Africa and South America climate change is predicted to exacerbate water stress significantly (Steffen et al. 2004).

nerability analysis (e.g., of global water resources) – rather than an isolated assessment of ecosystem impacts – as a most critical aspect is needed, also drawing from the conditions for successful improvement already achieved (Lomborg 2001). The assessment (e.g., of the impacts of land-use/cover change on the hydrological cycle) needs to synthesize the effects of systemic changes in the Earth System (Steffen et al. 2004; Pimentel et al. 2004), and it needs to explore the prospects for (e.g., improved water-use) efficiency and/or (e.g., water) productivity as well (Gleick 2003; Pimentel et al. 2004).

It has further been proposed to describe both the intended (direct) and unintentional (indirect) trade-offs using "spider diagrams" (which show the relative impacts of land-use and land-use/cover change on key ecosystem goods and services), and to address the challenge of quantifying these trade-offs in units meaningful to scientists, decision-makers and lay people alike (DeFries et al. 2004a,b). In a step further, it has been suggested to link impacts to typical trajectories of land change (i.e., including the feedback structure and causative mechanisms) and (arche)typical situations in the land-use transition (e.g., frontier situation, agricultural cropland consolidation, industrialization/urbanization, post-industrial economy) (Lambin et al. 2003; Mustard et al. 2004).

Addressing the need to quantify trade-offs (and drawing from a wide array of local case studies), intervention points along pathways of land change have been explored to mitigate impacts in contemporary agricultural frontier situations of dryland as well as humid forest ecosystems. In a comparative perspective, most important causative interactions (i.e., to be directly influenced) and most important feedbacks (i.e., to be enforced or turned around) have been identified, as well as the global-local interplays of causative factors for tropical deforestation (while national-local interplays have been found to be characteristic for dryland situations). This leaves some opportunities for interventions at multiple scales (see Chap. 7), given that no universal pattern among the myriad combinations of land use, ecosystem and society types appears, and hence no universal applications or mitigating policies work out (Geist et al. 2006).

An underlying assumption of the study was that a careful identification of the factors at work in a given location will be a prerequisite for getting the mix of mitigation measures right while minimizing the cost to local peoples' livelihood opportunities and other legitimate development objectives. Accurate, objective information is needed regarding the private and social costs and benefits of alternative land-use systems on which to base inevitably controversial decisions, and to help weigh up the difficult choices, a tool such as the matrix of the Alternatives to Slash-and-Burn (ASB) Programme is useful (see Sect. 7.2, message 8). In the ASB matrix, natural forest and the land-use systems that replace it are scored against different environmental, socio-economic and institutional criteria reflecting the objectives of different interest groups. To enable results to be compared across sites, the systems specific to each site are grouped according to broad categories, ranging from forests and agroforests to grasslands and pastures. The criteria may be adjusted to specific locations, but the matrix always comprises indicators for (a) two major global environmental concerns: carbon storage and biodiversity, (b) agronomic sustainability, assessed according to a range of soil, nutrient, and pest trends, (c) policy objectives: employment opportunities and economic growth, with the latter expressed in social prices (i.e., adjusted for trade policy distortions and capital market failures, but not for environmental externalities such as carbon sequestration), (d) smallholders' concerns: returns to their labor and land, their workload, food security for their family, and start-up costs of new systems or techniques, and (e) policy and institutional barriers to adoption by smallholders, including the availability of credit, and improved technology, and access to and the performance of input and product markets (Tomich et al. 1998, 2005).

Over the past ten years, or so, ASB researchers filled in this matrix for representative benchmark sites across the humid tropics. The social, political and economic factors at work at these sites vary greatly, as also does their current resource endowment. The sites range from the densely populated lowlands of the Indonesian island of Sumatra, through a region of varying population density and access to markets south of Yaoundé in Cameroon, to the remote forests of Acre State in the far west of the Brazilian Amazon, where settlement by small-scale farmers is relatively recent and forest is still plentiful. At each site, ASB researchers have evaluated land-use systems both as they are currently practiced and in the alternative forms that could be possible through policy, institutional and technological innovations. A key question addressed was whether the intensification of land use through technological innovation could reduce both poverty and deforestation (Tomich et al. 1998, 2005).

The matrix allows researchers, policy makers, environmentalists and others to identify and discuss trade-offs among the various objectives of different interest groups, and/or to discuss ways of promoting land-use systems that could provide a better balance among trade-offs without making any group worse off, but that still were not broadly adopted. The studies in Indonesia and Cameroon reveal the feasibility of a "middle path" of development involving smallholder agroforests and community forest management for timber and other products. In Brazil, small-scale managed forestry poses the same potential benefits. Such a path could deliver an attractive balance between environmental benefits and equitable economic growth. "Could" is the operative word, however, since whether or not this balance is struck in practice will depend on the ability of these countries to deliver the necessary policy and institutional innovations (Vosti et al. 2003).

An area that deserves more attention concerns how positive or negative impacts of land-use change are distributed among stakeholders: who is impacted by land-use change *versus* who makes land-use decisions? In that respect, agents of change may not necessarily be the same as the stakeholders most affected by change. This is especially the case under economic globalization where demand is geographically separated from supply of land resources. An urban consumer of tropical wood will be much less impacted by tropical deforestation than a slash-and-burn farmer earning a living from forest products. Distributional effects are therefore essential to understand interactions between land-use change and human development.

Chapter 5

Modeling Land-Use and Land-Cover Change

Peter H. Verburg · Kasper Kok · Robert Gilmore Pontius Jr. · A. Veldkamp

5.1 Introduction

The decade since the initiation of the Land-Use/Cover Change (LUCC) project in 1995 (see Chap. 1) has witnessed considerable advances in the field of modeling of land-use/cover change. The science plan of the project indicated that the major task would be the development of a new generation of land-use/cover change models capable of simulating the major socio-economic and biophysical driving forces of land-use and land-cover change. In addition, these models were supposed to be able to handle interactions at several spatial and temporal scales. Recent publications indicate that the LUCC science community has successfully met this challenge and a wide range of advanced models, aiming at different scales and research questions, is now available (Briassoulis 2000; Agarwal et al. 2001; Veldkamp and Lambin 2001; Parker et al. 2003; Nagendra et al. 2004; Veldkamp and Verburg 2004; Verburg et al. 2004b; Verburg and Veldkamp 2005). One of the most important observations that can be made examining the range of available land-use/cover change models is the wide variety of approaches and concepts underlying the models. This chapter intends to describe the variety of modeling approaches, discuss the strengths and weaknesses of current approaches and indicate the remaining challenges for the land-use science community. Not being able to discuss all individual models and approaches, we will focus on broad distinctions between approaches and discuss how modelers have dealt with a number of important aspects of the functioning of the land-use system. A land-use system is understood here as a type of land use with interrelated determining factors with strong functional relations with each other (see Fig. 1.2). These factors include a wide range of land-use influencing factors than can be biophysical, economic, social, cultural, political, or institutional. The discussion of modeling approaches in this chapter is illustrated with examples of models and results from selected research projects.

5.2 The Role of Models in Land-Use/Cover Change Research

Modeling involves the use of artificial representations of the interactions within the land-use system to explore its dynamics and possible future development. Modeling is one of the methods in the portfolio of techniques and approaches available to unravel the dynamics of the land-use system. Whereas descriptive and narrative approaches focus on mostly qualitative descriptions of the land-use system (see Chap. 3), models require a structural, mostly quantitative analysis. Gaps in knowledge become obvious during the model-building process and the sensitivity of land-use patterns to changes in key variables as well as to other variables can be tested. Sensitivity analysis can help to identify the most important mechanisms of change in a certain area that could not be identified from field observation. Such results may lead to new insights or guide further analysis of the land-use change processes. In this perspective, models are used as a learning tool to formalize knowledge. Since real-life experiments in land-use systems are difficult, computer models can be used as a computational laboratory in which the hypotheses about the processes of land-use change are tested. Finally, models can play a role in communication between researchers. One of the major difficulties in pluri-disciplinary research is to find ways to express oneself that are acceptable to all the disciplines involved, and that are free from the connotation of any or all of them. It is thus a major potential asset of models that they can be used to express phenomena and ideas in ways that can be understood in the same rigorous manner by practitioners of different disciplines (van der Leeuw 2004).

Apart from being a learning tool in unraveling the driving factors and system dynamics, land-use change models play an important role in exploring possible future developments in the land-use system. With a model the functioning of the system can be explored through "what-if" scenarios and the visualization of alternative land-use configurations that may be the result of policy decisions or developments in society as described in scenarios (see Chap. 6). These exploratory and projective capacities allow models to be used as a communication and learning environment for stakeholders involved in land-use decision making. Projections can be used as an early warning system for the effects of future land-use changes and pin-point hot-spots that are priority areas for in-depth analysis or policy intervention (see Chap. 7).

5.3 The Diversity of Modeling Approaches

The large diversity of modeling approaches that have evolved over the past years has challenged different authors to review and classify the different approaches. Such classification systems are mostly based on the dominant land-use change processes addressed by the model, the simulation technique used in the model or the underlying theory. For deforestation models, a general overview of models is provided by Lambin (1997), while Kaimowitz and Angelsen (1998) focus on deforestation models based on economic theory. Miller et al. (1999) and U.S. EPA (2000) present a review of integrated urban models, while Parker et al. (2003) and Bousquet and Le Page (2004) provide overviews of multi-agent modeling approaches. Lambin et al. (2000) review models for agricultural intensification, and Bockstael and Irwin (2000) review a number of land-use models in terms of economic theory foundations. Agarwal et al. (2001) review 19 models based on their spatial, temporal and human-choice complexity, while Briassoulis (2000) and Verburg et al. (2004b) give broad overviews of land-use models. More recent developments and new approaches are included in the special issues edited by the LUCC Focus 3 office (Veldkamp and Lambin 2001; Veldkamp and Verburg 2004; Verburg and Veldkamp 2005). The diversity of approaches can be explained by the wide range of research questions in which models are used as a tool, the different scales of application, ranging from the very local to the global extent, and the absence of an all-compassing theory of land-use change (see Chap. 1). In this chapter we do not aim at a classification of models or approaches. Instead, we will discuss a number of characteristics that can be used to make broad distinctions between the different approaches.

5.3.1 Spatial Versus Non-Spatial

An important first distinction between different model types is the distinction between spatial and non-spatial models. This distinction is of major importance when selecting a model type for a specific application since it largely determines the type of research questions the model may answer for that application. Spatial models aim at spatially explicit representations of land-use change at some level of spatial detail, in which land-use change is indicated for individual pixels in a raster or other spatial entities such as administrative units. This group of models is, therefore, able to explore spatial variation in land-use change and account for variation in the social and biophysical environment. A few examples of spatial models are well-known models such as the Conversion of Land Use and its Effects (CLUE) model, the SLEUTH model and GEOMOD (Pontius et al. 2001; Verburg et al. 2002; Goldstein et al. 2004).

The group of non-spatial models focuses on modeling the rate and magnitude of land-use change without specific attention for its spatial distribution. An example of such a model for the Sahel region (SALU) is presented by Stéphenne and Lambin (2001), while Evans et al. (2001) present a non-spatial model for deforestation in Altamira, part of the Amazon region. This parcel-level model calculates the utility of specific land-use activities to identify those land-uses that are most optimal at each time point, and labor is allocated to these activities based on the availability of household and wage labor. The model reports the proportion of the parcel in the following land-cover classes at each time point using a 1-year interval: mature forest, secondary successional forest, perennial crops, annual crops and pasture.

In spite of the large variety of modeling approaches it is possible to identify a common structure valid for a large number of spatially explicit land-use change models (except multi-agent models). In the model structure, a distinction is made between the calculation of the magnitude of change and the allocation of change (see Fig. 5.1). Both calculations are based on a set of driving factors, some steering the magnitude of change, while others only steer the location of change. Sometimes the same driving factor can influence both quantity and location of change. Based on the interpretation of one or more driving factors that are supposed to be determinants of the location of land-use change, a so-called suitability or preference map is created that indicates the suitability of a location for a specific land-use type relative to the suitability of other locations. The selection of the driving factors used in the model and its translation to a suitability map is one of the main components of a land-use model. A wide variety of approaches exist including: (a) rule-based systems based on either theory or expert knowledge; (b) suitability maps based on empirical analysis; and (c) transition rules dependent on the land uses in the neighborhood (e.g., cellular automata).

Besides the suitability map, the pattern of land-use change is also determined by the requirements for the different land-use types and competition among land uses. Therefore, in most models, a rule-based system is used to allocate the actual land-use changes based on the suitability map. These rules vary between using a simple cut-off value to select the locations with the highest suitability from the suitability map to dynamic modeling of competition between land uses based on land-use type specific characteristics.

Fig. 5.1.
Generalized model structure
of spatially explicit land-use
change models

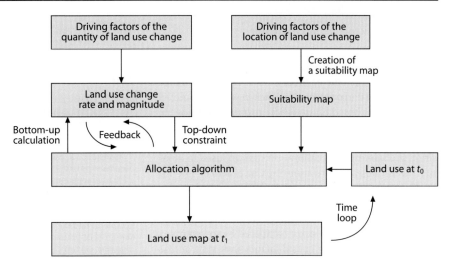

A wide variety of approaches is used to calculate the claims of different land uses for space. In a number of models a bottom-up approach is chosen in which the spatial dynamics and allocation rules determine the aggregated quantity of land-use/cover change. However, often a top-down approach is chosen in which the quantity of change is based on a set of driving factors. This quantity is used as a constraint in the actual allocation procedure. Alternatively, a hybrid approach is used in which the land-use requirements are influenced by a feedback from the allocation module to account for, e.g., land availability and changing land values.

5.3.2 Dynamic Versus Static

Apart from distinguishing models by their spatial representation, it is also possible to make a distinction between broad groups of models based on their temporal characteristics. The calculation of the coefficients of a regression model explaining the spatial distribution of land-use changes as a function of a number of hypothesized driving factors can be seen as a static model of land-use/cover change and is widely applied (Nelson and Hellerstein 1997; Chomitz and Thomas 2003; Overmars and Verburg 2005a). Although such regression models can be used to predict future land-use/cover change they often do not account for feedbacks and pathdependencies. Dynamic models give specific attention to the temporal dynamics of land-use systems, represented by the competition between land-uses, irreversibility of past changes leading to pathdependence in system evolution and fixed land-use change trajectories. Static models can be used to test our knowledge of the driving factors of land-use change, while dynamic models are used for projections of future land-use change or when trajectories of land-use change are studied. Examples of dynamic models include most multi-agent models and many spatially explicit models such as GEOMOD, CLUE and SLEUTH.

5.3.3 Descriptive Versus Prescriptive

Descriptive models aim at simulating the functioning of the land-use system and the explorative simulation of near future land-use patterns. Prescriptive models, in contrast, aim at the calculation of optimized land-use configurations that best match a set of goals and objectives. Descriptive models are based on the actual land-use system and dominant processes that lead to changes in this system. The model output provides insights in the functioning of the land-use system through testing hypothesis and the analysis of interactions between entities at a lower level that result in patterns at a higher level. Furthermore, this type of models is suitable to calculate projections of land-use/cover change for scenario conditions. Prescriptive models mostly include the actual land-use system as a constraint for more optimal land-use configurations (Lambin et al. 2000). The basic objective of most prescriptive or optimization models is that any parcel of land, given its attributes and its location, is modeled as being used in the way that best matches a series of defined objectives. Prescriptive models are especially useful for policy analysis as a spatial visualization of the land-use pattern that is the optimal solution based on preferred constraints and objectives (van Ittersum et al. 2004). However, prescriptive models do not provide insights in the actual land-use change trajectories and the intermediate conditions that might be needed to reach the optimized situation. Besides, many prescriptive models assume optimal economic behavior by the actors, and it is difficult to include non-optimal behavior of people, e.g., as result of differences in values, attitudes and cultures (Rabin 1998). While, at an aggregate level, these limitations are likely to be non-significant, they are more important as one looks at fine-scale land-use change processes and is interested in the diversity between actors (Lambin et al. 2000).

5.3.4   Deductive Versus Inductive

Another major difference between broad groups of land-use models is the role of theory. Most land-use change models rely on the inductive approach in which the model specification is based on statistical correlations between land change and a suite of explanatory variables that provide insight into this change. Multivariate statistics, transition probabilities and calibration runs are used to identify the variables and relationships in practice. The inductive approach has become extremely popular because of the absence of a strong, all-encompassing theory to explain land-use change to guide research. Theoretical developments in this field are hampered by the complex interplay of underlying causal factors that vary across temporal and spatial scales and organizational levels that may work directly or through longer causal routes and that may be associated with very different scientific realms, such as ecology, economy, sociology, or geography. Meta-analysis of case studies have indicated that the actual processes at work in a certain area are very much dependent on the context and specific conditions of an area or a site, also varying over time (see Chap. 3). The (inductive) interpretation of dominant land-use processes from a careful analysis of spatio-temporal patterns in collected data on land-use change is therefore often seen as the most straightforward technique to deal with this complexity. Different types of inductive models exist, ranging from models in which decision making by actors is specified as a set of decision rules and interactions based on observations (e.g., Parker et al. 2003) to models in which the relation between land-use pattern and the spatial variability in the socio-economic and biophysical environment is captured by statistical techniques, often regression (Verburg and Chen 2000; Geoghegan et al. 2001; Nelson et al. 2001), but also other empirical techniques such as neural networks (Pijanowski et al. 2002a, 2005). It should be noted that most studies are not purely inductive: theory and prior understanding of the decision-making process is used to select factors for inclusion and possibly suggest the functional form. However, the quantification of the relations is based on inductive methods.

The counterpoints to these studies are deductive land-use models based on theory that predicts pattern from process. Such process-led studies use theory to guide the characterization of land-use pattern relationships explored in the land-change model. The deductive approach is important to explore for several reasons. It structures the model around the critical human-environment relationships identified within the theory, and focuses attention on the data required to explore those relationships. In contrast, an inductive approach too frequently draws on readily available data (often proxy variables) that may not fully represent the processes at work,

---

Box 5.1. The Von Thünen model as a framework for understanding deforestation

Arild Angelsen · Norwegian University of Life Sciences, Ås, Norway

Johann Heinrich von Thünen (1783–1850) published in 1826 one of the first spatially explicit land-use models (Von Thünen 1966). In essence, von Thünen's *The Isolated State* is about how land rent – determined by distance from a centre – and perishability of agricultural produce shape the land uses.

Consider a simple model where land only has two uses: agriculture and forest. Agricultural production per hectare (ha) of the homogenous land is given (y), and the produce sold in a central market at a given price (p). The labor and capital required per ha are l and k, with inputs prices w (wage) and q (annual costs of capital). Finally, transport costs per km are denoted v and the distance from the centre d. This defines the land rent or profit of agriculture (per ha): $r = py - wl - qk - vd$. The rent declines with distance, and the agricultural frontier is where agricultural expansion is not profitable anymore, i.e., $r = 0$. Thus the frontier is at $d = (py - wl - qk)/v$.

This model yields several critical insights on the immediate causes of deforestation. Higher output prices (p) and technologies that increase yield (y) or reduce input costs (l, k) make expansion more attractive. Lower costs of capital (q) in the form of better access to credit and lower interest rates pull in the same direction. Reduced access costs (v) – new or better roads – also provide a great stimulus for deforestation. Higher wages (w) work in the opposite direction.

This simple framework has served as basis for a number of empirical investigations. A survey of more than 140 economic models of deforestation (Angelsen and Kaimowitz 1999) finds a broad consensus on three immediate causes of deforestation: higher agricultural prices, more/better roads, and low wages and shortage of off-farm employment (see Sect. 3.3.2). Although several extensions of this model are possible and provide more realistic descriptions, the key lesson from the von Thünen model remains that in many cases farmers or companies deforest because it is the most profitable alternative and they have the necessary means to do so.

---

forcing the analysis to sidestep the role of these processes. Pattern-led studies also tend to explore only those processes and concepts that are likely to explain the observed land-use pattern (Laney 2004). The classical example of a deductive land-use change model based on economic theory is the Von Thünen model (see Box 5.1). More recent deductive models of land-use change are presented by Angelsen (1999), who compares four different model specifications based on economic theory for agricultural expansion, and Walker and Solecki (2004) and Walker (2004) who develop theoretical models for deforestation and wetland conversion.

5.3.5   Agent-Based Versus Pixel-Based Representations

A final distinction between model types can be made based on the simulated objects. In many spatially explicit models, the unit of analysis is based on an area of land, either a polygon representing a field, plot or census track, or a pixel as part of a raster-based representation. Land-use changes are calculated for these spatial objects, directly resulting in maps that show the changes in land-use pattern. At the most local level, with the unit of analysis being a plot, field, or farm, the match with agents of land-use change, e.g., a farmer, is very good. Here the unit of analysis coincides with the level of decision-making. At higher organizational level, individual farmers or plot owners can no longer be represented explicitly and the simulations usually do not match with the units of decision making.

Another group of models use individual agents as units of simulation. Several characteristics define agents: they are autonomous, they share an environment through agent communication and interaction, and they make decisions that tie behavior to the environment. Such multi-agent systems give emphasis to the decision-making process of the agents and to the social organization and landscape in which these individuals are embedded. An agent is not necessarily an individual: an agent can represent any level of organization (a herd, a cohort, a village, etc.) (Parker et al. 2003; Bousquet and Le Page 2004). Disadvantages of using the agent as the basic unit of simulation is the difficulty to link agent behavior to the actual land areas (Rindfuss et al. 2002, 2004b) and to adequately represent spatial behavior. Therefore, both approaches of modeling have (dis)advantages, and the appropriate approach depends on the research questions and temporal and spatial extent of the model.

5.3.6   Global Versus Regional Models

Regional applications of land-use models vary in extent between local case studies of a few square kilometers to the country or continental level with resolutions varying between 50 m^2 to 1000 km^2. Numerous different models have been developed for these scales, and examples are given throughout this chapter. The situation is different for land-use/cover models that operate at the continental or global scale. Only few global models of land-use/cover change have been developed and those global model analyses are not typically aimed at investigating land-use/cover change issues *per se*, but, land-use/cover change can play an important role in analyses of climate change, biodiversity loss, agricultural production or world markets. Global models have thus addressed a range of land use-related questions. Early, well-known attempts at global modeling – most prominently the World3 model from the *Limits to Growth* study (Meadows et al. 1972) – were heavily criticized for being too aggregated to be meaningful and thin on empirical or theoretical support for presumed quantitative relationships among variables. A new generation of global Integrated Assessment Models (IAMs) was developed in the late 1980s and early 1990s primarily to assess the climate change issue and future scenarios for food and agricul-

ture (see Box 5.2). These models typically consist of linked sub-components representing population, economic activity leading to demand for agricultural products, technological and other factors that determine how these products are supplied; emissions of radiatively active gases associated with this production, resulting change in atmospheric composition and climate, and impacts of climate change on ecosystems and society (Steffen et al. 2004). IAMs rest on a much more rigorous framework than the earlier work. They have played prominent roles in recent assessments by the Intergovernmental Panel on Climate Change (e.g., IPCC 2000a), the Global Environment Outlook (UNEP 2002), and scenarios for future ecological changes being produced for the Millennium Ecosystem Assessment (see Chap. 6). They have also figured prominently in prospective agricultural studies, including those carried out for the International Food Policy Research Institute (Rosegrant et al. 2002) and the United Nations (Fischer et al. 2002). Another group of global models that address land-use dynamics are global economy models. Most of these models are equilibrium models, aiming to explain land allocation by demand-supply structures of the land-intensive sectors. Examples of such models are IMPACT (Rosegrant et al. 2002) and GTAP (Hsin et al. 2004). In these macro-economic models, land is usually allocated according to its relative economic return under different uses commonly achieved via a competitive market of land-intensive products. However, the geographic representation of the heterogeneity of production processes and land-use patterns is commonly not represented in these models. An exception to this is the IIASA-LUC model for China (Fischer and Sun 2001).

Overall, the land-use/cover modules of most global scale models have seen little development over the last decade. Although global land-cover change models still serve an important role as component of IAMs, e.g., in climate change assessments, only a few new models have been developed. Moreover, the central issue of global (climate) change that has guided model development over the past decade has lead to a focus on other elements of global models than on land-use/cover change. However, the recently completed Millennium Ecosystem Assessment (2005) has shown the grown interest of global modelers in land-use related processes and might well trigger a stronger effort in development of land-use/cover modules (see Chap. 7). The inherent coarse spatial resolution of global scale modeling poses many constraints to adequately capture the diversity in regional conditions of land-use/cover change and link transition rules to actual decision making. The most promising avenue for bridging the gap between these highly aggregated global models and the local case study literature appears to be in spatially explicit models of large world regions such as EURURALIS (Klijn et al. 2005).

Box 5.2. Global land-use modeling within an integrated assessment model

Bas Eickhout and *Tom Kram* · Netherlands Environmental Assessment Agency (MNP), Bilthoven, The Netherlands

IMAGE-2 (Integrated Model to Assess the Global Environment) is an integrated assessment model that simulates the environmental consequences of human activities worldwide (Alcamo et al. 1998; Eickhout et al. 2004). IMAGE-2 represents interactions between society, the biosphere and the climate system to assess environmental issues like climate change, biodiversity and human well being. The objective of the IMAGE-2 model is to explore the long-term dynamics of global environmental change, which requires a representation of how the world system could evolve. Future land-use, for example, is the result of interacting demographic, technological, economical, social, cultural and political forces. The model is designed to compare business-as-usual scenarios with specific mitigation and adaptation scenarios. The socio-economic and energy-use calculations are performed for 17 world regions. The atmospheric and ocean components are based on globally aggregated approaches. The land use and terrestrial-carbon cal-

culations are performed on a grid of 0.5 × 0.5 degrees, to mimic the detailed processes that occur and give insight in the consequences for ecosystems and agricultural activities.

IMAGE has been used in many studies to quantify different storylines and consequently played an important role in global assessments like the emission scenarios of the Intergovernmental Panel on Climate Change, the Global Environment Outlook of UNEP and the Millennium Ecosystem Assessment. In a recent application, the IMAGE model provides insight in economic and environmental consequences of four different trade liberalization scenarios. The main conclusion of this analysis is that liberalization can be helpful in gaining welfare; however, uncoordinated liberalization can lead to severe pressures on the environment. This conclusion is visualized in Fig. 5.2, in which the pressures on ecosystems by a scenario with full focus on liberalization are calculated (see Chap. 6).

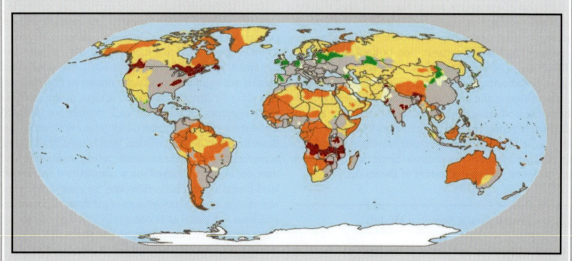

Fig. 5.2. Change in pressures on ecosystems between 2000 and 2030 for a full liberalization scenario. *Dark red* visualizes areas that will become agricultural land and *dark green* areas that will be abandoned from agricultural use. The *yellow-orange colors* visualize the increasing pressure on ecosystems through population density, economic activities and climate change

5.3.7   What's the Best Model?

There is no single approach that is clearly superior to model land-use/cover change. The choice of model is largely dependent on the research or policy questions that need to be answered, while issues of data availability might also play a role. No modeling approach is capable to answer all questions. Furthermore, the research questions may pose restrictions on the applicability and suitability of a particular model by its spatial and temporal scale and dominant land-use change process. For example, a spatially explicit cellular automata (CA) model may be well suited to explore urban growth dynamics, but is incapable of fully exploring the driving factors of agricultural transitions.

The wide selection of models and modeling approaches that has become available provides the re-

searcher with the opportunity to select the modeling approach that best fits the research questions and characteristics of the study area. In many cases, it may even be most appropriate to use a range of modeling approaches to study different aspects of the system under study. Box 5.3 provides an example of the application of different approaches in one study area. When models are used to improve and test our understanding of the driving factors of land-use change, the combined use of inductive and deductive modeling approaches may lead to complementary insights. Inductive techniques can be used to explore data sets and suggest possible driving factors that can be tested on causality in deductive modeling approaches. Such a combination leads to a more direct linkage between land-use change processes and observed patterns of change (Geoghegan et al. 1998).

Box 5.3. Land-use/cover change models in interdisciplinary research

Stephen J. Walsh · University of North Carolina, Chapel Hill, USA

The land-change science community has been active in modeling land-use and land-cover dynamics through a variety of approaches that integrate endogenous and exogenous factors, space and time scales, and feedbacks among people, place, and environment (Walsh and Crews-Meyer 2002). For instance, multi-level models or generalized linear mixed models are used to estimate the effects of farm-level variables on land-use/cover change patterns taking into account the contextual influences of community for each farm (Pan et al. 2004). Another modeling approach, based within complexity theory, uses cellular automata (CA) approaches by taking into account initial conditions, growth or transition rules, and neighborhood effects. Finally, agent-based models, also cast within the context of complexity theory and non-linear systems, consist of autonomous decision-making entities (agents), an environment through which agents interact, and rules that define the relationships between agents and their environment, as well as the sequence of actions in the model. Complexity theory conceives the world as consisting of self-organized systems, either reproducing their state (or stable state) through negative feedbacks with their environment, or moving along trajectories from one state to another as a result of positive feedbacks (Messina and Walsh 2001).

In the Northern Ecuadorian Amazon, for example, the greatest changes on the land are those created by agricultural colonists following in the wake of oil exploration who gained access on roads that made isolated areas accessible for development (see Sect. 3.3.3). Modeling approaches are needed that characterize dynamic and complex land-use/cover change patterns.

Among the more important findings from the use of the above models for assessing land-use/cover change in the frontier of the northern Ecuadorian Amazon are insights as follows: *(a)* rapid population growth caused substantial subdivision of plots, which in turn created a more complex and fragmented landscape; *(b)* key factors predicting landscape complexity are population size and composition, plot fragmentation, location of the plot relative to roads and towns, age of plot, soil quality, and topography; *(c)* family size and the number of males have direct effects on land clearing and use; *(d)* land use evolves over time with the family life-cycle and the duration on the plot; *(e)* technical assistance programs lead to more land in crops, less in pasture, and less total land cleared; *(f)* education level is important in determining the area in agriculture; *(g)* a direct correlation exists between distance to roads and towns and deforestation patterns; and *(h)* human settlement is affected by pattern-process feedbacks of land-use/cover change.

5.4 Spatial and Temporal Dimensions of Land-Use/Cover Change Modeling

In this section, we discuss the current capacities of land-use/cover change models to deal with the spatial and temporal dimensions of land-use systems. Issues related to the spatial and temporal dimensions have frequently been mentioned as a priority for land-use/cover change research and modeling, i.e., spatial scales, spatial interaction and autocorrelation, and temporal dynamics and feedbacks (Turner II et al. 1995; Lambin et al. 1999; McConnell and Moran 2001; van der Veen and Rotmans 2001; Veldkamp and Lambin 2001).

5.4.1 Spatial Scales and Level of Analysis

Scale is the spatial, temporal, quantitative, or analytic dimension used by scientists to measure and study objects and processes (Gibson et al. 2000). All scales have extent and resolution. For each process important to land-use and land-cover change, a range of spatial scales may be defined which has a significant influence on the land-use pattern (Meentemeyer 1989; Dovers 1995). Often, the range of spatial scales over which the driving forces and associated land-use change processes act correspond with levels of organization. Level refers to level of organization in a hierarchically organized system (e.g., individual, ecosystem, landscape or institution) and is characterized by its rank ordering in the hierarchical system. Many interactions and feedbacks between these processes occur at different levels of organization. A

clear understanding and representation of the functioning of the land-use system at different scales is therefore of prime importance for land-use/cover change modeling.

Differences in scientific discipline, tradition and research question have resulted in differences in the scales and levels that are addressed by the different land-use models. Coleman (1990) developed a framework that describes the interaction between the micro and macro level for social systems, which can be applied to land-use systems as well. Land-use change models are often based on remote sensing and GIS data at the regional (macro) level, while at the same time it is tried to explain these macro-level developments by specifying a micro-level mechanism. Figure 5.3, based on the work of Coleman (1990), depicts the relations between the macro and micro level. Macro-level analyses (pathway A) of land use are often based on empirical techniques, e.g., modeling of the spatial pattern of land cover derived from remote sensing. Pathway B explains the underlying process that lead to the different land-use patterns, e.g., the individual decisions in response to land-use policies. Together, these individual decisions lead to the changes in land-use pattern that cannot be predicted from a simple, linear combination of the individual behaviors. Therefore, explicit attention should be given to the interactions between agents and feedbacks in the decision-making process. Following this trajectory one can explain why differences in both micro and macro conditions lead to different land-use patterns. Linking macro-level analysis with micro-level dynamics could account for the differences in the modeling approach and/or method of analysis used.

Fig. 5.3.
Representation of the linkage
between micro-level and
macro-level research in
land-use change

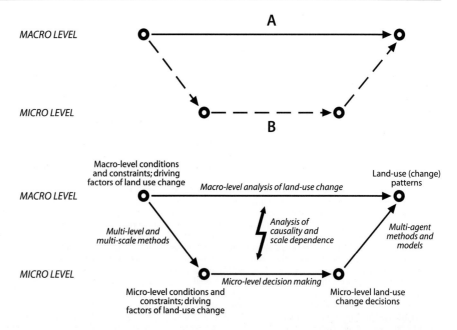

Most current models of land-use/cover change, however, follow trajectory A in Fig. 5.3, using relations between macro-scale variables to simulate land-use change. Many of these macro-level approaches are based on an analysis of the spatial structure of land use or the interactions between sectors of the economy. Examples of such models are the CLUE model (Verburg et al. 1999; Verburg and Veldkamp 2004), GEOMOD (Pontius et al. 2001), LOV (White and Engelen 2000) and LTM (Pijanowski et al. 2002a).

Micro-level modeling approaches based on microeconomic theory applied to simulate land-use changes have a long history. Most of these models start from the viewpoint of individual landowners who make land-use decisions with the objective to maximize expected returns or utility, and use economic theory to guide model development, including choice of functional form and explanatory variables (Ruben et al. 1998). The assumptions of behavior are valid for the micro level. This limits these models to applications that are able to discern all individuals. Difficulties arise from scaling these models, as they have primary been designed to work at the micro level. Jansen and Stoorvogel (1998) and Hijmans and Van Ittersum (1996) have shown the problems of scale that arise when this type of models are used at higher aggregation levels.

More recently, micro-level models have been developed that better address the scaling issues involved by explicitly following pathway B indicated in Fig. 5.3. These models are based on multi-agent systems. Multi-agent models simulate decision making by individual agents of land-use change, explicitly addressing interactions among individuals. The explicit attention for interactions between agents makes it possible for this type of models to simulate the emergent properties of systems. These

are properties at the macro scale that are not predictable from observing the micro units in isolation. If the decision rules of the agents are set such that they sufficiently look like human decision making, they can simulate behavior at the meso level of social organization, i.e., the behavior of heterogeneous groups of actors. Multi-agent based models of land-use/cover change are particularly well suited for representing complex spatial interactions under heterogeneous conditions and for modeling decentralized, autonomous decision making (Parker et al. 2003; Bousquet and Le Page 2004). Multi-agent systems are able to formalize decision-forming behavior of individual stakeholders, based on a theoretical argumentation. Most multi-agent models focus on either hypothetical or simplified situations to explore interactions between agents and between agents and the environment, rather than simulating landscape change at the regional level. An example of a multi-agent model application is given in Box 5.4. Other examples in land-use/cover change science include Huigen (2004), Deadman et al. (2004), Berger (2001), Castella et al. (2005), and the models reviewed by Parker et al. (2003) and Bousquet and Le Page (2004).

Often, the choice for a certain scale in land-use models is based on arbitrary, subjective reasons or scientific tradition (i.e., a micro- or macro-level perspective) and not reported explicitly (Watson 1978; Gibson et al. 2000). Models that rely on geographic data often use a regular grid to represent all data and processes. The resolution of analysis is determined by the measurement technique or data quality instead of the processes specified. Other approaches chose a specific level of analysis, e.g., the household level, which can be the level of the processes studied in the particular case study. For specific data sets optimal levels of analysis may exist where predictability is

---

Box 5.4. SLUDGE: an agent-based model to explore "edge-effect externalities"

Dawn Parker · George Mason University, Fairfax, USA

Rates of urban expansion in the U.S. are increasing, with urban areas characterized by decreasing density and perceived increasing rates of land-use fragmentation. Such patterns are often characterized as "urban sprawl" and are hypothesized to have broad negative impacts on humans and the environment. The Simulated Land Use Dependent on Edge-Effect Externalities (SLUDGE) model is a local-scale, spatially explicit agent-based model (ABM), designed to explore the hypothesis that "edge-effect externalities" – i.e., spatial conflicts which reduce the value of a given land use when a border is shared with a conflicting land use – contribute to fragmented patterns of land use at the urban-rural interface, and investigate the corollary implication that these resulting patterns are socially and economically inefficient (Parker and Meretsky 2004). An ABM was determined to be the best modeling technique to meet this goal, due to ABM's ability to model interdependencies between agent choices, impacts of heterogeneous land values, and feedbacks between macro-scale land-use and land-rent patterns and micro-scale drivers of agent land-use decisions.

SLUDGE is a hybrid ABM-Cellular Automata model that operates over an abstract cellular landscape and allocates each cell to the highest valued of two possible land uses: urban (white) and agricultural (black) – see Fig. 5.4 and Box 5.5. Payoffs for each land use depend on fixed returns to agriculture, transportation costs, externalities generated by nearest neighbors, and an endogenously determined rent for urban land. Land-use quantity, location, and pattern are iteratively and jointly determined through feedback from changing land values and surrounding land uses. The model has been used to demonstrate a series of "existence proofs": *(a)* that market equilibrium patterns of land use may be sub-optimally fragmented; *(b)* that possible Pareto improving rearrangement of land use may require coordination or bargaining between agents; *(c)* that conflicts between urban and residential land users lead to a more compact urban form; *(d)* that negative externalities from urban residents to agricultural producers lead to a socially inefficient expansion of urban land; and *(e)* that conflicts between urban land users can lead to fragmented patterns of urban development consistent with existing definitions of urban sprawl.

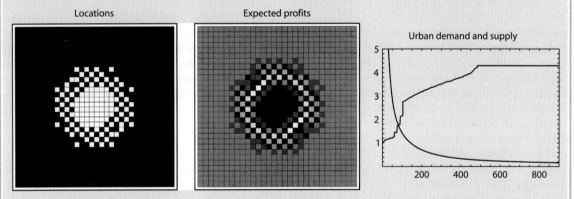

Locations   Expected profits   Urban demand and supply

Fig. 5.4. Illustration of a declining, concentric rent gradient, and an upward-sloping supply curve depending on the distance from the central market (*white* "location" cells are urban; *black* are agriculture)

---

highest (Veldkamp and Fresco 1997; Goodwin and Fahrig 1998; Verburg et al. 2003), but unfortunately there is no consistency on which level is optimal across land-use types or across spatial and temporal scales. Therefore, it is better not to use *a priori* levels of observation, but rather extract the appropriate levels for observation and modeling from a careful analysis of the land-use system and available data (Gardner 1998; O'Neill and King 1998).

Due to our limited capacities for the observation of land use, the extent and resolution chosen in land-use modeling are mostly linked. Studies at large spatial extent invariably have a relatively coarse resolution, due to our methods for observation, data analysis capacity and costs (see Sect. 2.5.2). This implies that features that can be observed in small regional case studies are generally not observable in studies for larger regions. On the other hand, due to their small extent, local studies often lack information about the context of the case study area that can be derived from the coarser scale data.

The importance of scale issues for land-use/cover change modelers can be summarized as follows:

- Land use is the result of multiple processes that act over different scales. At each scale different processes have a dominant influence on land use (Turner II et al. 1995).
- Aggregation of detailed scale processes does not straightforwardly lead to a proper representation of the higher-level process. Non-linearity, emergence and collective behavior cause scale dependency (Easterling 1997; Gibson et al. 2000).
- Our observations are bound by the extent and resolution of measurement that usually do not correspond to the level at which the processes operate. This causes observations to provide only a partial description of the whole multi-scale land-use system.

Although the importance of explicitly dealing with scale issues in land-use models is generally recognized, most existing models are only capable of performing an analysis at a single scale. Many models based on microeconomic assumptions tend to aggregate individual action but neglect the emergent properties of collective values and actions (Riebsame and Parton 1994). Approaches that implement multiple scales can be distinguished by the implementation of a multi-scale procedure in either the structure of the model or in the quantification of the driving factors. The latter approach acknowledges that different driving forces are important at different scales and it takes explicit account of the scale dependency of the quantitative relation between land use and its driving forces (see Chap. 3). Two approaches of quantifying the multi-scale relations between land use and driving forces are known. The first is based on data that are artificially gridded at multiple resolutions; at each individual resolution, the relations between land use and driving forces are statistically determined (Veldkamp and Fresco 1997; de Koning et al. 1998; Walsh et al. 1999; Verburg and Chen 2000; Walsh et al. 2001; Braimoh and Vlek 2004). The second approach uses multi-level statistics (Goldstein 1995) to explore the driving factors of land-use/ cover change over a range of scales. The first applications of multi-level statistics were used in the analysis of social science data of educational performances in schools (Aitkin et al. 1981). More recently, it was found that this technique is also useful for the analysis of land use, taking the hierarchical structure of land use into account. The land-use structure in Japan was analyzed taking different factors at different administrative levels into account using data for municipalities nested within prefectures (Hoshino 2001). A similar approach was followed by Colin Polsky (Polsky and Easterling III 2001; Polsky 2004) for the analysis of the land-use structure in the Great Plains of the United States of America. Also, in this study administrative units at different hierarchical levels were used. Others have used the micro-level organization of households nested within communities as a basis for a multi-level statistical analysis (Overmars and Verburg 2006; Pan and Bilsborrow 2005).

Besides accounting for the hierarchical structure in driving factors in empirical models, many dynamic models implement different scales within the structure of the model. In its simplest form the micro-level dynamics simulated by the model are constrained by conditions at a higher level. Other approaches allow interactions between levels or even use different, interacting models at different levels. Recently, the multi-scale, multi-model approach has been adopted as pivotal method in a number of larger-scale projects. Examples of multi-scale frameworks are ATEAM (Rounsevell et al. 2005) and EURURALIS (Klijn et al. 2005; Verburg et al. 2006). Within EURURALIS, global interactions determining the production and consumption characteristics of different regions are modeled by economic models at the global scale in connection with integrated assessment models to account for feedback of changes in climate. Land allocation at more detailed scales within the countries of the European Union is done by a spatially explicit land allocation model that accounts for variations in socio-economic, policy and biophysical location characteristics. This procedure was chosen since no single land-

use change model is able to account for both the local variation in driving factors as well as for the driving factors that operate at the global scale.

5.4.2   Spatial Autocorrelation and Spatial Interaction

Land-use patterns nearly always exhibit spatial autocorrelation. The explanation for this autocorrelation can be found, for a large part, in the clustered distribution of landscape features and gradients in environmental conditions that are important determinants of the land-use pattern. Another reason for spatially autocorrelated land-use patterns are the spatial interactions between land-use types themselves. Especially in the context of urban growth, neighborhood interactions are often addressed based on the notion that urban development can be conceived as a self-organizing system in which natural constraints and institutional controls (e.g., land-use policies) temper the way in which local decision-making processes produce macroscopic urban form. The importance of structural spatial dependencies is increasingly recognized by geographers and economists.

Different processes can explain the importance of neighborhood interactions. Simple mechanisms for economic interaction between locations are provided by the central place theory (Christaller 1933) that describes the uniform pattern of towns and cities in space as a function of the distance that consumers in the surrounding region travel to the nearest facilities. Spatial interaction among the location of facilities, residential areas and industries has been given more attention in the work of Krugman (Fujita et al. 1999; Krugman 1999). The spatial interactions are explained by a number of factors that either cause concentration of urban functions (centripetal forces: economies of scale, localized knowledge spillovers, thick labor markets) and others that lead to a spatial spread of urban functions (centrifugal forces: congestion, land rents, factor immobility etc.). Also, in agricultural landscapes spatial interaction may be an important determinant of land-use pattern, e.g., the adoption of particular farming technologies or cultivation patterns may be related to development in neighboring locations (Perz and Skole 2003; Polsky 2004).

Different approaches have been developed to address spatial autocorrelation in land-use patterns. In this chapter, we discuss how spatial autocorrelation is dealt with in statistical models and in cellular automata models.

Spatial autocorrelation in the residuals of a statistical model that relates land use to a set of supposed explanatory factors can be a problem for model estimation and interpretation (Anselin 1988; Overmars et al. 2003). Spatial statistical techniques have been developed to detect, correct and, empirically, quantify spatial dependencies (Bell and Bockstael 2000; Anselin 2002; Munroe et al. 2002). These advances in statistical techniques have allowed an increased incorporation of corrections for spatial autocorrelation in the residuals of regression equations in econometric models of land-use change. Implementation can be done through advanced measures of autocorrelation (Bell and Bockstael 2000; Walker et al. 2000; Brown et al. 2002; Munroe et al. 2002; Perz and Skole 2003). More often, simple measures of neighborhood composition, e.g., the area of the same land-use type in the neighborhood, are included as explanatory factors in regression models explaining land-use change (Geoghegan et al. 1997; Nelson and Hellerstein 1997; Munroe et al. 2002; Nelson et al. 2002; Verburg et al. 2004a). It should be noted that spatial autocorrelation in land-use patterns is scale dependent (Overmars et al. 2003). At an aggregate level residential areas are clustered, having a positive spatial autocorrelation. However, Irwin and Geoghegan (2001) found that, at the scale of individual parcels in the Patuxent watershed, there was evidence of a negative spatial interaction among developed parcels, implying that a developed land parcel "repels" neighboring development due to negative spatial externalities that are generated from development, e.g., congestion effects (Irwin and Bockstael 2002). The presence of such an effect implies that, *ceteris paribus*, a parcel's probability of development decreases as the amount of existing neighboring development increases. The existence of different causal processes at different scales means that spatial interactions should again be studied at multiple scales while relations found at a particular scale can only be used at that scale.

The most popular method to implement neighborhood interactions in dynamic land-use change models are cellular automata – see Box 5.5. Cellular automata (CA) were originally conceived by Ulam and Von Neumann in the 1940s to provide a formal framework for investigating the behavior of complex, extended systems (von Neumann 1966). In land-use models, cellular automata typically model the transition of a cell from one land use to another depending on the land use within the neighborhood of the cell. Cellular automata are used in almost all land-use change models for urban environments (White et al. 1997; Clarke and Gaydos 1998; Wu 1999; Ward et al. 2000; Jenerette and Wu 2001; Sui and Zeng 2001; Torrens and O'Sullivan 2001; Silva and Clarke 2002; Herold et al. 2003). Besides urbanization, CA-based models now also simulate other processes of land-use change, e.g., Messina and Walsh (2001) study land-use and land-cover dynamics in the Ecuadorian Amazon, an area where tropical forest is converted into agricultural land (see Box 5.3).

The definition of the transition rules of a CA model is the most essential part to obtain realistic simulations of land-use and land-cover change. Land-use change is the result of a complicated decision-making process; however, the transition rules of CA models are often defined on an *ad hoc* basis. Standardized methods to derive the transition rules are lacking. In an editorial on research priorities for CA and urban simulation, Torrens and O'Sullivan (2001) argue that urban CA models are now mostly technology, driven instead of really informing theories through the exploration of hypothetical ideas about urban dynamics.

Recently, different approaches have evolved to better match the transition rule set with reality. Sui and Zeng (2001) use historic conversions of land use to derive empirical evidence for the importance of the different factors and use multiple regression techniques to quantify the weights of the different factors within the transition rules. Other authors use advanced calibration methods for the model as a whole to fine-tune the coefficients of the transition rules based on a number of pattern and quantity measures (Clarke et al. 1996; Messina and Walsh 2001; Straatman et al. 2004). The main drawback of calibration techniques is formed by the huge set of parameters to be calibrated and consequently, the large amount of computing time. A good initial set of transition rules would be of great help to get these procedures on their way. Calibration of CA transition rules is complex due to the many interacting coefficients that do not necessarily yield unique solutions: different processes (rule sets) may lead to identical patterns. Calibration, therefore, does not always lead to new understandings of the relative importance of the different coefficients and is inappropriate for testing hypothesis concerning the underlying factors of urban development. The same argument holds for other methods that calibrate the transition rule set without explicating the relations used. Li and Yeh (2001, 2002) propose a method that overcomes the definition problem of the transition rules of a CA model by training artificial neural networks. However, neural networks do not give insight in the relations actually used in mod-

---

Box 5.5. Cellular Automata as a tool to incorporate positive spatial autocorrelation

Keith Clarke · University of California, Santa Barbara, USA

Cellular automata (CA) are simple models that can simulate both simple and complex behavior. They are defined by *(a)* a spatial grid or tessellation of cells; *(b)* a set of states which each grid cell can assume; *(c)* an initial state configuration of the whole grid; and *(d)* rules that define transitions between cell states based on neighboring states. With a long and detailed heritage in computer science, CA have been devised that can mimic almost any spatial or temporal configuration. The SLEUTH model is one of several proposed for modeling urban growth and land-use change (Clarke and Gaydos 1998). SLEUTH reads in the status of urban map layers at different time periods in the past, and uses them to "train" a complex CA. The model allows self modification that is changes are made to the behavior rules based on how fast or slow the system as a whole is growing. Over 35 SLEUTH applications to cities worldwide have been conducted, in a large variety of planning contexts such as to the growth of informal settlements in Africa and to urban encroachment on waste disposal sites in Brazil.

eling, leaving the user uninformed about the possible lack of causality in the relations that are used in the model. Also the method of Yang and Billings (2000a,b), that solves this inverse problem of cellular automata based on genetic algorithms has a number of drawbacks. This method is, at present, only operational for simple, binary patterns. Land-use patterns with multiple different land-use types are much more difficult to unravel.

The different possible specifications of transition rules and neighborhood lead to large uncertainties and potentials for error propagation in model simulations (Yeh and Li 2006). Novel methods to specify the transition rules in CA models, either based on theory or careful data analysis, are therefore urgently needed. A number of recent efforts may guide the way into this direction (Li and Yeh 2004; Verburg et al. 2004a; Caruso et al. 2005).

Besides the frequently addressed interaction between neighboring spatial entities, spatial interactions can also act over larger distances. A change in land use in the upstream part of a river may affect land use in the downstream part through sedimentation of eroded materials leading to a functional connectivity between the two areas. Another example of spatial connectivity is the migration of companies from one part of the country to another part when all available land area is occupied at the first location. Analysis of these interactions is essential to understand the spatial structure of land use. Globalization of the economy will cause these interactions to have a large spatial extent, leading to connectivity in land use between continents (see Sect. 4.3.3). One of the methods for implementing spatial interaction over larger distances is the use of network analysis. In many models, driving forces have been included that indicate travel times, distances or barriers to access markets, ports and other facilities. Often, models that are based on economic theory take travel costs to a market into account (Jones 1983; Chomitz and Gray 1996; Nelson et al. 2002). Most often, simple distance measures are used. However, it is also possible to use sophisticated techniques to calculate travel times/costs and use the results to explain the land-use structure. This type of calculations are included in combined urban-transportation models (Miller et al. 1999).

5.4.3 Temporal Dynamics: Trajectories of Change and Feedbacks

Changes in land use and land cover are often non-linear, feedbacks and thresholds often play an important role (Turner II 1997; Turner II et al. 2003a; Steffen et al. 2004). In this case, dynamic modeling and the subdivision of the simulation period into time steps becomes essential. Only then, land-use change analysis can account for the path dependency of system evolution, the possibility of multiple stable states, and multiple trajectories. Land-use change

cannot be simply explained as the equilibrium result of the present set of driving forces. In other words, land-use change may be dependent on initial conditions, and small, essentially random events may lead to very different outcomes, making prediction problematic (see Sect. 4.3.2). Exemplary is the effect of transportation infrastructure on the pattern of development. Road expansion and improvement not only lead to more development but may also lead to a different pattern through a reorganization of the market structure, which feeds back on infrastructure development (see Box 3.5). Thus, certain trajectories of land-use change may be the result of "lock in" that comes from systems that exhibit autocatalytic behavior.

In most deforestation and urbanization models, a one-way conversion from one land-use category to another category is assumed because of the focus on a single land-use conversion (Clarke and Gaydos 1998; Pontius and Pacheco 2004). However, in agricultural and semi-natural landscapes changes in land use are often reversible or cyclic and can be determined by the land-use history of a location leading to path dependence of the land-use change processes (see Box 5.6). Recent studies that have analyzed land-use change trajectories in more detail (Fox et al. 2000; Mertens and Lambin 2000; Nagendra et al. 2003; Geist and Lambin 2004) have confirmed that land-use and land-cover changes often exhibit high degrees of spatial and temporal complexity. This complexity arises from particular chains of events and sequences of causes and effects that lead to specific land-use changes ("pathways of land change"; see Sect. 3.5).

In a number of models, temporal dynamics are taken into account using initial land use as a criterion for the allowed changes. Cellular automata do this explicitly in the decision rules that determine the conversion probability. In the CLUE-S model (Verburg and Veldkamp 2004), a specific land-use conversion elasticity is given to each land-use type. This elasticity will cause some land-use types to be more reluctant to change (e.g., plantations of permanent crops) whereas others easily shift location (e.g., shifting cultivation). In the SLEUTH urban growth model (Clarke and Gaydos 1998) even more explicit functions to enforce temporal autocorrelation are implemented that also take the "age" of a new urban development centre into account. In the economic land-allocation model of the Patuxent Landscape Model (Irwin and Geoghegan 2001), the land-use conversion decision is posed as an optimal timing decision in which the landowner maximizes expected profits by choosing the optimal conversion time. That time is chosen so that the present discounted value of expected returns from converting the parcel to residential use is maximized.

Another source of temporal complexity in land-use modeling is the influence of feedbacks between land-use decisions and land-use change impacts. For a proper description of certain land-use types, e.g., long fallow systems, feedback processes such as nutrient depletion

Box 5.6. The impact of path dependence on landscape pattern

Non-linear changes in demand for arable land can have severe impacts on the landscape for a case study of the Czech Republic. The EURURALIS scenario study projected for one of the scenarios non-linear changes in the arable land area for most accession countries of the EU (Klijn et al. 2005). The CLUE-S land-use change model (Verburg et al. 2006) was used to allocate these changes spatially. The increase in the area of arable land during the 2000–2010 period comes at the cost of a decrease in natural area. After 2010, abandonment of arable land is expected. Part of the abandoned area will change into new natural areas. However, as can be seen in Fig. 5.5, nature does not return at the locations where it is lost during the first decade. During the first ten years, it is mainly the small patches of nature in the main agricultural area that are lost (see arrow in map b), while new nature develops on abandoned marginal lands, mostly adjacent to existing nature areas (arrow on map c). This pathway of change has important, irreversible consequences for the rural area and landscape diversity in different parts of the country (see Sect. 4.5).

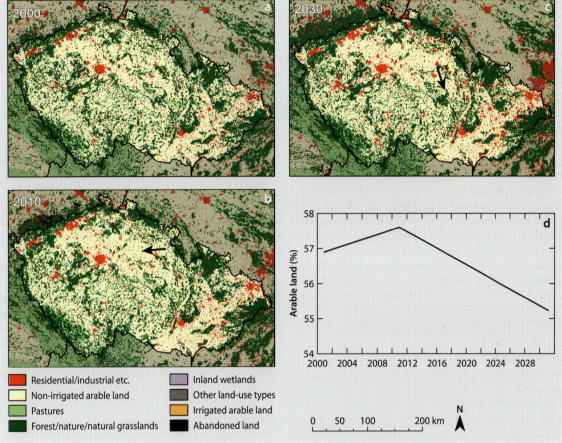

Fig. 5.5. Aggregate changes in arable land area (**d**) and resulting land-use patterns for 2000 (**a**), 2010 (**b**) and 2030 (**c**) in the Czech Republic for the "B1" or "Global Co-operation" scenario

upon prolonged use of agricultural land, should be implemented (van Noordwijk 2002; Verburg et al. 2004c). Other examples of feedback mechanisms influencing land use are climate change (Carvalho et al. 2004), hydrological change (Ducrot et al. 2004) and the interaction between urbanization and transportation structures (Miller et al. 2004).

The combination of temporal and spatial dynamics often causes complex, non-linear behavior. However, many land-use models are based on an extrapolation of the trend in land-use change through the use of a regression on this change (Mertens and Lambin 2000; Geoghegan et al. 2001; Schneider and Pontius 2001; Serneels and Lambin 2001b). This type of models is therefore less suitable for longer-term scenario analysis, as they are only valid within the range of the land-use changes on which they are based, which is usually in the order of one or two decades. The validity of the relations is also violated upon a change in competitive conditions between the land-use types, e.g., caused by a change in demand. This critique does not apply to all models based on statistical quantification. When these models are based on the analysis of the structure (pattern) of land use instead of the change in land use and are combined with dynamic modeling of competition between land-use types, they have a much wider range of applications.

Land-use change decisions are made at different time scales, some decisions are based on short term dynamics (such as daily weather fluctuations), others are based on long-term dynamics (e.g., climate change). Most land-use models use annual time steps in the calculations. This means that short-term dynamics are often ignored or, when they can have an additive effect, are aggregated to yearly changes. However, this aggregation can hamper the linkage with the actual decision making shorter time scales (Laney 2004). The need for multi-scale temporal models was acknowledged in transportation modeling, where short-term decisions depend on the daily activity schedules and unexpected events (Arentze and Timmermans 2000; Arentze et al. 2001). The link between this type of transportation models and land use is straightforward. If changes in the daily activity schedule are required on a regular basis, individuals will need to adjust their activity agenda or the factors affecting the agenda, for example, by relocation. Such a decision is a typical long-term decision, evolving from regular changes in short-term decisions.

Temporal complexity and feedback mechanisms still pose a major challenge to land-use/cover change modelers. These challenges not only include the development of well balanced approaches for adequately dealing with this complexity, also appropriate tools to validate predictions of path-dependent systems are needed. Box 5.7 discusses a recent development of such approaches.

Box 5.7. Path dependence and the validation of agent-based spatial models of land use

Dan Brown · University of Michigan, Ann Arbor, USA

Our contemporary understanding of land-change processes as complex adaptive systems requires (Lambin et al. 2003) validation methods that acknowledge unpredictability as a possible outcome. The methods of map comparison outlined in this chapter evaluate a model's predictive accuracy, i.e., its ability to produce land-use patterns that are highly correlated with the actual land-use pattern. However, in using a model for policy evaluation and scenario development, we are also concerned with its process accuracy. Importantly, the predictability of a real-world system that includes feedbacks and exhibits path-dependent behavior is necessarily limited. Yet, because we have only one real-world outcome to evaluate, there is a tendency to seek a model that matches that one outcome (i.e., map) very well. The possibility exists, therefore, for a model that matches the real-world processes well to fit the observed patterns less well than another model with less realistic processes. To balance these two potentially conflicting motivations, Brown et al. (2005) suggest an approach to validating stochastic models that recognizes the concept of the invariant region, i.e., the locations where land-use type is almost certain, and thus path independent; and the variant region, i.e., the area where land use depends on a particular series of events, and is thus path dependent. Researchers can use this approach to improve their ability to communicate how well their model performs against real-world patterns, including the cases in which it is relatively unlikely to predict well because of either path dependence or stochastic uncertainty.

5.5   Calibration and Validation of Land-Use/Cover Change Models

Rykiel (1996) defines validation as "a demonstration that a model within its domain of applicability possesses a satisfactory range of accuracy consistent with the intended application of the model". Model validation is therefore the process of measuring the agreement between the model prediction and independent data. If there is a good match, then the method used to make the prediction is said to be valid. It is crucial to distinguish between model calibration and model validation. Calibration refers to the process of creating a model such that it is consistent with the data used to create the model. It is essential that the available data set be split into two separate subsets, called calibration data and validation data.

Calibration and validation of land-use/cover change models has mostly not been given a lot of attention. This can be attributed to the difficulty of obtaining appropriate data for calibration and validation as well as to the lack of specific methods for calibration and validation of land-use/cover change models (Veldkamp and Lambin 2001; Walker 2003). In recent years, more methods have become available, either developed in other disciplines (e.g., Costanza 1989; Manel et al. 2001) or specifically for land-use/cover models (Pontius et al. 2004; Visser 2004).

Confusion about the validation of land-use/cover models can originate from lack of distinction between goodness-of-fit of calibration *versus* goodness-of-fit of validation. The goodness-of-fit of an empirical model, e.g., a regression model that relates supposed driving factors to observed land-use patterns, only measures the goodness of calibration, whereas the performance of the same regression model to predict land-use pattern from driving factors for another location or time period could be considered a validation. If there is a good fit between the model prediction and the validation data, then the model might be able to make accurate extrapolations to other spatial and temporal extents. If the characteristics and mechanisms of the other spatial and temporal extents are similar to the characteristics and mechanisms that existed during the calibration and validation phases, then the model should be able to extrapolate to other extents with a level of accuracy similar to the performance in the validation. If the model fails to attain a good fit in the validation, then one should have little confidence in the model's ability to extrapolate accurately to other extents. Obviously, we will not know *a priori* whether the mechanisms during the calibration and validation phases will continue into the extrapolation, because we can never know something until we have empirical data (Oreskes et al. 1994).

There are no agreed criteria among scientists concerning either the method or the level of what is considered a "good match" between validation data and a model prediction. Pontius has suggested that a reasonable minimum criterion would be that the agreement between the validation data and the prediction from a scientist's model should be better than the agreement between the validation data and the prediction from a null model (Pontius et al. 2004; Pontius and Malanson 2005). The prediction from a null model is the naïve prediction that one would make if one were not to create any model.

There are numerous mathematical methods to compare the patterns in maps. A common first step in computing the agreement between a validation map and a prediction map is to calculate the percent of pixels classified correctly. This statistic should be compared to the percent correct in the comparison between the validation data and the prediction from the null model. The percent correct is the most common statistic reported because it is simple to compute and is relatively intuitive to interpret. The disadvantage of the percent correct statistic is that it fails to capture patterns that are immediately obvious to the human eye (Hagen 2002). There are many other statistics available that compute pattern metrics to compare the patterns in two maps (Ritters et al. 1995). Whatever statistic is chosen, the scientist should be cognizant of two important components in the comparison of the pattern between two maps of a common categorical variable. These components are comparison in terms of quantity of each category and comparison in terms of location of each category. Comparison in terms of quantity considers whether the proportion of each category on one map is similar to the proportion of the corresponding category on the other map. Comparison in terms of location considers whether the position of each category on one map is similar to the position of the corresponding category on the other map. Pontius (2000) describes how to budget the agreement and disagreement for these two components.

None of the statistical measurements above match perfectly the human eye's ability to recognize patterns. Therefore, all of the methods of statistical map comparison should be complimented by a visual assessment, to see whether the selected statistic is measuring the characteristic that the scientist thinks is important. However, visual assessment can be influenced dramatically by subjective aspects of map production, such as selection of the color palette. Therefore, objective statistical measurement is essential to maintain scientific rigor.

Any measurement to compare two maps can be extremely sensitive to the scale of the analysis. Therefore, it is advisable to compute the measurement at various scales to examine the degree to which the results are sensitive to changes in scale (Costanza 1989). Increasingly, scientists are creating methods to examine how results are sensitive to scale. Pontius (2002) shows how to compute the components of agreement and disagreement in terms of quantity and location at multiple resolutions. Multiple scale analysis allows the statistics to conform more closely to the patterns and clusters that the human eye sees. Multiple-scale validation is also important because it allows the scientist to see whether the model makes predictions at scales that are relevant to the purpose of the model. Many models are calibrated and make predictions at fine resolutions that match the available data. For example, satellite data are commonly available in the form of 30-meter pixels. However, the relevant questions concerning land change may occur at coarser resolutions (see Sect. 2.4 and 2.5). For example, many global climate models operate at scales of 1 degree longitude by 1 degree latitude. Scientists are sometimes reluctant to change the scale of the raw data because any adjustments introduce additional artifacts into the data. Multiple resolution validation allows a scientist to see whether the 30-m resolution model performs sufficiently accurately at a resolution that is relevant to the purpose of the model.

When there is lack of fit between the data and the prediction, then the quality of both the model and the data should be examined and improvements should be made to which ever is found to be worse. If the data quality is high, then focus should be on improving the quality of the model and to make it consistent with the data. If the data quality is low, then the focus should be on getting better data, as working with inaccurate data might lead to the development of an inaccurate model.

It is not particularly useful to attempt to crown a model as valid, or to condemn a model as invalid based on the validation results. It is more useful to state carefully the degree to which a model is valid. Validation should measure the performance of a model in a manner that enables the scientist to know the level of trust that one should put in the model. Useful validation should also give the modeler information necessary to improve the model.

In many of the earlier land-use change model applications validation was lacking, mainly due to a lack of good data sets and appropriate validation techniques. The recent literature on validation of land-use/cover change and ecological models has provided an incentive to make validation part of the standard modeling procedure (Kok et al. 2001; Walker 2003; Pontius et al. 2004; Visser 2004; Brown et al. 2005; Pijanowski et al. 2005).

5.6 Conclusions

Over the past decade much progress has been made in the development of land-use/cover change modeling approaches. This is reflected in the large number of different models that have been developed for the local and regional scales. Besides offering a range of different techniques applicable to different scales and contexts, large progress has been made in linking different disciplinary perspectives.

Global-scale land-use/cover change models have seen less development recently. Although global land-cover change models still serve an important role as components of integrated assessment models, only few new models have been developed. It is urgently needed that different approaches for modeling land use/cover at the global scale are tested and validated to better equip the many environmental and social assessments at the global scale.

Substantial progress has been made on different aspects that are part of the modeling cycle – see Fig. 5.7. These include (a) areas of model validation, (b) participatory approaches to develop storylines of scenarios, (c) incorporation of biophysical and social impacts, and (d) the linkage between process-based and pattern, based descriptions of the driving forces.

In general, it can be concluded that the land change modeling community rapidly becomes more integrated, benefiting from the integration of different disciplines and system-based approaches. Examples include the use of techniques and methods developed in other disciplines that help to better develop simulation algorithms. Multi-level statistics, originating from educational research have already proven to be useful in the analysis of the hierarchical structure of land use. Hydraulic models may

---

Box 5.8. Integrating spatial and actor-based land-use/cover change research in the Philippines

Koen P. Overmars · Wageningen University and University of Leiden, The Netherlands

Different disciplines within land-use/cover change have different foci in their research and try to explain land-use changes from their own perspective and at different scales and levels. In this box, an example of research is given where process-oriented research methodology from the social sciences (Action-in-Context) is combined with system-based, pattern-oriented research originating from geography (CLUE approach). Action-in-Context is a conceptual approach for actor-oriented research, which investigates causal relations and options and motivations of farmers (de Groot 1992; Verburg et al. 2003). CLUE is a spatially explicit land-use change model based on a top-down approach of spatial analysis of land-use patterns (Verburg et al. 2002). The integration leads to the development of two different modeling approaches that help to explore scenarios of future land-use change: a multi-agent model that captures human-environment interactions at the community level (Huigen 2004), and a spatial model that combines the strengths of actor-oriented and geographical approaches at the watershed level.

Within the first modeling approach, a modeling toolbox is built that incorporates multi-agent techniques that can be used to explore the dynamics of land-use change in the area. The Action-in-Context methodology is used as a framework to study options and motivations of the actor, actor environment relations and the interactions and relations between actors. The result is a spatially explicit model that can deal with dynamic processes like migration, expansion of agricultural area and actor behavior.

In the second model, the actor-oriented research from project one, which provides information about causality and land-use change processes, is combined with empirical findings (multivariate statistics) to describe relations between land use and its explanatory factors at the watershed level (Overmars and Verburg 2005a). This information is incorporated in a CLUE model at the watershed level. The actor-oriented research provides the strong causal relations that explain the empirical findings. The resulting model simulates future scenarios in a spatially explicit manner – see Fig. 5.6.

Although both approaches and models have similar objectives, they provide insights at different levels and provide different and complementary types of information to better target interventions for sustainable development (see Sect. 7.4).

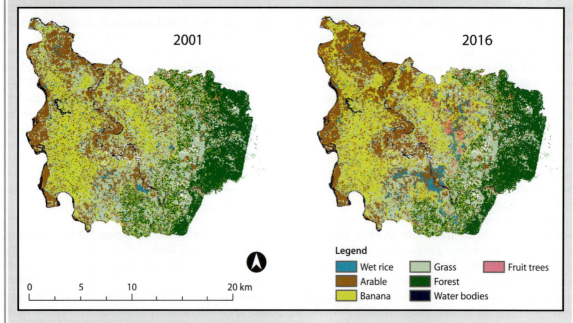

Fig. 5.6. Observed land-use map of 2001 and a simulated land-use map (CLUE model) of 2016

help to understand traffic congestion in linked urban-transportation models, while ecological models can give hints of dealing with the hierarchical organization of land use (Dale and Pearson 1999). This type of integration requires land-use researchers to move beyond their disciplinary traditions and attempt to develop overarching interdisciplinary theoretical frameworks that encompass integrated processes. A couple of research projects in land-use analysis has already shown that such approaches can result in innovative results, e.g., recent attempts to link social science research with geographical data (Geoghegan et al. 1998; Walsh et al. 1999; Mertens et al. 2000; Walker et al. 2000; Walsh and Crews-Meyer 2002; Overmars and Verburg 2005a). Such integration can strengthen both the modeling and narrative approaches – see Box 5.8.

In spite of the successes and progress in land-use/cover change modeling, challenges for further improvement remain. In the discussions among the approximately 100 participants of the LUCC Focus 3 workshop *Integrated assessment of the land system: the future of land use*, held in Amsterdam from 28 to 30 October 2004, several challenges were identified. Important for progress in the field of land-use/cover change modeling is the integration of the different components that are part of the actual modeling cycle, indicated by the arrows in Fig. 5.7. Solid lines indicate interactions that are accounted for in at least some of the approaches. Dashed lines indicate research frontiers that are currently given limited attention (see Chap. 8).

The research frontier for land-use/cover change modeling is two-fold. First, a number of methodological issues are not yet fully developed and need further attention. To name a few of the most pressing:

- Validation. Whereas progress has been made in the validation of spatially explicit land-use/cover change models, validation of multi-agent models needs specific attention, since no adequate measures are available yet (Brown et al. 2005). Furthermore, besides validation it is important to better develop methods to partition the error in model outcomes in different components (e.g., input data, model specification, quantity, location etc.). Such uncertainty analysis will guide further improvements to the most uncertain components of the model (Yeh and Li 2006).
- Linking process-, and pattern-based approaches to quantify land-use dynamics. Modelers still have difficulty to take stock of the information contained in qualitative land-use studies due to the difficulty to match organizational levels with spatial entities. Strong interaction of inductive and deductive approaches can certainly benefit the quantification of land-use change processes important to land-use/cover change models.
- Scaling issues. The multi-scale structure of the land-use system has always been an important item on the research agenda of the LUCC project. All land-use/cover change modelers have to deal with scale in some way: either through linking individual actors with institutions and spatial patterns within model building, or through the communication of model results across different scales to the stakeholders. Awareness of the pitfalls and challenges of these issues has increased during the last decade. However, in spite of the progress there is still a need for approaches and techniques that can deal more adequately with scaling issues. Therefore, understanding the interactions between and across scales will most likely remain at the research frontier of land-use/cover change modelers for the next decade.

Fig. 5.7.
Overview of the potential use of land-use/cover change models to support policy. *Source:* Kok et al. (2004)

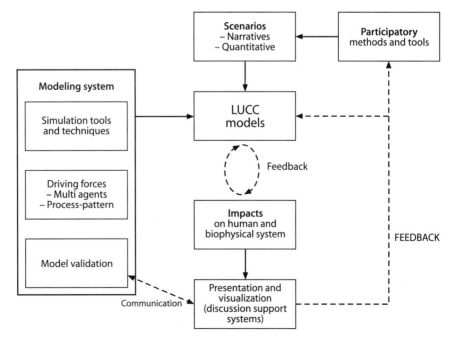

- Translation of qualitative storylines to quantitative input in land-use/cover change models. Although land-use/cover change models are frequently used for scenario simulations (see Chap. 6), the translation of qualitative storylines into quantitative model conditions is often done on an *ad-hoc* basis. Further development of tools and techniques will facilitate the use of models within scenario studies.

- Incorporation of human and environmental feedbacks in land-use/cover change models. System-based approaches focusing on the feedbacks between different components of the land-use system are needed to enhance our understanding of the role of land use in vulnerability and resilience of the human-environment system.

- Addressing urban-rural interactions. Few models explicitly address the interaction between urban and rural areas. The research communities focusing on urban dynamics and those focusing on agricultural or forest changes have developed separately, each using different techniques and model approaches. More recently these communities are better integrated and experiences with different modeling techniques are exchanged (Verburg and Veldkamp 2005). However, the challenges of addressing the spatial interactions between urban and rural areas remain (Irwin and Bockstael 2004). Large impacts are to be expected of these interactions both in developed countries through the emergence of multi-functional land uses in the rural hinterlands of cities, and in developing countries where unequal development between cities and rural areas is an important issue. The emergence of multi-functional land use requires a step beyond the common modeling approaches that represent land use by a single, not-overlapping, land-cover type in each spatial entity. Land-use change models for such regions should focus on the functionalities of land use and allow overlapping functions. This requires conceptual as well as practical innovations.

As indicated in Fig. 5.7, there are a couple of key feedbacks that have hardly been touched upon. These feedbacks are related to the use of models as policy support tool and relate to the communication of results of land-use/cover change modeling to the decision makers and other stakeholders (see Chap. 7). The use of land-use models in policy support is limited. This is partly due to the relatively recent development of full-fledged modeling approaches and the explorative phase of many models. The methodological challenges have been so overwhelming that the production of stakeholder, relevant outputs was not always given explicit attention. Furthermore, most validation exercises have indicated that uncertainty in land-use simulations is high. Such limitations are common for models of complex integrated systems and, although progress is made, some of these constraints are inherent to the study

of complex systems. Therefore, visualization issues and adequate presentation and communication of the results and associated validity are important issues that need to be given attention. Communicating the results of land-change models to the different stakeholders involves the use of the appropriate visualization techniques so that these will be better understood and better appreciated. This may also include the use of models in role-playing games (RPGs). Moreover, the feedbacks requires that the views of stakeholders (e.g., policy on land use) be translated to development scenarios that may be evaluated or considered in the land-use/cover change models, taking into account the scale and the decision variables involved. The model scenario simulations can be used as part of the decision- or discussion-support for stakeholders toward sustainable management and development.

One of the major constraints in using information from complex system analysis in policy is that not always the policy-relevant questions are addressed. Insufficient attention is given to an inventory of the questions of the policy makers due to the sometimes inward-looking attitude of scientists. Furthermore, frequently, a mismatch appears between the scale at which the stakeholders have an influence and the range of scales addressed by the models (Tomich et al. 2004a,b). In a study that analyzed the use of decision support systems by policy makers, Uran and Janssen (2003) found that decision support systems are not adequately used by the stakeholders they were developed for. Among the identified reasons, the lack of communication and feedback between developers and stakeholders is ranked as the main challenge to improve the use of such systems. Geertman and Stillwell (2004) reach similar recommendations after an inventory of planning support systems: the development of planning support systems should be an integral part of the planning procedure and context. The communication of model results should no longer be the final stage of a research project. Decision makers should become part of the process, instead of merely the end users. Furthermore, the authors stress the importance to present results to stakeholders such that these are tuned to the knowledge and skills of the stakeholders; the scientists should take seriously its users and leave them with the feeling that they have been taken seriously. Finally, it is indicated that the scientists should be aware of the fact that many stakeholders address issues from an interdisciplinary perspective. As a consequence, a presentation of research findings that intends to connect to people's way of thinking should address issues in an interdisciplinary manner, linking the spatial to the social, the environmental to the economic, and so forth. This provides another incentive to scientists in this field to adapt an interdisciplinary way of thinking.

Finally, it is important to make a distinction between actual decision support and discussion support. Presenting clear-cut solutions for land-use decisions to policy

---

Box 5.9. Multi-agent systems, companion modeling and land-use change

François Bousquet · CIRAD – Tere Green Cirad-Tera-Green, Montpellier, France

Since 1994, a group of researchers from different disciplines has developed complementary activities on the theme of multi-agent systems (MAS) simulations and renewable resource management, which are very often strongly related to land-use change issues:

- Development of abstract models, also called artificial societies, which help to understand the generic properties of interacting processes (Antona et al. 1998).
- Development of models applied to concrete and local problems to understand the dynamics of natural and renewable resources and their management (Mathevet et al. 2003).
- Development of a simulation platform (CORMAS, common-pool resources and multi-agent systems) (Bousquet et al. 1998).
- Development of a methodology, companion modeling, for the use for these MAS tools within the very wide framework of collective decision support.

The approach was named "companion modeling" because it is used as a tool in the mediation process (the social dimension of the companion) and it co-evolves with the social process (temporal and adaptive dimension). The companion modeling approach is an iterative process based on repetitive back-and-forth steps between model building and field activities by researchers. Changes in the model being used or developed may be introduced as new information or ideas from the field were obtained. This approach may also be combined with other methods of data collection and analysis during the modeling process. Intuitively, a MAS model is a role-playing game (RPG) simulated by the computer. Thus, we proposed (Bousquet et al. 1999) to set up RPGs, similar to the MAS model, with the objective of making real stakeholders play the game, allowing them:

- To understand the model and the difference between the model and the reality,
- To validate it or to propose modifications, and
- To be able to follow MAS simulations on the computer, and to propose scenarios.

The applications range from irrigated schemes in Senegal to upland agriculture in northern Vietnam. With the multiplication of application case studies in various places, many new scientific questions and technical issues emerged that were addressed by the research group.

---

makers often disregards the different opinions among stakeholders and the policy-making context. Projects that used land-use models as a tool to provoke and inform discussions among different stakeholders and policy makers have the potential to be more successful, as indicated by successful projects at the European level (Klijn et al. 2005; Verburg et al. 2006b; Rounsevell et al. 2005) and a number of local to regional projects based on the CORMAS model – see Box 5.9.

This chapter has shown that a large variety of concepts, approaches and techniques for land-use/cover change modeling are already available: combining the strengths of these concepts, approaches and techniques is the best concept for further progress in this field. Bundling of strengths of the multi-disciplinary land-use research community will help to better understand these complex systems and to better communicate with the stakeholders of land-use change. The large contribution of the land-use/cover change modeling community to land change science in general can mainly be attributed to the fact that modeling has provided an enormous incentive for researchers from different disciplines to work together on the same issues and actually formalize the interactions between system components. This process challenged many researchers to analyze the land-use system from different perspectives while focusing on a formal description of the system dynamics.

As is illustrated by the many successful projects and publications, land-use/cover change modeling has made an important contribution to land-use/cover change science in general, and will most likely continue to do so in the future.

Chapter 6

Searching for the Future of Land: Scenarios from the Local to Global Scale

Joseph Alcamo · Kasper Kok · Gerald Busch · Jörg A. Priess · Bas Eickhout · Mark Rounsevell · Dale S. Rothman
Maik Heistermann

6.1 Introduction

Much of the scientific research concerned with land-use and land-cover issues is motivated by questions related to global environmental change. For example, will deforestation continue, and if yes, where, and at what rate? How will demographic changes affect future land use and cover? How will economic growth influence future land use and cover? What will be the magnitude of emissions of greenhouse gases related to land use and cover? A common characteristic of these and other issues related to global environmental change is that they stimulate questions not only about past and present changes in land use and cover but also about their future changes (Brouwer and McCarl 2006). The main objective of this chapter is to summarize the state of understanding about the future of land. What are the range and predominant views of this future? What are the views on the global, continental, regional and local levels? We review what (we think) we know and don't know about the future of land by reviewing published scenarios from the global to local scale. Our aim is to identify the main messages of these scenarios especially relevant to global change issues, and to recommend how scenarios can be improved to better address the outstanding questions about global change and land use/cover.

In the first section of the chapter, we describe how scenario analysis is used as a convenient tool to envision the future of land use and cover. In the next section, we describe the main messages of large-scale scenarios and their insights into plausible global and continental-scale trends. We then review regional and local scenarios and discuss in particular current efforts to link these scenarios with the goals of different actors influencing local land-use change. Finally, we identify the shortcomings of current scenarios, and discuss how they might be improved.

6.2 Scenario Analysis: a Method for Anticipating the Future of Land

Although research on the future of land is clearly needed, the scientific community has been hesitant to take up this challenge – an understandable situation consider-

ing that the projection of land use/cover requires assumptions about future global vegetation (including future areas of cropland, forest and grassland) as well anticipating society's countless decisions on where to settle, where to build, where to grow its crops, and what lands to protect. Some researchers have found a partial solution to this challenge by developing scenarios of future land use and cover. Scenarios are plausible views of the future based on "if, then" assertions – If the specified conditions are met, then future land use and land cover will be realized in a particular way. Scenario analysis is the procedure by which scenarios are developed, compared, and evaluated. Scenario analysis does not eliminate the uncertainties about the future, but it does provide a means to represent current knowledge in the form of consistent, conditional statements about the future.

6.2.1 Qualitative Scenarios

There are a variety of ways of classifying land scenarios. One way is to distinguish between qualitative and quantitative scenarios. Qualitative scenarios describe possible futures in the form of words rather than numbers. They can take the form of images, diagrams, phrases, or outlines, but more commonly they are made up of narrative texts, called storylines. Qualitative scenarios have the advantage of being able to represent the views of several different stakeholders and experts at the same time. Another advantage is that well-written storylines can be an understandable and interesting way of communicating information about the future, at least as compared to dry tables of numbers or confusing graphs. A drawback is that, by definition, they do not satisfy a need for numerical information. For example, numerical estimates are needed of the future extent and type of forest land in order to compute the flux of carbon dioxide between the biosphere and atmosphere.

It is common now to develop qualitative scenarios through a participatory approach, meaning a set of procedures through which experts and stakeholders work together to develop the scenarios. Experts are individuals with expertise relevant to the scenario exercise, and stakeholders are individuals or organizations with a special interest in the outcomes of the scenarios. Of course, it is not always easy to distinguish between experts and stakeholders. While there is a variety of different participatory approaches, they typically include a scenario panel made up of stakeholders and experts that develop the basic ideas of the qualitative scenarios at a series of intensive meetings. Between meetings, a secretariat prepares input to the scenarios and elaborates storylines. The SAS (story and simulation) procedure is a participatory approach used to develop both qualitative and quantitative scenarios (Alcamo 2001). Here, storylines are outlined and refined at scenario panel meetings, and between meetings, a secretariat works with modeling teams to quantify the scenarios (see Chap. 5). A key feature of this approach is that the qualitative and quantitative scenarios are developed hand-in-hand through a series of iterations.

6.2.2   Quantitative Scenarios

Quantitative scenarios are usually computed by formalized computer models and provide numerical information in the form of tables, graphs and maps. A disadvantage is that their exactness implies that we know more about the future than we actually do. Another disadvantage is that the models used to compute quantitative scenarios embed many assumptions about the future. These models tend to represent a limited point of view about how the world works (as compared to qualitative scenarios) and therefore provide a narrow view of the future. Furthermore, because not all processes of land-use change can be modeled, by definition, quantitative scenarios omit these processes. An additional drawback is that the basics of modeling are difficult for the non-specialist to understand.

There are also advantages of producing quantitative scenarios based on models. Model developers point out that their assumptions about the world are clearly written down in the form of model equations, inputs and coefficients. Although these are not easily understandable to non-experts, the assumptions are at least documented and usually more transparent than the undocumented and unspoken assumptions behind qualitative scenarios. Another advantage of quantitative scenarios based on models is that these models are often published in the scientific literature and have therefore received some degree of scientific scrutiny. The types of models used for computing future land use and cover are presented in Chap. 5 and some of the main techniques used in the models are presented in Box 6.1.

Since there are convincing arguments for using either qualitative or quantitative scenarios, a popular current approach is to use a combination of both. All of the global scenarios presented later, and some of the regional scenarios, are combined qualitative and quantitative scenarios.

Box 6.1.   Main approaches to modeling future land use and cover

Rule based models / cellular automata models. Models usually based on cellular automata (CA) or similar techniques, operating at various spatial-temporal scales. Note that the original CAs operate in a homogenous environment and the states of cells depend only on the states of their neighbors, while CAs used in land-use models operate in heterogeneous environments and can also take into account external driving forces such as changes in climate or product markets.

Empirical/statistical models. Both economists and natural scientists employ this category of models, although usually with quite distinct sets of explaining variables or drivers of land-use change. These models are typically based on regression techniques using linear or logistic assumptions. The models can be either static (using regression output as final product) or dynamic (using regression output as suitability maps in a dynamic allocation procedure).

Agent-based models. These models are usually based on an available agent-simulation library such as SWARM or COR-MAS. They are applied to a broad range of themes (deforestation, agriculture, urban growth) and often as part of a participatory scenario-building approach. These models are usually used to build local or regional scenarios in which agents represent people, households, or social/ethnic groups.

Macro-economic models. These models are built on general or partial equilibrium sets of macro-economic equations, in which land is not considered in a spatially explicit way, but is usually represented as a production factor. The heterogeneity of land is either ignored, or accounted for by different productivities or yield functions.

Land use accounting models. These models use a spread-sheet program to keep track of the assumptions of a scenario and their consequences on land use/cover. Linear relationships are sometimes used to compute future land use/cover as a function of changing driving forces.

6.3   Global and Continental Scenarios

6.3.1   Methodological Issues

Independent of their type, all scenarios require a coherent set of assumptions for the driving forces of future land use/cover. The driving forces typically used by scenario developers include demographic changes, economic growth and technological development (see Box 6.2). The preparation of these input data is a major undertaking because a large number of internally consistent driving forces must be specified ("internally consistent" is used here to mean driving forces that have consistent trends according to the knowledge of the scenario developer or the assumptions of the scenario). An example of the large effort needed to specify driving forces for global ecosystem scenarios is given by Nelson et al. (2005). A common strategy for maintaining the internal consistency of driving forces is to first develop storylines, as mentioned above, that provide a logic for the many different assumptions about future changes in population and other drivers. This approach is used in the Environmental Outlook Report (GEO) of UNEP (2004a) and the Special Report on Emissions (SRES) of the Intergovernmental Panel on Climate Change (IPCC 2000a).

While there are many different ways to model land changes, only two of these have been used to develop global scenarios because of data deficiencies, scaling mismatches, or long preparation and run time. The two approaches are land use accounting models (Kemp-Benedict et al. 2002) and rule-based/cellular automata models (Alcamo et al. 1998; Eickhout et al. 2005; IMAGE-Team 2001) (see Box 6.1 and Chap. 5).

Figures 6.1 through 6.3 show outcomes of selected global scenarios based on these modeling approaches. Included are scenarios from GEO (UNEP 2002, 2004a), SRES (IPCC 2000a), and the Global Scenarios Group (Gallopin et al. 1997; Gallopin and Raskin 2002; Raskin et al. 2002). We note that comparing scenarios produced with different methods and by different groups raises some methodological problems that should be kept in mind throughout this chapter. For example:

- The classification of land use/cover is not uniform.
- Different estimates of initial areal coverage for particular land-cover types are used.
- Different methods (qualitative or quantitative) are used for developing scenarios.

6.3.2 Global Scenario Results

Most global scenarios show very dynamic changes in agricultural land (see Fig. 6.1) caused by the trade-off between food supply and demand as moderated by international trade. Changes in demand for agricultural land are driven by changes in population, income, food preferences and commodity prices, while supply is driven by agricultural management, fertilizer input, soil degradation, and climate-related changes in the biophysical suitability of land for agricultural production.

Scenarios with a greater extent of agricultural land (see Fig. 6.1) result from assumptions about high population growth rates together with low but steady economic growth which combine to stimulate large increases in food demand. At the same time, assumed slower rates of technological progress lead to slow to negligible increases in crop yield. These combined effects lead to a sizable expansion (up to 40%) of agricultural land between 1995 and 2100 – see Fig. 6.1. The majority of scenarios show a growth in agricultural land during this period. The scenarios with a smaller extent of agricultural land have lower population assumptions leading to smaller food demands, while higher economic growth stimulates technological progress leading to rapid increases in crop yields. The sum of these effects is lower demand for agricultural land, with the lowest scenario showing a decline of more than 20% in the global area of agricultural land. Such large changes could have an important effect on the magnitude of greenhouse gas emissions, release of nutrients and other trace substances to aquatic ecosystems, and other large-scale impacts on the Earth System (see Chap. 4).

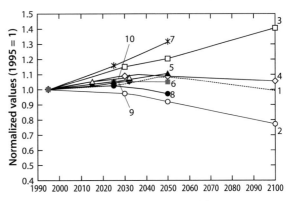

Fig. 6.1. Global scenarios of agricultural land from 1995 to 2100. *Sources:* Scenarios 1, 2, 3, 4: IPCC-SRES scenarios "A1", "A2", "B1", "B2" (IPCC 2000a,b) computed with IMAGE model (IMAGE-Team 2001). Scenarios 5, 6, 7, 8: Scenarios of Global Scenario Group "Market Forces", "Policy Reform", "Fortress World", "Great Transition" computed by PoleStar model (Kemp-Benedict et al. 2002). Scenarios 9, 10: "GEO-3" scenarios (UNEP 2004a) "Markets First", "Policy First" computed with PoleStar model. "Agricultural land" comprises the land-cover classes "Agricultural Land" and "Extensive Grassland" within the IPCC-SRES scenarios computed by the IMAGE model, and is the sum of "Cropland" and "Grazing Land" in the remaining scenarios

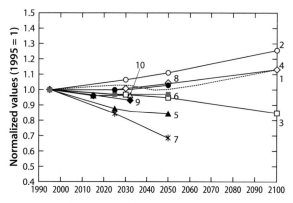

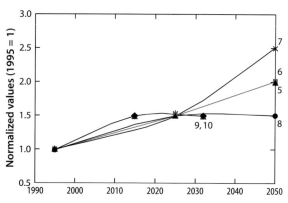

Fig. 6.2. Global scenarios of forest land from 1995 to 2100. The key to scenario numbers is the same as in Fig. 6.1. "Forest land" is defined as the sum of "Carbon Plantations", "Regrowth Forest", "Boreal Forest", "Cool Conifer Forest", "Temperate Mixed Forest", "Temperate Deciduous Forest", "Warm Mixed Forest", and "Tropical Forest" within the SRES scenarios computed by the IMAGE model. For the remaining scenarios forest land is the sum of "Natural Forest" and "Plantation"

Fig. 6.3. Global scenarios of urban land from 1995 to 2050. *Sources:* Scenarios 5, 6, 7, 8: Scenarios of Global Scenario Group "Market Forces", "Policy Reform", "Fortress World", "Great Transition" computed by PoleStar model (Kemp-Benedict et al. 2002). Scenarios 9, 10: "GEO-3" scenarios (UNEP 2004a) "Markets First", "Policy First" computed with PoleStar model

One of the key uncertainties in these scenarios is the question of how the world's population will be fed in the future, i.e., will food come from the intensification of agricultural land, that is, by boosting crop yields with increasing fertilizer, irrigation and other inputs, or from extensification, by expanding the area of cultivated land? How much food will be provided by imports, and conversely, how much agricultural production will be exported? The scenarios presented in Fig. 6.1 assumed various degrees of extensification, intensification and world food trade and their wide range reflects the uncertainties of these factors.

The global forest scenarios largely mirror the agricultural scenarios (see Fig. 6.2), and illustrate both the positive and negative aspects of existing scenarios. On one hand the forest scenarios are a valuable illustration of the connection between agricultural trends and the future tempo of global deforestation or afforestation. On the other hand, these scenarios imply that forest trends are driven almost exclusively by cropland expansion or contraction. They deal only superficially with driving forces such as global trade in forest products and the establishment of future forest plantations to sequester carbon from the atmosphere. Global scenarios in general need to incorporate many more of the actual driving forces of land-use/cover change and in a more realistic way (see Chap. 3).

There are very few published global scenarios of changes in urban area (see Fig. 6.3), and these give a limited view of urban developments. All show a steep increase over the next decade, with about half estimating a stabilization of urban areas by 2025. Stabilization, however, occurs only after urban areas are about 50% larger than their 1995 area. The remaining few scenarios show urban area still expanding at a linear or exponential rate in 2050. The set of scenarios in 2050 shows an increase from 1.5 to 2.5 over the extent of urban land in 1995. These estimates are based

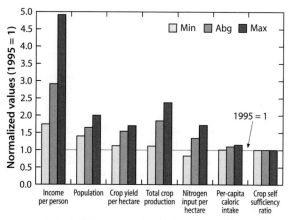

Fig. 6.4. Drivers of global scenarios of land use and cover from 1995 to 2050

on the multiplication of estimates of current urban space requirements per person (for different world regions) times the future trend in urban population (Kemp-Benedict et al. 2002). Hence, they do not account for changing spatial requirements of settlement areas.

Figure 6.4 presents the assumptions of some important drivers of the global scenarios. These are global averages of the values assumed for various world regions. The driver with largest relative increase is income, and this affects the change in agricultural area, particularly through increases in per capita food consumption. Income growth also influences the assumption for nitrogen fertilizer input and other variables in some scenarios. Assumptions about population growth affect the total crop production (per capita caloric uptake multiplied by population). Note that the assumed growth of population is modest compared to the growth of income. The increase in total crop production (assumed or computed across all scenarios) is partly satisfied on new agricultural land and partly by augmenting production on existing land (we return to this issue later). Crop yield increases from 10 to 70% between 1995 and 2050 depending on the scenario, primarily because of an increase of 20 to 70% in the amount of nitrogen fertilizer applied per hectare, and partly because of favorable changes in climate. The global average caloric intake does not significantly increase, although most scenarios assume a marked increase in food consumption in developing parts of the world.

We note that driving forces in the global and other scenarios described in this paper are almost always assumed to be external factors that drive land-use changes. In reality, not only is land-use change driven by external factors, but land-use change in turn feeds back to these external factors. For example, migrants escaping a threatening political or economic situation outside of a region could be major agents of changes within a particular region and could eventually cause a depletion of suitable agricultural land which in turn could dampen the migration rate into the region. Including feedbacks to driving forces is an important task for scenario developers and is further discussed in Sect. 6.6.

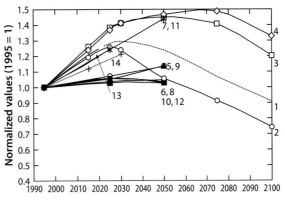

Fig. 6.5. Scenarios of agricultural land in Africa from 1995 to 2100. *Sources:* Scenarios 1, 2, 3, 4: IPCC-SRES scenarios "A1", "A2", "B1", "B2" (IPCC 2000a,b) computed with IMAGE model (IMAGE-Team 2001). Scenarios 5, 6, 7, 8: Scenarios of Global Scenario Group "Market Forces", "Policy Reform", "Fortress World", "Great Transition" computed by PoleStar model (Kemp-Benedict et al. 2002). Scenarios 9, 10, 11, 12: "GEO-3" scenarios (UNEP 2004a) "Markets First", "Policy First", "Security First", and "Sustainability First" computed with PoleStar model. Scenario 13 refers to the "Reference Scenario" of the OECD "Environmental Outlook" study computed by Pole-Star model (Kemp-Benedict et al. 2002). Scenario 14 addresses the "Reference Scenario" of the FAO "Agriculture towards 2015/30" study. "Agricultural land" is defined as in Fig. 6.1

6.3.3   African Scenario Results

The same tools and approaches used to develop global scenarios have been applied to continental-scale scenarios. To illustrate the differences between trends in developing and developed parts of the world, we review scenarios for Africa and Europe. By comparing these regions we also show the consequences of increasing food demand (Africa) and stabilizing food demand (Europe) on future land use/cover.

The scenarios we review for Africa come from the same references as the global scenarios with the addition of the FAO *Agriculture towards 2015/2030* study (FAO 2000b) and the OECD *Environmental Outlook* study (OECD 2001). To interpret these scenarios it is useful to examine results for different time periods. Focusing on trends from 1995 to 2025, almost all scenarios indicate a continuous expansion of agricultural land, with an intermediate estimate of 25% and a range from 0 to 45% – see Fig. 6.5. By comparison, the actual net expansion of agricultural land between 1980 and 1995 was only about 2%. The scenarios, however, take into account the additional agricultural land needed to satisfy both a growing population and a higher per capita food demand arising from accelerating economic growth rates. In addition, some scenarios include large areal demands for biofuel crops as a possible future strategy to reduce greenhouse gas emissions.

Between 2025 and 2050, the scenarios begin to take on more distinctive trends. The higher scenarios show an expansion of agricultural land from 1995 to 2050 of about 40 to 60%, reflecting the assumption of higher population growth (compared to other scenarios) and

slower diffusion of technology which hinders Africa from benefiting from advances in agricultural technology. The lower scenarios result from assuming lower population and a vigorous exchange of information, technology, and products across borders which leads to higher economic efficiency of agricultural production and higher crop yields. Comparing 2050 to 1995, there is a net increase in agricultural land in all but a few of the scenarios.

Expanding the time horizon to 2100 (see Fig. 6.5) reveals clearly defined turning points at which the trend in agricultural land changes its direction between 2010 and 2050. These turning points occur in several different scenarios and correspond to an eventual slowing of food demand and technological catch-up in Africa which accelerates improvements in crop yield. The net effect is a shift from expanding to contracting agricultural land. The fact that these turning points are apparent only after several decades illustrates the importance of considering the long term trend of land-use/cover change.

According to most scenarios, the expansion of agricultural land causes a continuing reduction in African forested land up to 2025 (see Fig. 6.6) which is likely to have ongoing consequences on biodiversity, water resources, climate and other aspects of Africa's environment. Although the scenarios indicate a continuation of deforestation, they also show a slowing of the rate of deforestation. As compared to a rate of 0.8% per year from 1980 to 1995 (FAO 1999, 2003), the scenarios show a rate of 0.2 to 0.7% per year between 1995 and 2025 (with tropical deforestation rates in the 1980s and 1990s estimated to be about 20 to 30% lower

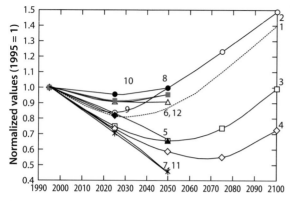

Fig. 6.6. Scenarios of forest land in Africa from 1995 to 2100. The key to scenario numbers is the same as in Fig. 6.5, except scenarios 13, and 14 which do not contain forest-land cover. "Forest land" is defined as in Fig. 6.2

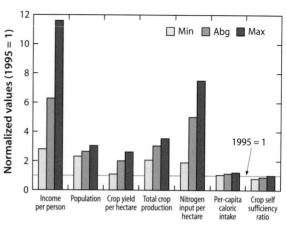

Fig. 6.7. Drivers of scenarios of land use and cover in Africa from 1995 to 2050

than these estimates; see Sect. 2.3.1). However, the scenarios may in general underestimate deforestation because they do not include a comprehensive description of the many causes of changing forest land (see Chap. 3).

After 2025, the slowing and eventual reversal of agricultural expansion also results in a further slowing and reversal of deforestation (see Fig. 6.6). Some scenarios even show a significant expansion of forested area by 2100 relative to 1995. This raises interesting questions, e.g., if the pressure of expanding cropland is alleviated, can deforestation be reversed within this time frame? (see Box 6.3). In particular, is it ecologically feasible for tropical forest ecosystems to re-establish themselves within a few decades as in these scenarios? And, what are the consequences of this reversal on terrestrial biodiversity, the global water cycle and other aspects of the Earth System? By stimulating such questions, scenario analysis provides a useful input to the research agenda of Earth Systems science.

The assumptions for the drivers of the African land scenarios are depicted in Fig. 6.7. As in the global case, income grows much faster than population. Average income growth is about a factor of 6 between 1995 and 2050. Yet, this very large growth in income does not translate into a similarly large increase in caloric intake (10 to 30% during the same period, depending on the scenario). Apparently, the scenarios assume that it is the quality rather than quantity of food that is lacking in Africa. While the average scenario assumes a population increase of a factor of 2.6, total crop production increases by a factor of 3, so food production is assumed to more than keep up with the population. Only for the lowest scenarios does the increase in population exceed the increase in crop production. In these cases, an increase in imported food partly compensates for the production gap.

Crop yield grows by an average factor of 2, stimulated by the factor of 4 increase of nitrogen fertilizer input per hectare. Increasing yields make it possible to gain part of the new crop production on existing agricultural land.

Box 6.3. Is a quick reversal of deforestation feasible?

The African scenarios indicate that a slowing and reversal of agricultural land expansion could halt deforestation and lead to re-establishment of the tropical forest within a few decades. Is this realistic? In principle, the answer is, yes, with respect to both biomass accumulation and spatial coverage (e.g., Achard et al. 2002, 2004; IPCC 2000b; Otsamo et al. 1997; Rudel et al. 2005; Silver et al. 2000). In terms of plant biomass and soil carbon, a forest may require longer to recover, from a few decades to a century (Silver et al. 2000). The rate of re- or afforestation at a given site depends on climatic conditions, soil fertility, seed dispersal and in case of managed forests and plantations also management options. Silver et al. (2000) also found that on average tree biomass accumulated fastest on abandoned agricultural land as compared to other types of abandoned land. On the other hand, agricultural land is often abandoned because of soil degradation associated with decreased productivity. In this case Zanne and Chapman (2001) found that the renewal of biomass will take longer than on abandoned agricultural land with soils in good condition. Under any circumstances the restoration of tree biodiversity and forest structure may need a much longer period of time, while other types of biota (insects, herbaceous plants, fungi) may require shorter or longer periods of time to recover, or may not be able to recover at all (as in the case of large mammals requiring large undisturbed habitats) (see Sect. 4.5).

Regarding the rate of deforestation as compared to afforestation, several of the scenarios for Africa imply that the tempo of these two processes are of the same order of magnitude. By comparison, Rudel et al. (2005) found that observed tropical deforestation is on the average twice as rapid as re- and afforestation, based on a relatively small number of studies of individual countries.

To sum up, some but not all aspects of a tropical forest may be fairly rapidly re-established after the pressures of deforestation are released.

The value of the food self-sufficiency ratio (production divided by production plus imports minus exports) is currently approximately 0.9, indicating that Africa is a net importer of food. As shown in Fig. 6.7, this ratio will decrease about 10% between 1995 and 2050 across all scenarios, indicating a deepening dependence of Africa on food imports.

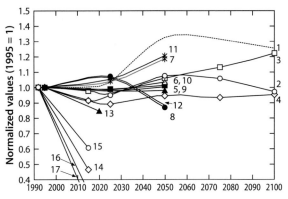

Fig. 6.8. Scenarios of agricultural land in Europe from 1995 to 2100. *Sources:* Scenarios 1, 2, 3, 4: IPCC-SRES scenarios "A1", "A2", "B1", "B2" (IPCC 2000a,b) computed with IMAGE model (IMAGE-Team 2001). Scenarios 5, 6, 7, 8: Scenarios of Global Scenario Group "Market Forces", "Policy Reform", "Fortress World", "Great Transition" computed by PoleStar model (Kemp-Benedict et al. 2002). Scenarios 9, 10, 11, 12: "GEO-3" scenarios (UNEP 2004a) "Markets First", "Policy First", "Security First", and "Sustainability First" computed with PoleStar model. Scenario 13 addresses the OECD Environmental Outlook "Reference Scenario" computed by PoleStar model (Kemp-Benedict et al. 2002). Scenarios 14, 15, 16, 17: WRR scenarios "Nature and Landscape", "Regional Development", "Free Markets and Free Trade", and "Environmental Protection"

6.3.4 European Scenario Results

The European scenarios we review here are the same as the global scenarios with the addition of the following studies: *Ground for Choices* (WRR 1992), the OECD *Environmental Outlook* (OECD 2001), and the EURURALIS study (Klijn et al. 2005). The available set of scenarios of Europe's agricultural land give a wide range of views (see Fig. 6.8). The lower boundary is set by the *Ground for Choices* study (WRR 1992) which estimated the impact of steadily decreasing agricultural subsidies up to 2015 and used an optimization approach for agricultural production and labor costs. As a result, these scenarios show a 35 to 80% shrinkage in agricultural land relative to 1995. A more typical result is given by the IPCC-SRES scenarios as applied in the EURURALIS Project (see Box 6.4) which indicate a decrease of around 3 to 6% between 1995 and 2030 in the 25 countries of the European Union.

At the opposite extreme, the highest IPCC-SRES scenario suggests that expanding the export of agricultural commodities from Europe could result in a 35% expansion of agricultural land (relative to 1995). The scenarios in-between do not show large changes up to 2025. Afterwards, however, they exhibit a wide range of different trends and views about the future. The fact that most scenarios begin to diverge only after 2025 is another illustration of the importance of incorporating a longer time horizon for studies of future land-use and cover. Some agricultural scenarios show a change in direction but this occurs later than in the African scenarios.

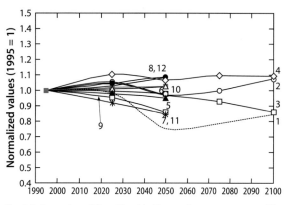

Fig. 6.9. Scenarios of forest land in Europe from 1995 to 2100. The key to scenario numbers is the same as in Fig. 6.8, except the scenarios 13 to 17 which do not contain forest-land cover. "Forest land" is defined as in Fig. 6.2

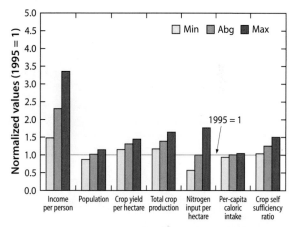

Fig. 6.10. Drivers of scenarios of land use and cover in Europe from 1995 to 2050

Similar to the agricultural scenarios, the forest scenarios do not show large changes up to 2025, but sharply diverge afterwards (see Fig. 6.9). Some long-term scenarios show a reversal in the trend of decreasing forest area at mid-century in response to declining agricultural land area. The rate of reforestation is slower here than in the African forest scenarios (see Fig. 6.6), and may be feasible because of the heavy management of Europe's forests.

Estimates of future forest coverage in most studies are computed in the same way as in the global and African scenarios in that changes in forest area only mirror changes in agricultural area. Most forest scenarios neglect the factors that determine the extent of forest area in Europe such as policies for nature protection and landscape preservation, forest-management practices, and trade in wood products. (An exception are the EURURALIS scenarios shown in Box 6.4 which examine European land-use policies in detail and computed ongoing abandonment of agricultural land and an increase in "natural land" which is likely to include new forest areas). Another

Box 6.4. European scenarios (2000–2030) from the EURURALIS project

Peter H. Verburg · Dept. of Environmental Sciences, Wageningen University, Wageningen, The Netherlands

EURURALIS was sponsored by the Netherlands as part of its chairmanship of the European Union in 2004 with the aim to analyze potential land-use/cover change in Europe (Klijn et al. 2005). Four scenarios were evaluated based on the IPCC SRES global storylines. A number of models were used to translate the scenarios into high resolution assessments of changes for the 25 countries of the European Union. Global economic and integrated assessment models (GTAP and IMAGE) were used to calculate changes in demand for agricultural areas at the national level, while a spatially explicit land-use model (CLUE-S) was used to translate these demands into land-use patterns (van Meijl et al. 2006).

Table 6.1 shows the area of the 25 member states of the European Union (EU-25) facing urbanization, agricultural land abandonment, and/or new "natural land". The maps in Fig. 6.11 illustrate how the incorporation of spatial policies results in very different land-use patterns (1×1 km²) for southern France. In the B2 scenario (Regional Communities), the "Less Favored Areas" (shaded areas in 2000 map which indicate areas of low productiv-

ity; see Sect. 6.4.4) are maintained leading to incentives for continuation of arable agriculture, thus slowing land abandonment in these areas. In the B1 scenario (Global Cooperation), the Less Favored Areas are only incentives for managed grasslands, which leads to an almost complete disappearance of agriculture in these areas. Thus, patterns of land-use change are very different, although the overall percentage of change is similar.

Table 6.1. Change in land use between 2000 and 2030 (as percentage of total land area of EU-25)

	A1	A2	B1	B2
Urban land	2.4	1.4	1.3	0.4
Agricultural land abandoned	6.4	2.5	6.3	5.2
"Natural land"	2.1	0.6	4.6	3.2

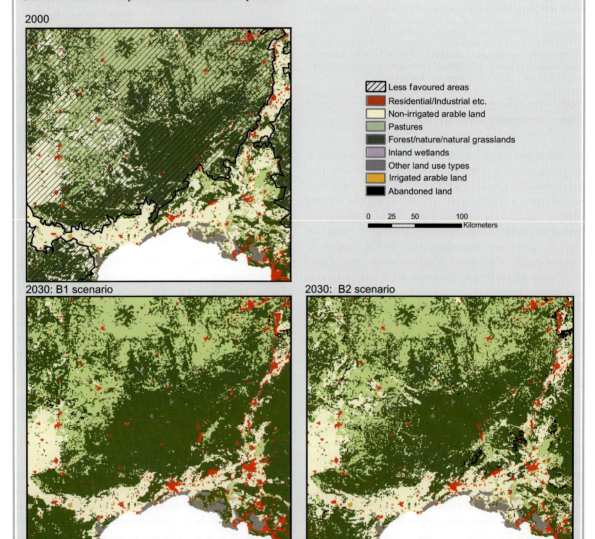

2000

Less favoured areas
Residential/Industrial etc.
Non-irrigated arable land
Pastures
Forest/nature/natural grasslands
Inland wetlands
Other land use types
Irrigated arable land
Abandoned land

0   25   50   100
Kilometers

2030: B1 scenario   2030: B2 scenario

Fig. 6.11. Different land-use patterns (1×1 km²) for southern France as a result of the incorporation of spatial policies

deficit is that forest scenarios of Europe and other regions usually do not distinguish between primary and secondary forests which have dissimilar roles in the regulation of the water cycle, the support of species, and other global change relevant processes.

The assumed rate of change of driving forces in Europe (see Fig. 6.10) are more moderate than for Africa (see Fig. 6.7). This applies in general to developed *versus* developing regions in existing scenarios and reflects the thinking that Europe and other industrialized parts of the world will materially develop much less in the coming decades than Africa and other developing regions. Perhaps, this is a too narrow view of the future, since it is imaginable that various social, economic or political events could narrow or widen the gap in growth between developed and developing countries.

Population growth assumptions range from a small decrease to a small increase, while income growth ranges from a factor of 1.5 to 3.3 from 1995 to 2050 (for the various scenarios). In the case of Europe (as other industrialized world regions) the increase in income does not translate into an increase in caloric intake since this is already at its saturation level. Crop yields modestly increase because of improved agricultural management, and because of increased fertilizer input in some scenarios. The average scenario assumes that nitrogen fertilizer input remains constant, while the lowest assumes a decrease of 30% and the highest an increase of 50% between 1995 and 2050. Europe is currently a net food import area (self-sufficiency ratio = 0.95) but the ratio will increase according to the scenarios by an average factor of 1.2 between 1995 and 2050, thus making Europe a net exporter of food products.

6.4 Regional and Local Scenarios

6.4.1 Methodological Issues

The variety and number of regional and local land-use scenarios is much larger than global scenarios. This variety is caused primarily by the much wider range of place-specific questions that are being addressed and place-specific factors determining land use and cover. Other causes are methodological problems mentioned earlier and varying availability of reliable data.

On one hand, regional studies of future land use have objectives similar to that of global studies in that they also offer insight into the consequences of current actions and uncertainties of the future and thus support more informed and rational decision-making. On the other hand, while global studies tend to focus on producing scenarios, regional studies often concentrate on developing tools for direct decision support because in principle land-use change can be steered by local stakeholders (Peterson et al. 2003).

Regional scenarios also differ from global scenarios with respect to the basic questions they address. Whereas global scenarios tend to ask how much land-use change will take place, regional scenarios tend to address where it will take place. Although Lambin et al. (2000) suggest that the magnitude of change might be more informative than its location, most regional scenario studies have in practice focused on the location of change and have employed spatially explicit models to map this change. A typical procedure is to, first, develop storylines that specify the trends of socio-economic, environmental and institutional variables determining land use, as well as the resulting direction or even order of magnitude of land-use change. Quantitative models are, then, used to allocate where the land-use change will take place, consistent with the trends specified in the storyline.

The typical drivers included in regional and local scenarios are similar to those used in global scenarios but, of course, are described in much greater detail. In comparison to global scenarios, regional and local storylines often include governance issues, technology, and changes in the social system. These translate into similar quantitative drivers, although data on social issues are often limited and economic drivers (income, trade, subsidies, prices) dominate. The location of change is determined by a range of factors, including biophysical (for example topography, soil, and/or precipitation), demographic (population, accessibility), and socio-economic (land tenure, education level). The determining mix of factors depends on local characteristics. In Brazil, for example, the distance of development to road is very often the most important factor, boosted by the launch of the *Avança Brasil* which involves very high investments for road paving (e.g., Alves 2001b; Laurance et al. 2001) (see Fig. 7.3). By comparison, European scenarios would not be complete without including the effects of the Common Agricultural Policy (CAP), while many studies single out soil characteristics as the main determinant of land use (e.g., Bakker et al. 2005).

Although the diversity of drivers is high, population is the single most frequently mentioned driving force, both in determining quantity and location of change (e.g., Kok 2004). Published land-use scenarios, however, still tend to simplify the impacts of population because of lack of data, despite a strong plea that population will hardly ever be the key single driver (see Chap. 3). Recently, more complex measurements of accessibility (Verburg et al. 2004d), income and education level are being included in land-use models.

In the following paragraphs, we review a small selection of the many regional and local scenarios that have been developed. To minimize the problems of interpreting scenarios based on different methodologies, we review only the subset of scenarios which fulfill one or more of the following conditions: *(a)* they are embedded in regional and/or global developments (e.g., scenarios produced by the Millennium Ecosystem Assessment or EURURALIS); *(b)* they were developed using a single framework/methodology applied at different locations (e.g., scenarios based on the CLUE, SLEUTH, or Environment Explorer models); *(c)* they have employed a proven methodology such as the cellular automata approach; and/or *(d)* they are considered "archetypal" scenarios for a particular location.

6.4.2   Results from Regional and Local Scenarios

While most global/continental scenarios have a long perspective (usually up to 2050, some up to 2100), most regional/local scenarios are short term (usually up to 2015, some up to 2025). However, there are exceptions as we will see later. Short-term scenarios tend to be extrapolations of current trends, while long-term scenarios are usually derived from a top-down, multi-scale methodology and incorporate non-linear system changes and feedbacks. We begin with a review of short term regional scenarios.

The picture that emerges from many short term studies is not encouraging from the perspective of environmental change. In Latin America, the vast majority of scenarios indicate that deforestation will continue unabated, although there are exceptions (e.g., Fearnside 2003). Examples of regional deforestation scenarios are given in Box 6.5. Growing populations, expanding economies and increasing urbanization characterize the situation in Southeast Asia (Roetter et al. 2005). The few available regional scenarios for Africa (e.g., Thornton et al. 2003) suggest that further increases in population and income will change dietary preferences and boost food demand. Since increasing food demand cannot be easily covered by boosting crop productivity and imports,

agricultural land will greatly expand. This is consistent with the results of most continental-scale African scenarios (see Fig. 6.5) which indicate a strong expansion of agricultural land over the coming few decades. However, as noted above, the continental scenarios show a slowing of this expansion and its eventual reversal over a longer time period.

In North America, the focus of land research has traditionally been on monitoring current land-use/cover change and describing historical changes, thus gaining understanding of the current patterns of land use and important (historical) drivers of change. Recently, however, the emphasis has shifted to scenario development. Examples are the work of spatial economists (e.g., Irwin and Bockstael 2002); the use of agent-based models in the SLUCE project (Spatial Land Use Change and Ecological Effects at the Rural-Urban Interface; see Brown et al. 2004); and the applications of the urban growth model SLEUTH (Clarke and Gaydos 1998). Land-use research is coordinated in a number of research programs, notably NASA's Land Cover Land Use Change Program (Gutman et al. 2004); the Human-Environment Regional Observatories (HERO); and the U.S. Global Change Research Program Element, Land-Use/Land-Cover Change (U.S. CCSP/SGCR 2003) with a particular emphasis on the future impact of climate change on crop productivity. It is to be expected that the number of land scenarios will increase rapidly in the near future.

Short term scenarios of European regions have analyzed the impact of the recent expansion of the European Union from 15 to 25 countries (e.g., Kohler 2004) and of the Common Agricultural Policy of the European Union (Topp and Mitchell 2003; ACCELERATES 2004). These scenarios indicate a continuation of urban-

---

Box 6.5. Scenarios of deforestation in Latin America (2000–2010)

The quantitative scenarios of deforestation in Latin America depicted below were derived through a multi-step procedure. First, qualitative storylines for Latin America were written based on information and requests from experts and decision makers ("Business as Usual", "Market Liberalization", "Sustainability"). The storylines were then quantified using FAOSTAT data. Finally, these data were input to the CLUE model (Verburg et al. 1999) which produced quantitative estimates of deforestation (Kok and Veldkamp 2000; Kok and Winograd 2002).

Figure 6.12 shows that deforestation rates remain high between 2000 and 2010. Although national level rates are lower in Central America than in the Brazilian Amazon, local rates (e.g., the Atlantic Coast of Costa Rica) are as high. The "Sustainable" scenario was formulated at the request of national policy makers and is a normative scenario. Despite the strong interest in a scenario with a reversal of deforestation, the quantification of this scenario indicated that deforestation is likely to continue in the short run in Costa Rica and Panama. During quantification it was assumed that sustainability measures (e.g., institutionalization of national parks, and changes in dietary patterns) only occur when the economy grows fast and human well-being is increased. But higher income and well-being also stimulate a higher demand for beef which leads to an expansion of grazing land, and hence to continuing deforestation. Moreover, the sustain-

ability scenario was not considered feasible by experts and decision makers involved in the scenario studies because it assumed that current trends of land-use policies, dietary patterns, and crop yield could be reversed within the next decade.

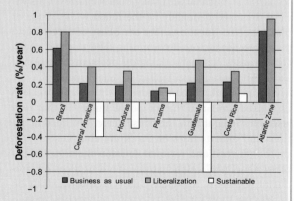

Fig. 6.12. Deforestation rate scenarios in Latin America under three scenarios

ization and land abandonment, together with further land and water-quality degradation.

One set of long-term studies of local land-use changes has focused on potential changes in agricultural areas up to 2100. For example, as a result of climate change the corn and wheat belts in North America may shift northward, reducing U.S. production of these crops and increasing their production in Canada (IPCC 1997). These studies analyze potential impacts on land use, but do not provide an integrated view of land-use changes incorporating socio-economic developments.

Other long-term studies focus on downscaling and applying global scenarios to the regional and local scale. Many of these studies have downscaled the IPCC SRES scenarios (IPCC 2000a). These include the work of the ATEAM project (Rounsevell et al. 2005, see Box 6.6 and Fig. 6.13) and the EURURALIS project mentioned earlier (Klijn et al. 2005; see Box 6.4). Other examples are the application of SLEUTH in the U.S. (Solecki and Oliveri 2004); land-use scenarios for the Netherlands (Kuhlman et al. 2006; De Nijs et al. 2004); and a local landscape study in Norfolk, England (Dockerty et al. 2005).

An important characteristic of regional and local scenarios is that they sometimes show solutions to global change problems that are overlooked by the coarse resolution of global scenarios. For example, local policies may effectively slow down deforestation in Brazil (Fearnside 2003), and crop-farming can be replaced by fish-farming in flooded areas in the Netherlands (White et al. 2004). Such local solutions could have a global impact if they can be propagated throughout the world.

6.4.3   Results from Urban Scenarios

The analysis of spatial developments in urban areas has proceeded separately from the regional and local studies mentioned above, and merits a separate discussion. The most common approach used for producing urban scenarios is cellular automata modeling because of its flexibility in handling rules that determine changes in urban areas. Other approaches include the land transformation model of Pijanowski et al. (2002a) and the agent-based model of Brown et al. (2004).

Up to now, urban scenarios have concentrated on future expansion of urban land, an important issue in both developed and developing countries. Over the last decades, urban populations in developed countries have been moving from dense, compact urban centers to new low-density urban areas on the outskirts of present cities. Meanwhile, a combination of high population growth and lack of (urban) planning has led to a large expansion of urban land in many developing countries. One of the main messages of urban scenarios is that urban land will continue to expand at many different locations. Some scenario studies (e.g., Pijanowski et al. 2002b) also suggest that the expansion of urban area may lead to a greater-than-proportional loss in fertile farmland (new urban areas not only occupy the best agricultural lands but also attract industry and infrastructure that claim an additional share of former rural land). These changes are of particular importance since they are usually irreversible over a long time period.

Scenario analysis has also shown that urban sprawl, and its opposite "compact growth", could lead to many different plausible spatial patterns of urban growth. The recent EURURALIS project (Klijn et al. 2005) considered different variants of sprawl- and compact-type growth in European cities (see Table 6.2) and found that factors such as local city planning policies have an important effect on the particular spatial pattern resulting from sprawl or compact growth. The EURURALIS scenarios also indicated that urbanization rates are likely to remain high until 2030 under the downscaled assumptions of the four IPCC-SRES scenarios (IPCC 2000a) (see Table 6.2). Solecki and Oliveri (2004) reached similar conclusions for the New York Metropolitan Region by downscaling two of the same four IPCC-SRES scenarios.

6.4.4   Results from Multi-Scale Scenarios

The close connection between future land use on the global and regional scales argues for the development of integrated global-regional land-use scenarios. The Millennium Ecosystem Assessment (MA) took first steps in this direction by constructing parallel global and regional land-use scenarios as part of their multi-scale assessment of ecosystem services (Millennium Ecosystem Assessment

Table 6.2.
Assumptions for characteristics of urban growth in the EU-25 between 2000 and 2030 from EURURALIS Project

	A1	A2	B1	B2
Type of urban growth	Sprawled	Sprawled	Compact	Compact
Large cities	No restrictions	No restrictions	Designated areas only	Designated areas only
Provincial towns	No incentives or restrictions	No incentives or restrictions	Designated areas only	Designated areas only
Small villages	Proliferation of second houses	Decrease in land abandonment regions	Designated areas only	Maintain size and structure

Scenarios are downscaled urban versions of the IPCC SRES (IPCC 2000a) storylines; see Box 6.4 for explanations of B1 and B2.

Box 6.6. Downscaling the ATEAM scenarios of land-use change – Bioenergy crops in the British Isles

Nicolas Dendoncker · Catholic University of Louvain, Louvain-la-Neuve, Belgium

The four ATEAM scenarios of land-use change (see Fig. 6.13) (Rounsevell et al. 2005), initially obtained at a resolution of 10 min, were further downscaled to a spatial resolution of 250 m following the methodology proposed by Dendoncker et al. (2006). The 250-m grids allow the representation of one land-use type per grid cell. While downscaling faces a number of methodological issues, the resulting data sets may serve as useful inputs to subsequent applications of the scenarios. Downscaling also allows for better visualization of the land-use patterns, which is not visible at the 10' resolution, when land use shares (in %) are represented.

Figure 6.13 shows a strong reduction of cropland areas in the four ATEAM scenarios. Grassland areas also have a tendency to decline in all scenarios. Generally speaking, it is projected that there will be a replacement of agricultural land used for food production by areas devoted to the production of bioenergy. In the British Isles (see Fig. 6.14), this is especially striking in scenario B2, which projects that large areas will be planted with bioenergy crops in 2050. Bioenergy crops can be as diverse as willow plantations, sugar beet or oilseed rape and are often presented as an important alternative source of energy in the context of climate change (see Sect. 4.2.4).

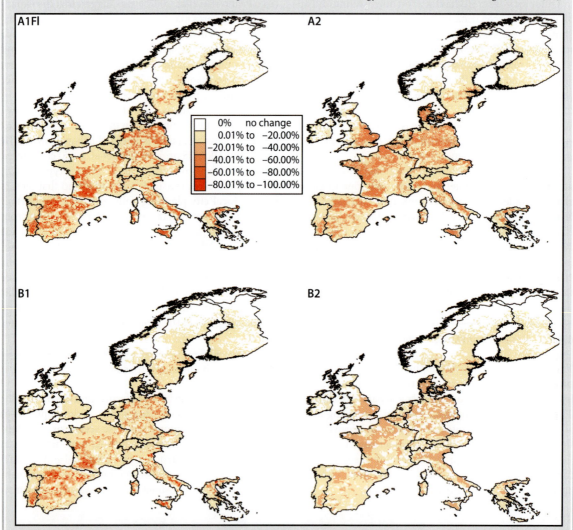

Fig. 6.13. Change (difference in % of each cell) in cropland area for food production by 2080 compared with the baseline for four storylines (*A1Fl, A2, B1, B2*) with climate calculated by HadCM3 – ATEAM project; see Box 6.4 for explanations

2003, 2005). The MA effort provides experience on how to set up a multi-scale scenario exercise. Figure 6.15 shows two different multi-scale organizational structures used in the MA, a fully hierarchically nested design (southern Africa) and a partly nested design (Portugal). Two parallel scenario exercises were conducted. On the global level, a

global scenario team developed four scenarios, which can be described by two axes of uncertainty (global *versus* regional development, and proactive *versus* reactive actions relative to environmental degradation). To drive the scenarios, a set of global driving forces with country-scale resolution was selected. On the regional level, different regional

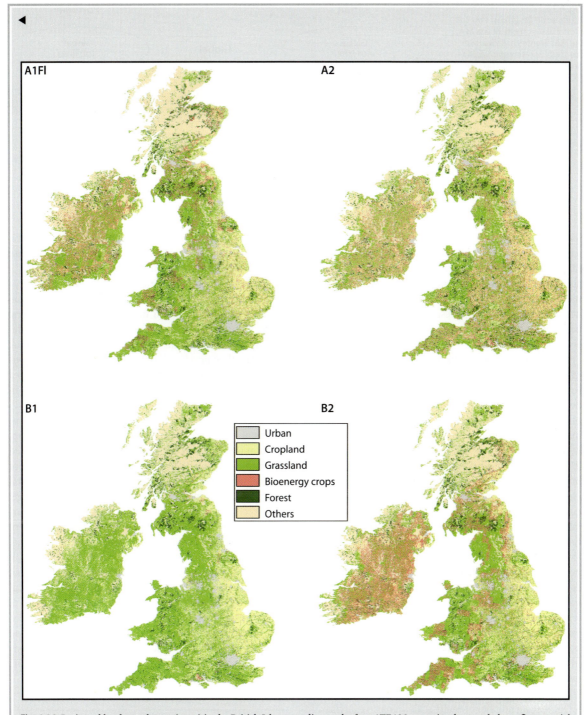

Fig. 6.14. Projected land-use change (2050) in the British Isles according to the four ATEAM scenarios downscaled to a finer spatial resolution showing the importance of bioenergy crops

scenario teams developed regional scenarios using the driving forces from the global scenario exercise as one of many inputs to their scenarios. While the global scenario exercise provided input to the regional scenarios, the regional scenarios were completed too late to provide feedback to the global scenarios.

Experience from the Portugal scenario exercise illustrates the difficulty in harmonizing regional and global scenarios. The global scenario "Global Orchestration" reflects a world of economic optimism in which farming areas are mostly located where production is highest and most efficient. When translated to Portugal by the regional scenario

Fig. 6.15.
Multi-scale designs of two sub-global assessments of the MA. *SafMA:* Southern Africa Millennium Assessment; *SADC:* Southern Africa Development Community

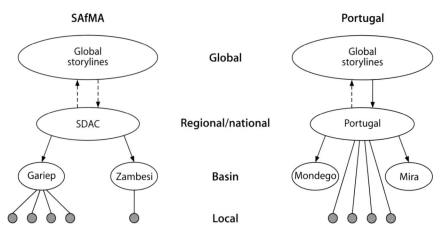

team, this scenario described a future in which regional agriculture is abandoned and replaced by oak forests, rural population migrates to cities and the expansion of uncultivated land leads to greater biodiversity. While international stakeholders consider "Global Orchestration" as a desirable scenario, Portuguese policy-makers had the opposite view because of the loss of rural employment and economic activity.

The Visions project (Rotmans et al. 2000) is another example of multi-scale scenarios, this time at the pan-European and local scales. Scenarios were first developed independently at the two scales and then mapped onto each other. Local scenarios tended to be generally positive and include local solutions to future challenges because of the multi-scale design (which encourages broad global and local thinking) and because of the involvement of stakeholders (who were interested in local solutions). In the Green Heart region in the Netherlands, for instance, agricultural entrepreneurs exploit more frequent extreme rainfall events and flooding by shifting their future focus to fish farming (White et al. 2004). In a subsequent project (MedAction; De Groot and Rotmans 2004) the three European scenarios were translated to fit land-use issues (Kok et al. 2003) and were downscaled to the Mediterranean region (Kok and Rothman 2003). Again, local scenarios tended to be a mix of higher-level changes and local innovative solutions. In the Guadalentín in Spain, water-transport networks are projected to sustain agriculture, while in the Agri Valley in southern Italy ecotourism is integrated with small-scale agriculture (Kok and Patel 2003).

The MA and Visions scenario exercises are just two of an increasing number of multi-scale scenario exercises. As mentioned earlier, many groups are downscaling global scenarios from the Intergovernmental Panel on Climate Change (IPCC 2000a), the Millennium Ecosystem Assessment (2003) and the Global Environmental Outlook of UNEP (2004a). One point of view is that downscaling a limited set of global scenarios is better than a "bottom-up" approach in which stakeholders help to develop local scenarios, in that downscaling provides a common, consistent framework for scenarios at many different locales and

regions (e.g., time horizon, time steps, categories of driving forces, definitions of land-use terms). Thus, it makes the scenarios from these places more comparable.

Another point of view is that global downscaling limits the creativity and diversity of regional scenarios. An example of this can be found in a number of downscaling efforts in Europe. The "Less Favored Areas" (LFA) are defined as agricultural areas that are economically marginal. Therefore, they provide a useful spatial indicator of non-optimal production areas (Rounsevell et al. 2005). This idea was implemented in a similar fashion in several studies – in the ATEAM studies (Rounsevell et al. 2005), in EURURALIS (see Box 6.4), in applications of the Land Use Scanner (Kuhlman et al. 2005) and in applications of the Environment Explorer (De Nijs et al. 2004). All these studies downscaled continental or global scenarios and used the LFA concept as a means to make the effect of the Common Agricultural Policy spatially explicit. Because spatial policies strongly and directly affect land-use patterns, these similarities carried over in the resulting land-use maps. The influence of the continental or global scenarios might be overly strong, thus weakening the local and regional signals. Based on the authors' experience, regional scenario exercises that emphasize stakeholder participation tend to stress local and regional factors and produce more diverse results.

To sum up, the multi-scale approach seems to be a promising method to standardize and harmonize local, regional and global studies, but it has only recently been given adequate attention. Many more studies are needed before any final conclusion on its usefulness can be drawn.

6.5   Main Findings of Scenarios

Although the scientific community is only beginning to study the future of land (Brouwer and McCarl 2006), the existing set of scenarios offers interesting insights to researchers. These scenarios range from the global/continental to regional/local and take the form of qualitative storylines and/or quantitative model output. The set of existing scenarios cover a wide variety of possible driving forces up to 2100.

They present "not implausible" futures of land use without making assertions about the probabilities of these futures.

There are some notable differences between global and regional scenarios. The published global scenarios have been based on only two modeling approaches, i.e., accounting and rule-based/cellular automata models, while the regional scenarios have used a wider variety of approaches. The global scenarios tend to be more expert driven, and cover a smaller set of potential futures than the regional scenarios. Global scenarios tend to be long term, while regional scenarios tend to be short term. Most of the global scenarios derived up to now mostly follow a few archetypical ideas of coming developments such as the continuation of current globalization trends or the reversal of globalization and collapse of international cooperation. Regional scenarios, because of their focus on smaller and more specific localities or regions, have tended to be more stakeholder driven. For these reasons, they also encompass a larger variety of views of the future, including the potential influence of local policy and institutions. However, it is usually difficult for developers of regional scenarios to set the physical/political boundaries of their scenarios, whereas developers of global scenarios do not have this problem. Global scenarios, by nature, focus on international, large-scale solutions to undesirable global change, while regional scenarios illustrate local solutions that may be overlooked by the coarse resolution of global scenarios.

Taken together, current land scenarios support the idea that fine, local spatial patterns of land-use change tend to be determined by local factors (e.g., city planning policies, local recreational preferences or topography), while the overriding forces for change come from outside drivers (e.g., world food trade, or society-wide changes in food preferences). This perspective is implicit in many scenarios and has an important influence on their results. The validity of these assumptions should be checked with empirical data (see Chap. 3).

The diversity of regional and local land-use scenarios makes it difficult to summarize their main findings. However, in their diversity may lie their strength in that regional and local scenarios provide a rich variety of different bottom-up views of the future. Nevertheless, constraining the range of regional and local scenarios by downscaling them from global scenarios has the advantage of making local land-use scenarios more consistent and comparable. The relative benefits and costs of these two approaches must be further discussed. It may even be possible to link global and regional scenarios in a way so that both gain from the other (see Sect. 6.6).

6.5.1   Changes in Extent of Urban Land

Scenarios have been developed for both the sum of global/continental changes in urban area, as well as for changes in the area of individual cities. The published scenarios of both types indicate a continuing increase in urban area over the decade 2000–2010, but some scenarios show a stabilization of global urban area by 2025. We remind the reader that scenarios are "if-then" propositions of what could occur given certain assumptions, and that different population, economic, and other assumptions could lead to scenarios of decreasing urban area. Nevertheless, for the range of assumptions adopted in the literature, urban area shows a global increase over at least the coming decade.

Regional and local scenarios also show that urbanization could lead to many different fine-scale patterns of land use in metropolitan areas. Some scenarios also show that fertile agricultural land could disappear at a faster rate than the expansion of urban area because of the additional infrastructure and other land requirements of the urban population.

6.5.2   Changes in Extent of Agricultural Land

The focus of most scenarios is on changing agricultural land, probably because agriculture is so important in terms of spatial outreach, ecosystem impact and for its political economy. Many scenarios emphasize the link between deforestation and agricultural land. The great majority of both regional and global scenarios indicate an expansion of agricultural land over the next decade, with the biggest changes occurring in the tropics. But many global scenarios also show turning points at which the trend in agricultural land changes its direction some time between 2010 and 2050. Many African scenarios point to an eventual slowing of population growth and technological catch-up which accelerates improvements in crop yields. The net effect is a shift from expanding to contracting agricultural land. If realized, this reversal in trends could relieve some of the pressure on existing unmanaged natural land and have positive consequences for biodiversity.

Although turning points are not implausible, up to now they have only been generated as a consequence of the input assumptions of scenarios and hence require empirical validation. Indeed, both scenarios and models require more rigorous descriptions of the future impacts of increasing food demand and depletion of suitable agricultural land. Another key uncertainty has to do with the way in which future food demand will be satisfied, i.e., will it be by expanding agricultural land, by intensification of existing land, or by world food trade? Much more research work is needed on this issue so that agricultural scenarios can capture a fuller range of possible futures.

6.5.3   Changes in Extent of Forest Land

The majority of regional scenarios indicate a continued rapid deforestation in many parts of Africa and Latin America over the next decade. Most global scenarios also show this short-term trend, but in addition suggest an eventual slowing of deforestation after a few decades as a result of the slowing of agricultural land expansion. This has important implications for carbon dioxide fluxes and other global change processes. Some scenarios for Africa even show a relatively rapid reversal of deforestation which raises the interesting question, whether it is ecologically feasible for tropical forest ecosystems to re-establish themselves within a few decades suggested by these scenarios?

Large-scale forest scenarios tend to mirror agricultural scenarios in that forest-land coverage is determined mostly (in the scenarios) by the expansion or contraction of agricultural land. This, of course, is an exaggerated simplification of reality, and future scenarios must take into account other factors that influence forest land such as conventional management practices (e.g., wood extraction), unconventional management practices (e.g., plantations for carbon sequestration), and protected areas of forests. Moreover, most existing global and regional scenarios do not distinguish between primary and secondary forests, which play different roles in the regulation of the water cycle, the support of species, and other global change processes.

6.5.4   Consequences for the Earth System

Taken together, the set of published scenarios imply that major changes in the Earth's land cover over the next decades are not implausible. These changes have large implications for the global water system (through modification of moisture and energy fluxes), for the rate of climate change (through changes in various climatic processes and in emissions of methane, nitrous oxide and other greenhouse gases), for biodiversity (through impacts on the integrity of habitats), for the global carbon cycle (through modifications in terrestrial carbon fluxes), and for other aspects of the Earth System (see Chap. 4).

6.6   Towards Better Land Scenarios

Although existing scenarios have served the needs of different audiences from local farmers to global policy makers, we have pointed out in the previous text that there are substantial opportunities for improvement. What direction should these improvements take? We suggest the goal of improvements should be to enhance the following four characteristics of scenarios. (This list builds on the three criteria (salience, credibility, legitimacy) for quality control of integrated assessment, presented by Jill Jäger at the Workshop on "Scenarios of the Future, the Future of Scenarios", Kassel, Germany, July 2002) (see Chap. 7):

- Relevance. Is the scenario relevant to its audience? Are the particular needs of the potential users addressed? The range of audiences for land scenarios is very wide,

extending from the community interested in global change processes (and land-use/cover change, in particular), to the concern of regional planners about local land-use changes.
- Credibility. Is the scenario plausible to its principal audience and developers? Are the statements and causal relationships consistent with existing information? Are the assumptions about the causal relationships underlying the qualitative scenarios (mental models) or quantitative scenarios (formalized models) transparent? Is the scientific rigor and methods used to develop the scenarios acceptable? Is the credibility of scenario developers high enough?
- Legitimacy. Does the scenario reflect points of view that are perceived to be fair by scenario users, or does the scenario promote particular beliefs, values or agendas? Was the process for developing scenarios perceived to be fair? Are the process and results adequately documented? (These factors are also important to the credibility of scenarios.)
- Creativity. Do the scenarios provoke new, creative thinking? Do they challenge current views about the future? (If this challenge is justified). Do they inform their audience about the implications of uncertainty?

The following paragraphs propose a range of actions for producing better scenarios by enhancing these characteristics.

6.6.1   Expand the Scope of Scenarios

While existing scenarios cover some of the basic dynamics of changing land use and cover, they still incorporate only a small fraction of the processes determining these dynamics. An important way to improve the credibility and relevance of scenarios would be to expand their scope to include more land-use/cover processes. By including more processes, the scenarios will gain scientific credibility because they are more likely to capture the driving forces and dynamics that will determine future land-use/cover changes. Likewise, covering more processes will make the scenarios more relevant to a wider range of scientific and policy users.

In the following paragraphs, we recommend six priorities for expanding the scope of scenarios.

- Describe in more detail the factors determining the extent of future agricultural land. As noted earlier in this chapter, most land scenarios focus on agricultural land because of its manifold importance. However, most of these scenarios are based on simplified assumptions about future farm management, crop yield and other factors that will determine the extent of future agricultural land. The credibility and relevance of agricultural land scenarios would be enhanced, if scenario builders provided a more detailed rationale for future trends in these factors. In particular, scenario builders should draw on either conceptual or formalized models to estimate future productivity of crop and grasslands, the future importance of new crops such as bioenergy plants, and the trade-off between future agricultural intensification and extensification.

- Give more attention to non-agricultural land. While the current focus of scenarios on agricultural land is understandable, neglecting other types of land results in an incomplete picture of future land use and cover. Land cover with natural vegetation (forests, grasslands) are often treated in scenarios as remnant land-cover classes (areas not needed for other purposes). Hence, greater attention should be given to future changes of non-agricultural land (forest, grassland, urban). In addition, more attention should be given to realistically representing competition between land-cover types, since many future policy interventions affect the availability of land (conservation of nature, carbon plantations, livelihood of rural areas, renewable energy etc.).

- Incorporate more detail about driving forces. Most land scenarios are driven by assumptions about external factors such as population, economic growth, and technological development. Although these factors are usually prescribed *ad hoc*, the reality is that they are affected by a host of other factors. The realism of land scenarios, and thereby their credibility and relevance, would be enhanced by including more detail and realism about future trends in these driving forces. Examples are:
  a the effect of social and cultural attitudes on food consumption, on land-use practices (e.g., farming systems), and on the priority given to the conservation of natural resources;
  b the impact of labor, capital and global food trade on agricultural production;
  c the effect of traditions and practices of land tenure on land-use patterns;
  d the effect of shifts of population from rural areas to urban or *vice versa*.

- Incorporate feedbacks into driving forces. In reality, not only is land use driven by external factors, but land-use change in turn feed back to these external factors. An example of such a feedback was given in Sect. 6.3.2. A key task for scenario developers is to incorporate the feedback from land-use change to external drivers, drawing on new knowledge about these feedbacks. This task can be achieved by modifying the models used to generate the scenarios. One way to modify the models would be to convert external drivers into internal variables in the model. Another way is to insert a switch in the model that indicates when "unrealistic" land-use change is computed. This switch would then send a signal to automatically modify the external drivers so that more "realistic" land-use change is computed.

- Include extreme events and changes in their periodicity. It is generally understood that flooding, fire and other extreme events have a profound but transient impact on land use and land cover (e.g., Kauffman 2004; Kok and Winograd 2002; Cochrane et al. 1999). At the same time, a single event usually does not have a persistent effect on land cover over the scale of several years, because vegetation and ecosystems tend to re-establish themselves after such events. However, it is also observed that recurrent extreme events can have an important influence on permanent land cover (e.g., Nepstad et al. 2004; van Noordwijk et al. 2004; Sorrensen 2004; Correia et al. 1999). One example is the role of periodic brush fires in determining the vegetation in chaparral landscapes. Hence, rather than including single extreme events in scenarios, it would be more consistent with current thinking to include a change in periodicity of extreme events (if appropriate for the setting of the scenarios). Including extreme events in this way could make scenarios more thought-provoking and thereby enhance their creativity.

- Inform stakeholders about the limitations of models. A challenge related to the limited scope of models is the communication problem that arises when stakeholders specify that a land scenario has 15 driving forces, but the model used to quantify the scenarios can only handle 5 of these driving forces. This is just one of the many mismatches that typically occur between the mental models of stakeholders and the simpler formalized models used for quantification of scenarios. This mismatch takes away from the consistency and credibility of the scenarios. In this case, a partial solution is simple: the model teams should inform stakeholders about the limitations of the models at an early stage of scenario development. The stakeholders then have the option of taking into account these limitations. Another option is to use simple, flexible models that can be adjusted quickly to the specifications of stakeholders during a scenario exercise.

6.6.2 Use Participatory Approaches to Scenario Development

We believe that the relevance, legitimacy and creativity of scenarios can be enhanced by developing them in partnership with stakeholders (i.e., individuals or organizations with a special interest in the outcomes of the scenarios). This is called the participatory approach to scenario development, as described earlier in the chapter. Typical of this approach is the use of a scenario panel consisting of stakeholders and experts to carry out the core work of scenario development.

How does the participatory approach enhance the relevance, legitimacy and creativity of scenarios? By including some of the potential users of the scenarios in the scenario panel (the stakeholders), the scenarios have a higher chance of addressing relevant policy questions. Since these stakeholders represent the different interest groups concerned with scenario outcomes, their participation also enhances the legitimacy of the scenarios. The participatory approach can also produce more creative scenarios because the wide range of views represented on the scenario panel often lead to new combinations of views about the future that are incorporated into less conventional and more creative scenarios.

However, a key to making scenarios more relevant, legitimate and creative is to ensure that the scenario panel is made up of a wide and representative group of stakeholders and experts. Otherwise, the scenario panel may be perceived as being biased towards one interest or another, thus undermining the credibility and legitimacy of the scenarios they produce. Moreover, a scenario panel with biased views will also narrow the scope and creativity of the scenarios they generate.

6.6.3   Improve the Transparency and Documentation of Scenarios

In this paragraph, we return to the question of how to maximize the credibility of scenarios. Sometimes credibility is associated with likelihood (the more likely a scenario, the higher its credibility), but this does not always hold for scenarios for two reasons. First, information about the likelihood of a scenario is usually not available. (For example, the authors of the IPCC emission scenarios explicitly advise scenario users that no likelihood should be assigned to the different scenarios; IPCC 2000a). Second, even unlikely scenarios can serve a useful purpose, as in the case of low-probability scenarios of accidents in nuclear power plants which are useful for developing accident contingency plans. Hence, the credibility of a scenario is not always related to its likelihood.

As an alternative, we believe that the credibility of a scenario can be associated with its internal logic, consistency and coherence. That is, the more logical, consistent and coherent the scenario, the higher its credibility. In turn, this logic, consistency and coherence must be transparent through the clear documentation of a scenario's basic assumptions, internal structure, and driving forces. This is a special challenge for qualitative scenarios because they are usually expressions of the complex mental models of stakeholders. To make the assumptions behind these scenarios more transparent, it may be possible to use well-established techniques of "soft systems research" that formalize human thinking and decision processes (e.g., Fishwick and Luker 1991; Checkland 1981). Another possible approach is to use spatial and/or historical analogs of the events in a scenario. In the case that models are used to generate scenarios, the credibility of the scenario can be enhanced by documenting the model and its assumptions in peer-reviewed scientific literature.

6.6.4   Build Interactive Scenarios

Another approach to increase the *credibility* of scenarios is to build interactive scenarios. This type of scenarios would increase the credibility of scenarios in general, because they provide a more realistic representation of the driving forces of scenarios.

Under this procedure, the time horizon of the scenario exercise (say 2005 to 2100) would be divided into smaller intervals (e.g., 2005 to 2020, 2020 to 2050, and 2050 to 2100). Rather than specifying driving forces over the entire time horizon as is usually done, the driving forces would be specified only for the first time interval. The next step would be to evaluate the consequences of these driving forces on land use/cover for the first time interval (either with a model or with storylines). The results of the first interval would then be used to set the starting conditions for the second interval. For example, if agricultural land in a study region is depleted by the end of the first scenario interval, this information could be used to assume a higher rate of migration from rural to urban areas in the second interval. In effect, the scenario developers would interact with the scenario itself, and would specify the feedback from land use to driving forces Rather than being specified only one time at the beginning of the scenarios, the driving forces would interact and be modified by the dynamics of the scenario.

A disadvantage of this method is the large effort it requires. We also note that the idea of interactive scenario development resembles the procedures of strategic gaming and policy exercises applied earlier to environmental and other problems (Checkland 1981; Fishwick and Luker 1991; Toth 1988, 1995).

6.6.5   Broaden the Realm of Application of Global Scenarios

An obvious way to increase the relevance of scenarios is to develop them for addressing a wider range of scientific and policy questions. Most existing global land scenarios were developed for analyzing climate change issues such as the emissions of land-related greenhouse gases or the flux of carbon dioxide between the atmosphere and biosphere. As a result, they have a bias towards processes important to climate change and this limits their relevance to other issues. Global scenarios could also be developed for analyzing other important issues such as the consequences of trade liberalization, or the planning of nature corridors for increasing the connectivity of protected areas. Land scenarios could also contribute to strategies for achieving the land-related Millennium Development Goals (such as the goal to reduce world hunger) and for analyzing the implementation of the terrestrial aspects of the Convention on Biodiversity (e.g., Leemans 1999). These applications will require an extension of the driving forces and processes covered by the scenarios.

6.6.6 Develop Multi-Scale Scenarios

In this paragraph we recommend developing multi-scale scenarios as a way of enhancing the credibility and relevance of scenarios in general. We noted earlier that existing global and regional scenarios tend to provide different kinds of information. Global scenarios provide a comprehensive picture of the implications of large-scale driving forces on land-use and cover change, while regional scenarios provide a more detailed representation of land-use/cover changes which can be related more realistically to biogeochemical processes such as soil degradation, changes in hydrology and land processes leading to emissions of greenhouse gases. Both types of scenarios lack a measure of credibility and relevance because they cannot capture the view of the others, and would gain credibility and relevance if they could be linked.

In the text, we referred to various efforts at developing multi-scale scenarios. A possible linkage would be to use global scenarios for setting boundary conditions and constraints for regional scenarios, e.g., the demands of global food markets or the implementation of national/international nature conservation goals. In the other direction, regional scenarios covering different parts of the world could provide input that is difficult to capture at the global scale. Some examples are the impact of land-related institutions (farming associations or regional planning organizations) on land-use change, visions of regional development pathways, the influence of cultural background on land-use practices, and attitudes towards nature protection.

6.6.7 Improve the Representation of Socio-Economic Behavior in Scenarios

Here we recommend increasing the credibility and creativity of scenarios by improving the representation of socio-economic behavior in scenarios, especially by applying agent-based modeling. Agent-based models have been used for simulations at the local and regional scale and have a high potential for use in the development of land scenarios at all scales (see Chap. 5). They provide a method to improve and formalize (in the sense of making more transparent and traceable) important social processes in scenarios, and thereby will increase the credibility of scenarios. For example, agent-based models can provide insight into interactions between actors relevant to land-use change such as between farming groups and the local government. Such approaches may also allow scenarios to incorporate the types of feedback processes that are currently poorly represented (as discussed above). This includes, in particular, processes that relate to policy-making and institutional responses to emerging environmental problems. By providing a platform for representing different ideas and policy responses, agent-based modeling can also help produce more creative scenarios. However, much work has to be done to enable the use of agent-based modeling or its results on the global level.

6.7 Conclusions

Summing up, although we are only in the early stages of analyzing the future state of land use and land cover on Earth, we have already learned much from existing scenarios. One clear message of the scenarios of particular importance to global change is that current land-use/cover patterns are not static. Indeed, major changes in the Earth's land cover over the next several decades, including trend reversals, are not implausible. The fact that some scenarios only begin to show distinctive trends after two or three decades also implies that a long-term view is needed to better anticipate the future of land.

Although we have not evaluated the impacts of potential changes in land use and cover, we believe that the scale of changes shown in the scenarios could have large implications on the Earth System. For that reason alone, we should devote greater effort to understanding the future of land.

Chapter 7

Linking Land-Change Science and Policy:
Current Lessons and Future Integration

Robin S. Reid · Thomas P. Tomich · Jianchu Xu · Helmut Geist · Alexander Mather · Ruth S. DeFries · Jianguo Liu
Diogenes Alves · Babatunde Agbola · Eric F. Lambin · Abha Chabbra · Tom Veldkamp · Kasper Kok
Meine van Noordwijk · David Thomas · Cheryl Palm · Peter H. Verburg

7.1   Introduction

Human use of the land and oceans is at the center of some of the most complicated and pressing problems faced by policy makers around the world today (e.g., DeFries et al. 2004b; Platt 2004; Millennium Ecosystem Assessment 2005). For the terrestrial biosphere, our need to balance current human needs and longer-term environmental sustainability often involves consideration of the way we use ecosystem goods and services produced by the land. Land-use is at the center of these trade-offs because changes in land use often enhance the share of energy, water and nutrients devoted to human needs but decrease the share available for other species and ecosystem functions. Problems as far ranging as improving human health or ensuring adequate food production cannot be solved unless policy makers understand how their policies alter land use and how altered land use affects ecosystem functions. For example, public health policy that adequately accounts for the future spread of mosquitoes that carry *Plasmodium* or malaria in the tropics often requires an understanding of the interplay between land use and climate (Lines 1995) (see Chap. 4). In China, agricultural policy makers are using a recent assessment of cropland area to create policies that ensure there will be enough land to meet China's rapidly growing demand for food, feedgrains, and raw materials that is driven by rapid economic growth (Welch and Pannell 1982; Yang and Li 2000; Ho and Lin 2004; Lin and Ho 2005), although it is not clear that other ecosystem services will be maintained in this process.

While policy makers must understand land use to address certain pressing policy issues, policy can also cause changes in land use. Some policies, such as those creating protected areas, directly affect land use, while others affect land-based activities like agriculture or forestry. But other policies, not intended to affect land use, can have profound but indirect impacts, particularly by influencing the underlying causes of that change. These include sectoral policies, like agricultural price policies, trade policy, and public investments in infrastructure, and macroeconomic policies, like exchange rates and monetary policy that influence interest rates and credit availability – see Fig. 7.1. For example, in Amazonia, developing road infrastructure within the framework of

Fig. 7.1.
Types of policies that affect land use from those directly affecting land use (land-use policies, in *front*), to those related to land-based activities (*middle*) and those indirectly affecting land use (*back*) (Mather 2006c)

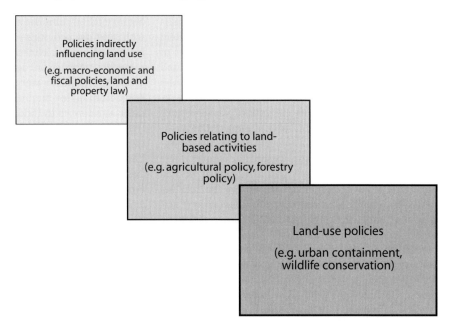

large-scale development programs has created a potent avenue for deforestation: 90% of all deforestation in the 1991–1997 period was observed within 100 km of major roads opened during the 1970s (Alves 2002b). Land use on humid forest uplands in Southeast Asia has changed rapidly in response to (or sometimes in spite of) sectoral and land policies regulating resettlement, land tenure and agricultural prices (Tomich et al. 2004c) and regional integration (Krumm and Kharas 2004). In East Africa and Central Asia, implementation of policy that privatizes land ownership in rangelands now causes rapid landscape fragmentation and expansion of cultivation and fencing (Rutten 1992; Williams 1996; Reid et al. 2005). Indeed, in drylands around the world, privatization of common property and public lands, public sector development projects, diffusion of agricultural technologies and chemical inputs, and market liberalization can trigger rapid intensification of land use with concomitant environmental problems in some cases (Beresford et al. 2001; Geist and Lambin 2004). European, U.S., and Japanese production subsidies and trade barriers distort world markets for agricultural products. This affects how farmers in both the former and the latter countries choose to use their land.

It is thus critical that good information about the causes and consequences of land-use change reach policy makers so that they can create more effective policies and understand policy impacts (Goetz et al. 2004). We are beginning to see cases around the world where lessons from land-change science are being used to revise old policies and create new ones. Information as simple as land-use maps can clarify land-management issues in indelible ways. International meetings to discuss global environmental policy matters often start with a presentation by a prominent scientist showing a map or graphic that originated within land-change science. New land-use research sometimes includes policy makers from the outset so that problems they face are the point of departure for the scientific process (Tomich et al. 2004a; Reid et al. 2005). As discussed below, some elegant ways of demonstrating the trade-offs between human needs and environmental sustainability are being used to address local and national policy concerns.

This chapter will examine interactions between land-change science and policy by first describing the key, credible lessons from the science of land-use that can be relevant to policy. We will then explore specific examples where land-change science is already part of the policy process. Finally, we will suggest how we can improve the links between land-change science and policy. Integration of science and policy will first be addressed by describing some of the needs and perceptions of policy makers. We will then describe some ways in which land-use scientists can better address those needs, using a conceptual framework that addresses three key characteristics of the type of science that successfully links with policy makers: science that is credible, salient and legitimate (Cash et al. 2003).

7.2 Key Public Policy Lessons from Land-Change Science

Over the last decade, land-change science has contributed strongly to our understanding of where, when, how fast and why people change their use of the land (see Chap. 2 and 3). We now have a credible and reliable science of land use. Here we discuss the information from that science that we think is most important to improving policy, with a focus on lessons that generally apply across the globe. Many of these lessons, however, are specific to regions, and we thus also present policy interventions suggested by different authors for specific regions. We define policy makers broadly as those land managers and political leaders who affect how land is used from very local levels in communities to national and international-level policy makers. We structure this section around nine straightforward statements about what we have learned; these are key messages to policy makers, meant to promote sustainable land use.

Message 1

> Some types of land use are more sustainable than others; this often depends how simple or diverse the land-use activity is.

Sustainable land use refers to the use of land resources to produce goods and services in such a way that, over the long term, the natural resource base is not damaged, and that future human needs can be met. The time horizon of the concept covers several generations. For various reasons, broad trends in agriculture run toward intensification and specialization at the plot level, often (but not inevitably) culminating in "monocultures" associated with land-use activities of much simpler structure and lower biodiversity richness than "polycultures". Consider a specific comparison: agricultural systems established at the humid forest margins following slash-and-burn range from highly biodiverse systems such as rubber or cacao agroforests in Indonesia and West Africa, respectively, to systems with much lower biodiversity like pastures in the Amazon or cassava plantations in Indonesia. The sustainability of these varied systems was measured and compared through the Alternatives to Slash-and-Burn (ASB) Programme according to three types of criteria: (a) environmental – carbon stocks and above- and belowground biodiversity; (b) agronomic – soil structure and biology, nutrient balances, and pests; and (c) socio-economic – returns to land and labor, implications for household food security, capital constraints arising from levels of investment required and years to positive cash flows, as well as an array of other policy, social and institutional indicators. The studies have revealed the feasibility of a "middle path" of development that delivers an attractive balance between environmental benefits and equitable economic growth. The Sumatran rubber agroforests and their cocoa and fruit counterparts in Cameroon contain about 25–50% of the carbon stocks of the natural forest (Palm et al. 2005). The biodiversity in these forests, though not as high as in natural forest, are far higher than those in monocrop tree plantations, short term fallows, or annual cropping systems (Gillison 2005). It is also interesting to note that there are many types of tree-based systems with similar levels of C storage but drastically different profitability and hence attractiveness to farmers (Gockowski et al. 2001). Agronomic criteria show moderate to high levels of sustainability in agroforests with pests and potentially negative nutrient balances as the main issues of concern, depending on the specific systems assessed (Hairiah et al. 2005). Simple tree crop systems (monoculture plantations) often experience problems of soil structure (compaction), besides problems with crop protection. Crop/fallow systems vary greatly in their effect on agronomic sustainability. The long fallow systems with low cropping intensity in Indonesia and Cameroon (traditional slash-and-burn shifting cultivation systems) are sustainable, but unimproved short fallow systems with intensified cropping have detrimental effects on soil structure, nutrient balance, and crop health; these also produce very low returns to labor. Continuous annual cropping, as with cassava in Indonesia, is often, but not always, problematic in the forest margins of the humid tropics. Pastures, particularly with improved management practices, tend to have a medium level of impact on the natural resource base, though impacts on global environmental issues (biodiversity and greenhouse gas emissions) may be quite large (see Chap. 4). A tool developed for analyzing these trade-offs in the tropical forest margins, the ASB matrix, is discussed under message 8 below.

In African dry forests and savannas, grazing can maintain the diversity of native plants, birds and butterflies more than in croplands (Soderstrom et al. 2003). Under-grazing has even been implicated in loss of plant diversity from grasslands across the world (e.g., Milchunas et al. 1988), as has over-grazing. In Africa and Europe, there are more native species in croplands with more complex features like hedgerows and woodlots (even in large trees) than in less complex landscapes with few of these features (Reid et al. 1997; Wilson et al. 1997; Soderstrom et al. 2001). However, complex, agricultural landscapes do not usually support large-bodied wild animals with large home ranges; farmers exterminate these species earlier in the process of clearing land. The diversity of small species (birds, insects) can be quite high on pastures, prompting European policies to preserve cattle pastures because of their high biodiversity. These examples suggest that agricultural land use can be compatible with biodiversity and other ecosystem services, which contribute to the nexus of agricultural biodiversity, dietary diversity and human health and nutrition, but this is far from always the case. This is an obvious place for policy to influence conservation of biodiversity, but the ability to influence land use outcomes depends greatly on public finance and administrative capacity. While elaborate land-management schemes can be implemented through land-use planning and incentive schemes in Europe and the United States of America, such approaches are problematic across most of the developing world.

Message 2

> Single factor causes are rare, but the range of "syndromes" (combinations of causes) is not infinite; some specific combinations account for a significant share of land-use change.

Although expressed in manifold ways, there are few, important causes of land-use change, that often work together in concert. And these can work in unexpected ways. For example, population growth sometimes causes land-use change and sometimes does not. But when "population" comes in as an explanatory variable, it is less fertility increase than migration, mainly in-migration to a given location or site. This phenomenon shows up in all major meta-analytical studies done under the umbrella of the Land-Use/Cover Change (LUCC) project (see Chap. 3). Moreover, even in the face of land scarcity and human population growth, agriculture and land use can stagnate. In addition, the location of growth is important. For example, farming land contracted and forests expanded in Europe at the same time that human populations were on the rise, because populations grew chiefly in the cities, not the countryside. Massive productivity increases and economic transformation (from agrarian to industrial) allowed support of larger populations with less agricultural land. Sectoral and macro-economic policies (e.g., price policies for agricultural inputs and outputs, infrastructure investments, land tenure and taxation policies, reforestation programs, and natural resources policies regulating exploitation of forests, minerals, and petroleum), are significant causes of land-use change, and thus are a set of levers held by policy makers that can influence either sustainable or unsustainable paths of land use. It is important to realize that while these policies interact to cause change, they also are aimed at a wide range of objectives, of which sustainable land use often is not the primary goal.

Policy makers will be more successful if they understand the underlying causes of land-use change (institutions, policies, population) as well as the proximate causes (logging, cultivation) that presently receive most attention in policy debates. Furthermore, effective policies need to account for the multiple and often interacting causes of land-use change, as highlighted in Chap. 3 – see Fig. 7.2. Lifestyle choices and shifting consumption patterns of goods and services are affecting land-use choices all over the world. For example, land users in the Yellowstone ecosystem, United States, are shifting from ranching to construction of leisure homes (Hanson et al. 2002), while semi-nomadic herders in Africa and Central Asia are choosing to settle to access schools and better health care (Rutten 1992; Blench 2000). In the most populous countries of the world (United States, India and China), economic integration and globalization, modified by national land policies, also strongly affect how and where people use the land.

In drylands and humid tropical forests, similar broad classes of factors underlie deforestation and desertification including: human population dynamics, market integration, urbanization, technological change (e.g., introduction of technical irrigation or new crop varieties), governance (e.g., corruption), changes in property rights, public attitudes and beliefs, individual household behaviors, and sometimes climate (Geist and Lambin 2002, 2004). While the factors to be considered may be similar, the main causes of change are not the same for humid and arid. For example, links to global markets are much more important for humid forests, while local drivers are more important for arid lands (Geist et al. 2006) – see Tables 7.1 and 7.2. The broad analytical "similarity" here relates to the large bundles of variables, but the scale (global, local) of the driving forces must be understood in context.

Message 3

Underlying causes, originating far from where land is actually changing, often drive local changes in the land.

With economic liberalization and globalization, people increasingly choose how they use the land on the basis of influences originating outside their communities, and this has major implications for transitions to sustainability (Lambin and Geist 2003a; Geist et al. 2006). Actually, agents of change become increasingly disconnected spatially from major stakeholders of these changes. However, the resulting change is almost always in response to a combination of local and global causes, leading to some uncertainty in likely outcomes. For example, even if local communities in East Africa can both reduce poverty and conserve wildlife through local land-use initiatives, these efforts will be unsustainable if they continually collide with inappropriate land-use policies (like subsidies that encourage crop cultivation) at the national level. In this case, local civil society groups that promote pastoral human rights are well aware of this need and act both locally and nationally in a synergistic fashion to agitate for change (Reid et al. 2005). Thus, working locally to sustain local land-use systems will likely succeed more quickly and maintain gains longer if national policies support rather than hinder local efforts – see Fig. 7.2.

Table 7.1. Driving forces of tropical deforestation by scale of influence

	All factors (range in %)	Demographic factors (%)[a]	Economic factors (%)	Technological factors (%)	Policy and institutional factors (%)	Cultural or sociopolitical factors (%)
	N = 152 cases	n = 93	n = 123	n = 107	n = 119	n = 101
Local	2 – 88	88	2	23	4	16
National	1 – 14	1	14	3	2	7
Global	0 – 1	–	1	–	–	–
Several scales: global-local interplays	11 – 94	11	82	74	94	77

[a] 6 cases of unspecified population pressure could not be attributed to scales. *Source:* Geist et al. (2006), p. 64.

Table 7.2. Driving forces of desertification by scale of influence

	All factors (range in %)	Demographic factors (%)[a]	Economic factors (%)	Technological factors (%)	Policy and institutional factors (%)	Cultural or sociopolitical factors (%)	Climatic factors (%)[a]
	N = 132	n = 73	n = 79	n = 91	n = 86	n = 55	n = 114
Local	12 – 29	23	18	29	12	16	–
National	4 – 20	–	13	–	20	4	–
Global	4 – 12	–	4	–	6	–	12
Several scales: national-local interplays	29 – 80	29	66	71	63	80	60

[a] 35 demography-driven and 32 climate-driven cases could not be attributed to scales. *Source:* Geist et al. (2006), p. 65.

Conversely, such local initiatives have little scope for success if adverse national policies and international market forces are ignored.

The liberalization of trade, and the opening up of new areas to national and international markets, can have several effects. One is to expand the scale of production and extent of monoculture of a particular commodity with possible effects on biodiversity. Another is for the production of particular commodities to be concentrated in particular areas, where they enjoy geographical advantages in environmental or other terms. This concentration could, in turn, yield economic benefits from positive spillovers through a concentration of knowledge, service provision, and marketing facilities (e.g., fertilizers, glasshouse heating, etc.). Some local "dis-benefits" might result from the former, and some local benefits from the latter. The significance both of distant causes and of national responses can be long-lasting. In the latter part of the 19th century, an episode of globalization involving the opening up of the American Prairies and the export of cheap grain from there had a major effect on European farming. Some countries, such as France, provided protection for their farmers, in the form of import tariffs. Others, such as Denmark and the Netherlands, encouraged diversification into the production of commodities in which there was less competition. Others again, such as Britain, took a *laissez-faire* approach, and left farmers fully exposed to competition. The effects of this episode, and especially of the differing responses, are still evident a century later.

Message 4

A finite set of pathways can be used to develop policy-relevant land-use scenarios that are relevant to different regions of the world.

A pathway is a particular set of events that together describe how land use changes in particular area, which is different from but related to the actual cause of the change described in message 2 (see Chap. 3). One obvious pathway is the opening up of a forest "frontier" by constructing a road, that results in conversion of native vegetation to cropland or pastures. To develop information on pathways that will be useful for policy development and land management in particular places, we must account for historical land-use patterns, climatic, economic and ecological constraints on land use, what causes change, how different causes act together (synergies), and how resulting land-use activities feed back to affect these causes. Once we have a basic functional understanding of these pathways, it will be clearer what policy interventions will and will not promote sustainable land use in specific cases. Understanding these pathways can also help land managers and policy makers anticipate changes and cope with uncertainty (see Chap. 6).

Message 5

Drivers can work together to create rapid land-cover change and unexpected land degradation; policy-oriented research should focus on these "hot spots" of rapid change and degradation.

Land cover changes faster in some locations than others around the globe (see Chap. 2). For example, deforestation mostly takes place at the edge of large forest areas and in conjunction with major investments in transportation networks and other infrastructure (e.g., the "arc of deforestation" in the Amazon Basin; Pacheco 2006c). At the national level, land use is changing more rapidly in transitional economies in post-socialist countries like China and Russia (Hill 1994; Kondrashov 2001) because of a rapid shift in property rights, decollectivization, decentralization and a collapse of employment opportunities in the non-agricultural sectors (Sturgeon and Sikor 2004).

Migration, education and land-tenure changes can together cause rapid changes in land use. In China and Kenya, for example, strong migration has expanded settlement and land use around and inside protected areas with surprising rapidity in the last 30 years (Liu et al. 2001, 2003a; Lamprey and Reid 2004). But additional social changes, through education and changes in land tenure, caused large cohesive families to split into smaller single family units at the same time. Migration and social change working together caused an explosion of household growth and settlement, with strong consequences for wildlife habitat in both cases. Careful analysis of these situations needs to be made quickly, and policy needs to focus on weakening synergistic causes that degrade the land. This could be done, in the Kenyan example, through new land use and access policy that allow secure land ownership but also supports the mobility of livestock herds and wildlife, particularly in times of stress during droughts.

Several parts of the world are not adequately represented in the available data sets (see Chap. 2), so it is possible that rapid change is occurring in locations where data are poor. Data on changes in drylands and mountains are the most incomplete of all types of change, because satellite imagery of these regions is difficult to interpret and we are largely unable to distinguish human-induced trends from large, climate-driven interannual variability in vegetation cover. Rapid land-cover changes that are still poorly documented at the global scale include, for example, changes in the (sub)tropical dry forests (e.g., miombo forests in southern Africa and chaco forests in South America); forest-cover changes caused by fires and insect damage; drainage or other changes in wetlands; soil degradation in croplands and changes in the extent and productive capacity of pastoral lands (Lambin et al. 2003). It is also possible that ecological impacts of change are large even in places where land-use change is slow, as in the case of depletion of wild mammals through hunting for bushmeat. These exceptions and gaps in our knowledge suggest that researchers should not focus solely on areas of readily-detected change in land cover.

Message 6

> Mobility and flexibility often are critical to sustainable land use.

Long-fallow, rotational shifting cultivation ("swidden agriculture") is one well-documented example of how mobility and flexibility underpin the sustainability of extensive smallholder systems; if these attributes are lost, such systems may collapse. Similarly, policies that support mobile lifestyles and flexible livelihood strategies can allow pastures to "rest" seasonally and thus curb overgrazing. Pastoral land use, all over the world, is shrinking as farmers push further into marginal lands and herders settle more often around infrastructure for water, health and education (Ellis and Swift 1988; Niamir-Fuller 1999). Access to large and diverse landscapes is critical to maintaining productivity of livestock in pastoral systems and reducing vulnerability of pastoral families, particularly

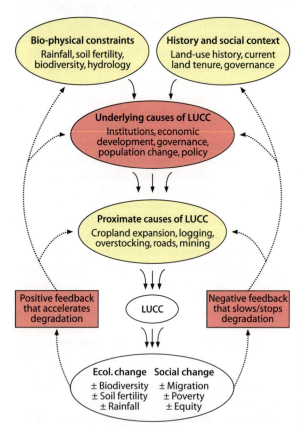

Fig. 7.2. Conceptual model showing where, during the processes of land-use change, national-level policy is likely to have the most impacts on land use (in *red*) or where intervention will be more difficult (*yellow*). Local policy will more easily impact the proximate causes of change; however, unless the underlying causes are addressed at the same time, local action may not be sustainable

during drought. For example, in a traditional system of transhumance, Sahel herders migrate large distances, following seasonally varying rainfall, to find greener pastures and full water holes. Recent privatization and sale of pieces of pastoral rangelands by pastoral peoples has been aptly termed "selling wealth to buy poverty" (Rutten 1992). In other agricultural systems, shifting cultivators and mountain farmers use mobility as a strategy to access resources over time. Policies need to provide mobile services to mobile communities to allow them good health care and educational opportunities while they, for example, move livestock to seasonal pastures.

Message 7

> Specific "entry points" exist where revised or new policies can improve land-use practices; it is possible to restore lands degraded by inappropriate land use, but sometimes the line between degradation and sustainability is fine.

Some policies support sustainable land use, while others do not. We will focus on the latter first. Policy can intervene to weaken some of the underlying causes of this unsustainable land-use change by revising perverse policies or generating new policy – see Fig. 7.2. In humid forests, much deforestation is caused by poor governance and perverse subsidies (like tax-breaks and low-interest loans) that encourage farmers to settle in forests (see Chap. 3). Some of these policy instruments are easier for policy makers to manipulate than others (such as trade or macro-economic policies), and thus can be the first places for policy action.

Policy can be targeted to weaken the positive feedbacks that accelerate unsustainable changes in land use and strengthen negative feedbacks that slow change (Lambin and Geist 2003b) – see Fig. 7.2. For example, good communication of the location and speed of land-use changes to policy makers can allow them to react in a timely manner to particularly fast or unexpected changes, or to start a protracted policy discussion in anticipation of future changes. In Brazil, for example, deforestation over the entire Amazon is monitored each year, so that changes can be detected and acted upon when there is the political will (INPE 2000; Alves 2001a). In Kenya, scientists have collected information on changes in land use and wildlife populations for over 40 years that highlight hot spots of change and other areas where coexistence of livestock and wildlife is sustainable (Said 2003). The key here is communication of information in a way that is useful to policy makers and engagement of policy makers often and early in the scientific analysis process. But, of course, while better information often is a necessary ingredient to improved policy, it is by no means sufficient. Typically, there are conflicts among the interests of particular groups within society regarding land-use priorities and between the broader public interest and narrow private interests in land-use outcomes. To be effective, land-use science and policy studies must also consider these contending interests and the balance of political power.

Good land management and appropriate policies can help farmers and herders avoid land degradation or restore degraded ecosystems. From all over the world, there are examples of farmers who use sustainable land-use practices, even in the face of growing human population density, when the institutions are appropriate, social networks are strong, access to markets and technologies is good and they have strategies to reduce risk (Schweik et al. 1997; Gray and Kevane 2001; Turner and Williams 2002; Dietz et al. 2003; Gibson et al. 2003; Tiffen 2003; Laney 2002). This has led some to suggest, for example, that "more people means less erosion" (Tiffen et al. 1994a), but "more people" can also lead to more erosion and less water conservation (Kates and Haarman 1991, 1992). Or even "more people and more forest" but "less livelihood security and poor environmental services" if the institutions, policies, markets and livelihood options are not in place. In one case study in Yunnan, southwest China, for example, forest cover increased at the expense of decreasing farmland and farmers' access to forest resources. However, monoculture reforestation with pine has caused both biophysical and socio-economic consequences, including negative effects on rural livelihoods (Xu et al. 2005a).

There can be a delicate tip point between trajectories ensuring innovation/restoration and those that cause degradation/deforestation, as demonstrated at a very local scale for the southern Yucatán (Klepeis and Turner 2001; Bray et al. 2004; Turner et al. 2004). This implies that scientists need to help policy makers monitor the effects of policy instruments, so that unexpected effects can be countered before degradation starts, or to model the probable effects of different policy instruments before they are deployed.

Message 8

> Land use that combines poverty reduction and nature conservation is rare, but new efforts exist to evaluate these often opposing goals more clearly, and monitor progress towards them.

Certainly there are examples where misguided policy, poor governance, and outright corruption undermine both conservation and development objectives; tropical forests are a well-documented case in point (e.g., Repetto and Gillis 1988). In these "lose-lose" cases, there may be opportunities to make incremental gains for people and nature through policy reform and better governance (Panayotou 1993). Unfortunately, though, there are few cases where a single type of land use achieves development without some sacrifice of conservation values of natural systems (where it is commercially viable, ecotourism is one such "win-win").

However, as already suggested, much depends on the point of reference and the trajectory of change. In the humid tropics, no forest-derived land use can match the global environmental values of natural forest – see Table 7.3 for the case of Sumatra. On the other hand, restoration of "degraded" tropical landscapes may provide a rare win-win opportunity, where restoration of ecological function and environmental services also could create livelihood opportunities if poor people are involved appropriately (Tomich et al. 2005). More common are situations where farmers can expand land use and improve their incomes (a win situation), while losing only part of the ecological services provided by a landscape (a small loss situation; DeFries et al. 2004b). The ASB matrix (Tomich et al. 2005; Palm et al. 2005) provides an approach to assessing the trade-offs and complementarities between losses of certain ecological services of global importance such as carbon stocks, which affect central functions of the climate system, and gains in the production of food, fiber and feeds to support local communities and national economic development. Tools like this allow identification of innovative policies and institutions needed to balance both sets of goals. The matrix also provides a basis for policy makers and stakeholders to assess trade-offs comparing among different land-use systems (and choices) regarding environmental and development goals (see Sect. 4.8). There is a new effort, supported by many land-change scientists and institutions, to develop a first-ever Climate, Community and Biodiversity (CCB) standards for different land-use practices. These standards are a public/private partnership seeking to recognize land users if they sequester carbon, conserve biodiversity and reduce poverty at the same time.

Empirical evidence shows that labor-intensive technological progress like new irrigation techniques often facilitates intensification on existing agricultural areas and, at the same time, has the potential to increase rural incomes. The increase in productivity on existing land leads labor-constrained households to allocate less time to land clearing and land expansion into upland areas and, in that way, has the potential to conserve forest cover on more marginal land; see, for example, Müller and Zeller (2002) in Vietnam, Pender et al. (2001) in Honduras, Pender et al. (2004) in Shively and Martinez et al. (2001) in the Philippines. In these cases, the key aspects underlying the win-win outcome are first that the technology is suited only to existing agricultural land (so it does not create incentives for conversion of wild lands) and second that it is labor-intensive as well as profitable (thereby inducing households to shift labor out of deforestation activities).

Another win-win example involves the edible mushroom, matsutake or pine mushroom (*Tricholoma* spp.), prized in Japan since ancient times. Recent dramatic increases in price and demand for these mushrooms have encouraged Tibetan collectors to shift from logging to collecting mushrooms for income generation, reviving cus-

Table 7.3. ASB matrix of trade-offs at the agriculture/forest margin in Sumatra

Land use	Global environment		Agronomic sustainability	National policymakers' concerns		Smallholders		
	Carbon sequestration	Biodiversity	Plot-level production sustainability	Potential profitability at social prices	Employment	Adoptability by smallholders — Production incentives at private prices	Smallholders' concerns — Household food security	
	Aboveground, time-averaged (Mt ha^{-1})	Aboveground, species per standard plot	Overall rating	Returns to land (US$ ha^{-1})	Average labor input (days ha^{-1} yr^{-1})	Returns to labor (US$ day^{-1})		Entitlement paths
Natural forest	306	120	1	0	0	0		NA
Community-based forest management	120	100	1	5	0.2 – 0.4	4.77		Food and income
Commercial logging	94	90	0.5	1080	31	0.78		Income
Rubber agroforest	66–79	90	0.5	0.7–878	111 – 150	1.67–2.25		Income
Oil palm monoculture	62	25	0.5	114	108	4.74		Income
Upland rice/bush fallow rotation	37	45	0.5	–62	15 – 25	1.47		Food
Continuous cassava degrading to *Imperata*	2	15	0	60	98 – 104	1.78		Food and income

Source: Geist et al. (2006), p. 66.

tomary institutions which manage forest habitats (alpine oak and pine forest) and regulate access to mushroom harvest in a particular place. This is a multi-million dollar trade for local people (Yeh 2000; Xu and Salas 2003). The key to the apparent case of a "win-win" possibility here would seem to be that the market for these mushrooms has dramatically increased the value of maintaining natural forests and diverted labor that would have gone into clearing forests.

Message 9

> Thorough understanding of key actors and local situations is important for the design of appropriate and successful policy interventions.

The importance of recognizing and understanding different actors has been widely recognized (see Chap. 5), for example, in the rapidly changing Brazilian Amazon (Alves 2001a; Mahar 2002; Walker 2004). The recognition of the different actors and social groups in this very large and diverse region is crucial for land-use policies because these need to recognize large regional differences in land use, demography and economics (Alves 2001a), and also because different groups have distinct social behaviors, land-use practices, and (often competing) interests. This is particularly important for two of the most important land-use policies for the Brazilian Amazon – Forest Code and Ecological-Economic Zoning – where, in some cases, the failure to identify the different actors and social groups has already affected policy formulation and its effectiveness (Alves 2001a; Mahar 2002).

There also is a need to understand the political ideology of the policy makers and politicians as well as the policy-making process. For example, large-scale rubber planting manifested state power during the socialist collective period in China. Rubber monocultures were introduced in marginal climatic zones. These large-scale settlement projects were viewed as part of the state's strategy to supply industrial raw materials in the national interest for political security through self-sufficiency during China's collective period. The outcomes, however, were inefficient (both technically and economically) as well as damaging to the environment (Xu et al. 2005b).

Using a framework developed within the Land-Use/Cover Change (LUCC) project, Geist and colleagues explored the type of actors involved in different regions of the world and at different scales in drylands and humid forests (Geist et al. 2006). They found that we need to discover and apply locally adapted methods and solutions and these need to be revised continually to maintain sustainable land uses. For example, for desertification problems, it is much more effective to identify and focus on individual problem areas or hot spots of desertification than to raise a general alarm since it is unlikely comprehensive evidence will be available (see Chap. 2).

There is increasing recognition of the critical role that community involvement can play in managing land-cover change. For example, the "tragedy of the commons" holds that open access to communal land causes overgrazing and land degradation (Dietz et al. 2003; Gibson et al. 2003). A synthesis of case studies throughout the world's drylands revealed that a more appropriate notion may be the "tragedy of enclosure" (Geist 1999a), which describes, for example, the loss of land for herders when other land uses encroach on grazing lands (Geist 2005). Case studies across the world have now clearly demonstrated that no single type of ownership, whether private, community or government, is by itself an automatic guarantee of effective management. When community management boundaries are well defined, legitimate, and effectively enforced, the social capital generated through community involvement can be very effective in promoting sustainable development and conservation over the long term, especially at local or regional scales (Nagendra 2006).

7.3 Influence of Land-Change Science on Policy: Some Successes and Failures

Clearly, several of these messages from land-change science may be broadly useful for policy research and analysis. However, producing credible scientific results is only one pre-requisite for establishing strong links between science and policy. Successful links always require scientists to listen to what policy makers need, to understand some of the processes and constraints to how policy actually is "made", to create new scientific designs and data needed to address these needs, and actively engage stakeholders with different points of view. Here, we ask: are there examples where credible land-change science is already salient and legitimate, and thus already part of the policy process? By *salient*, we mean information that is immediately relevant and useful to policy makers; *legitimate* information is unbiased in its creation and both fair and reasonably comprehensive in its treatment of opposing views and interests (Cash et al. 2003).

The different worldviews of researchers and policy makers create a cultural gap preventing adequate use of research (Neilson 2001) and adequate understanding of the needs of policy makers. These two groups have contrasting values and expectations and are rewarded for different behaviors. Scientists produce knowledge and often are rewarded for the number and profile of their technical publications; any activity that takes them away from these tasks may limit their chance of career advancement. Scientists are also rewarded for training students, but rarely for working with land managers and policy makers, except for those working in "boundary" organizations whose goals are to link research and policy (Cash et al. 2003). Ideally, in the arena of land-use issues, success for a policy maker lies in using policy instruments to maintain or improve land-management practices (Crewe and Young 2002), by responding to the needs of those who appoint them or their constituents. (In reality, policy makers will be responding to a range of interests and influences.)

An understanding of the policy development process provides scientists with an appreciation of places where they may engage and influence the process. The rational actor model, pioneered by Lasswell in the 1950s, portrayed the policy making process as a linear, non-iterative process, where policy makers rationally consider information on alternative options and then decide how to move forward. Few policies are actually created this way (Allison 1971); rather policy making is a complex interplay among political interests and competing discourses by multiple actors (Crewe and Young 2002). The key point is that scientists need to understand how organizational processes, bureaucratic politics, and other real-world phenomena (for example, corruption, bureaucracy, local politics) both open and foreclose opportunities for science to influence policy and its outcomes.

Scientists and policy makers also create and use different types of knowledge. Scientists (and local communities) tend to create and use process-based knowledge even including indigenous knowledge (Xu et al. 2005c), while policy makers use "rules of thumb" (M. van Noordwijk, personal observation). In addition, scientists often choose their areas of interest based on a subjective selection of "interesting cases" that may be of limited interest to politicians. Scientists also often focus too much on the creation of policy rather than on the implementation of policy, where local politics influence outcomes decisively (Grindle 1980).

What determines if policy makers use credible science in decision making? Scientific information that attains a balance of credibility, salience and legitimacy is most likely to effectively influence policy (Cash et al. 2003). Perhaps first and foremost, this information must address issues of sufficient importance (i.e., salience) to capture the attention of policy makers at the appropriate level (Tomich et al. 2004a). Salient research assesses the benefits and costs of different policy options or provides a solution to the problem. Participatory approaches and pilot demonstrations of solutions are particularly effective, and increase legitimacy (Court and Young 2003). Similarly, non-participatory approaches can be quite ineffective (Mahar 2002). Also crucial are strong communication links through informal and formal networks between researchers and policy makers that promote trust, openness, and legitimacy (Court and Young 2003). It is important for both researchers and policy makers to recognize each other's constraints in producing and using information (Crewe and Young 2002). Policy makers must realize that scientific knowledge is influenced by the values and beliefs of the scientists themselves, how-

ever strenuously they try to be objective. Scientists must realize that power relations within politics will likely affect the ability of policy makers to use the information they provide.

Researchers most often influence policy when they work with individuals or organizations who focus on the task of crossing the boundary of communication between researchers and policy makers (Cash et al. 2003), thereby improving saliency and legitimacy. These individuals or organizations promote active, interactive and inclusive communication between scientists and policy makers, translate information so the two groups understand each other, and mediate any misunderstandings between them (Cash et al. 2003). Civil society can often fill this role. Individual scientists, trusted by communities and policy makers alike, sometimes communicate among different actors in the policy process. These boundary-crossing activities – communication, translation and mediation – require real investments of time and energy by scientists (Guston 2001). This requires additional resources and is not a natural component of scientific inquiry.

But what is the evidence that some of the products of land-change science have influenced the policy dialogue at the international level? Similarly, are land-use scientists responding to the needs of policy makers? No formal assessment of this two-way translation exists, but it is easy to see some of the principles articulated above at work. The climate change assessments by the IPCC (Intergovernmental Panel for Climate Change), which included input from land-use scientists, were highly credible because they included an unprecedented range of scientific research. They were also salient and relevant for policy makers because the assessments appeared when the issue of climate change became a global public concern. Governmental involvement and the UN Framework on Climate Change (UNFCC) provided links between the scientists in the IPCC and policy makers. The Millennium Ecosystem Assessment involves many land-change scientists and has been designed to respond to the articulated need for policy advice at the global level for the future management of ecosystems worldwide (Millennium Ecosystem Assessment 2003, 2005), and thus includes land-use issues. These initiatives (and institutions like IGBP, IHDP and LUCC) are helping scientists to listen better to policy needs and to get their science directly to policy makers in appropriate forms. It also appears that land-change science is having an impact through individuals who act as "translators", bringing credible science into the public policy arena. The quantitative evidence of impact at all of these levels is weak, but qualitative evidence is abundant.

Qualitative impacts of land-change science on policy also abound at the local or national levels. In Brazil, research linking roads and deforestation (Reis et al. 2001; Alves 2002b; Soares-Filho et al. 2004) had significant impacts, along with other information, on the formulation of policies to curb or contain forest clearing in the

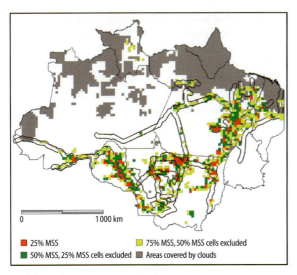

Fig. 7.3. Distribution of deforestation along road corridors in the Brazilian Amazon, showing areas with 25%, 50% and 75% deforestation (*colors*) and the areas within 25 km from the nearest road (*gray line*). Cells covered by a minimum of 50% of clouds also shown in gray (Alves 2002a). Information like this was used by scientists to show policy makers that road construction is linked to deforestation (see Chap. 3, Box 3.5)

Brazilian Amazon – see Fig. 7.3. This knowledge has led development banks and agencies to change their lending policies for road development projects in the Amazon (Redwood III 2002). It also motivated the Brazilian Federal government to establish public panels to discuss the paving of an important road link between Central Brazil and a major port on the Amazon River under the so called "Avança Brasil" development program.

In Nigeria, land-change research on urbanization has raised the profile of important issues of land-use change by providing credible information on the proximate causes, rates and locations of urbanization. In some ways, this research increased the saliency of the issue of urbanization by popularizing and disseminating research results to the public. Land-change science, because of its connection to high profile climate change research, has high political visibility in the government and NGO sectors, and has helped re-invigorate institutional support for urban planning.

Another example from East Africa uses the principles of establishing trust, strengthening researcher-policy networks, initiating research with a strong communication strategy, and establishing a network of research policy "translators" (Reid et al. 2005). This research team evaluates the trade-offs and complementarities inherent in different land-use practices in promoting pastoral welfare and conserving wildlife, goals often addressed by entirely different sectors of the government and donor communities. One key to this approach is identification (and re-identification over time) of the salient, policy-relevant issues for research with local communitiy members and leaders and also with national-level research and management institutions. Legitimacy was

established by including and addressing the wide-ranging concerns of different actors (individuals, institutions) that focus on agricultural development, land-use planning, water resources and wildlife conservation. The centerpiece of the communication strategy revolved around a group of researcher-community members, whose role was to establish legitimacy and guarantee saliency of the research, and to develop and strengthen researcher – policy maker links at the local and national levels. Another effective strategy was for the core research – communication team to act as a convenor and catalyst for other national and international researchers working in the same ecosystems to communicate with local and national policy makers. Specific activities to strengthen these links include feedback workshops with researcher and community members, meetings with policy makers to revise policy acts on wildlife and pastoral development, grants to international students to report their PhD results back to communities and discuss policy and management options, and meetings for researcher-policy maker discussions of salient issues. However, like most projects of this kind, no formal evaluation of the impacts of research on policy has been attempted.

The Krui people in Lampung Province, Indonesia, and their scientific colleagues on the ASB team together successfully reformed government policy that was set to violate their land tenure and appropriate their land for logging and conversion to an oil palm estate. They achieved this first by creating a credible and legitimate assessment of the social, ecological and economic benefits of their traditional agroforestry practices, so that government planners no longer classified their lands as "empty". Local groups were able to speak with conviction about the value of the way they used the land when policy makers visited their land, persuading policy makers to recognize the value of their lifeways. Six months after these visits and a report to the Ministry of Forestry, the Indonesian government reversed their appropriation policy (Tomich and Lewis 2001).

In China, political discourse (Brown 1995), technology advance (Welch and Pannell 1982), as well as a national land-use survey (Smil 1995) have aroused the Chinese state's concern about land use and food security. As a result, the state has implemented a very strict policy to maintain enough agricultural land to feed the population, a total of arable land area of no less than 1.28 million km^2 in China. The government is reclaiming land in northern China to compensate farmers for the loss of agricultural land mainly along the coast and in southern China due to urbanization and infrastructure development in the last two decades (Yang and Li 2000) which, in turn, paradoxically causes further land degradation and desertification in some cases.

Another very clear example of land-use scientists working to directly influence land policy involves panda conservation in Wolong Nature Reserve in China. Loss of high-quality panda habitat was faster after the reserve was set up (1974–1997) than before the reserve's establishment (1965–1974; Liu et al. 2001). This was due to a rapid increase in human population and an even faster jump in the number of households (Liu et al. 2003a), thus greatly expanding human settlement and other human activities (e.g., fuelwood collection and agriculture). This type of information helped the government develop and implement a set of new initiatives. The initiatives include: *(a)* establishment of an eco-hydropower plant to reduce fuelwood cutting, *(b)* direct payments (approximately $100–150 per household per year, or approximately 20% of average household income) to local communities to monitor natural forests and prevent illegal harvesting of trees, and *(c)* a grain-to-green program where farmers are given tree seedlings to plant in their fields and are paid (in the forms of grain and cash) for the amount of land they convert back to forest (Feng et al. 2006). Although the second and third programs are nation-wide in response to the 1998 major floods in China, their implementation in Wolong is mainly for panda habitat restoration and financial support for adjacent areas outside Wolong has been much less than that inside Wolong. Many suggestions based on the Wolong study (Liu et al. 2003b) are also being seriously considered for improving the entire nature reserve system in China because many of the reserves (almost 2 000 in total) are faced with similar challenges as Wolong. There are at least three reasons for this success: *(a)* the issue that the scientists tackled was high-profile or salient both within China and on the world stage, *(b)* the scientific team worked closely with policy makers, and *(c)* governance structures in China allow policy makers to enact policy quickly.

In Europe, policy makers initiated or funded several applications of land-use models to answer specific questions. For example, the EURURALIS project aims to develop an interactive, user-friendly meta-model to catalyze a balanced discussion about the future of the rural areas in 25 European countries from the perspective of sustainable land use in the coming decades – see Box 5.6. The project team interacts closely with the policy advisory group of the Dutch Ministry of Agriculture, Nature management and Food Quality (ANF), and the results will be discussed by the 25 nations. This work raised the profile and attention given by policy makers to land-use issues, but it is unclear if the results will be used to revise policy (Verburg et al. 2006b).

In Costa Rica, a team of scientists worked with policymakers to develop models that allow them to assess the environmental and socio-economic impacts of land-use/cover change, commissioned by the World Bank (Kok and Veldkamp 2001; Kok and Winograd 2002). Translation and communication of results between scientists and policy makers was one of the big challenges of this integrated team. Scientists presented land-cover change maps, showing hot spots of change, but policy makers

wanted piecharts and graphics of appropriate and inappropriate land uses. From the scientists' perspectives, this means the crucial information of the specific locations and rates of land-use change is lost in this translation. However, good progress is being made because policy makers now pay a good deal of attention to land-use change issues and they recognize the value of making future projections of land use.

7.4 How Can Land-Change Science Be More Useful in the Policy Process?

Despite these successes, why doesn't land-change science have more impact on policy now? How can science have more influence on land-use policy in the future? There are some clues from the research of those who have worked in this area and attempted to understand how research influences policy outside of land-change science (Garrett 1998; Sutton 1999; Court and Young 2003). All assessments admit that our understanding of these impacts is "thin" and better, more formal assessments need to be made. Despite this, there are some clear ways that land-change science could be more useful to policy makers. In thinking about this, scientists must understand that there is little chance for science to control policy outcomes. Rather, the key is for scientists to link their work to social/political processes and use this linkage to set more "salient" research priorities that will have a better chance of affecting those processes (van Noordwijk et al. 2001).

First, scientists need to listen to understand policy makers care the most about. Understanding needs and beliefs will allow scientists to design their research so that it is truly relevant and salient to policy makers. In this discussion of science and policy, we focus on scientists and policy makers, but it is particularly critical to include the viewpoints of the land users themselves throughout the process. One way to do this is to transform the current, relatively *ad hoc* information collection by land-use scientists, that may (or may not) have policy implications, into more purposeful land-change policy research that aims to be useful to policy makers (Tomich 1999; Tomich et al. 2004a) and land users. Policy research starts with a clear definition of a policy research problem, including assessment of policy objectives and the impact of existing policies, identification of relevant policy instruments, and establishing working relationships with policy makers who have influence over those policy instruments. One of the first steps for researchers seeking to embark on policy-relevant research is to listen to the questions that policy makers ask (Tomich et al. 2004a):

- Who cares? Who loses? Does anybody win? Are the negative (or positive) effects big enough to capture the attention of local people or of policy makers?

- So what? Is it a policy problem? Would action serve one or more public policy objectives?
- What can be done? Do we know enough to act? Will it work? What are the risks? What will it cost?

Once scientists listen to questions posed by policy makers and land users, they will be able to frame salient, appropriate and useful policy research questions. They will then be able to design their research to collect the most effective data to address the policy problem, which will depend, in part, on where the problem is in the policy issue cycle (Tomich et al. 2004a) – see Fig. 7.4. With a new issue, scientists need to focus on establishing if the issue is a problem, using process-based research that establishes cause and effect. This is where much of global land-change science has focused in the last decade, since much of our understanding of connections between land use and the environment, for example, is relatively new. Some of the land-change science at the local and national levels now focuses further along the cycle, on how big the problem is, what to do about it (mitigation or adaptation options) and how to monitor progress on addressing the problem. Towards the end of the issue cycle, after stakeholders have a broad understanding of the problem and have reached consensus on the need and way to act, then research is likely to have the most impact if it develops cheap, replicable and credible indicators (that will stand up under legal scrutiny, for example) for use in monitoring and enforcement.

In most cases, however, it simply is naïve to expect that better information alone will lead to better public policy and land use. Typically, the most that can be hoped for is that policy research can support the efforts of certain policy makers, politicians and others who share a commitment to core long-term land-use policy objectives such as reducing deforestation or combating desertification. Without links to influential individuals, prospects for constructive impact of policy research are severely limited. This also means that, if there are beneficial policy changes, these influential individuals – not researchers – deserve the credit (Tomich 1999).

Policy makers – especially in democratic societies – often want to maximize votes and agreement in short election cycles. Where opportunities exist, scientists and policy-makers alike need to put additional emphasis on win-win situations that deliver both short-term benefits for politicians and long-term conservation of natural resources. Of course, this political calculus applies even in non-democratic societies.

Another issue is that land-use problems often occur at landscape and regional scales. This creates a problem in policy, because, particularly in developing countries, there are few institutions that naturally operate at this scale: many function locally and internationally, but not in the "missing middle" (Tomich et al. 2004a). In general, collective action is more difficult when more people or institutions are involved, they are in different locations, and they speak different languages: they are substantially heterogeneous. Thus, in the missing middle, action is problematic and institutions are weaker, making policy action doubly difficult.

Scientists, land users and policy makers may find that it is useful to work directly with boundary or "translator" organizations whose goals are to bring the best of scientific information into policy (Cash et al. 2003; Soberon 2004). These organizations (or individuals) can mediate when scientists and policy makers (and other stakehold-

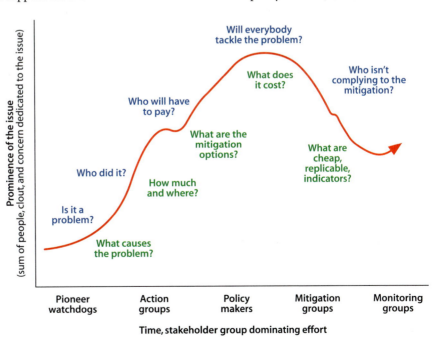

Fig. 7.4.
The "issue cycle" in a democracy showing, over time, the groups who focus on an environmental problem (*x*-axis), how prominent the problem is in public discourse (*y*-axis), the appropriate policy questions (*blue, above line*) and scientific questions (*green text, below line*; adapted from Tomich et al. 2004b)

ers) have different constraints and goals, and when they differ on what kind of information is credible and useful. They can ensure that communication is active, iterative and inclusive, thus strengthening the legitimacy of these interactions. They can also help scientists understand what policy makers and land users need and how the policy process works. It is helpful if key individuals in these boundary organizations are accountable to the scientists, land users and policy makers.

Land-use scientists need to work closely with policy makers and land users to identify – and in many cases develop, test, and validate – workable policy levers that effectively influence the rate and patterns of land-use change (Tomich et al. 2004c). There are, of course, policy instruments that are relatively easy to manipulate (at least technocratically, if not politically) and that have powerful effects on land use and land-use change. Examples include exchange rates and interest rates; price, trade, and marketing policies; and public expenditures for infrastructure (Tomich et al. 2004a). While it is important to recognize that finance ministries are far more powerful than others concerned with land use (e.g., agriculture, environment), they also have much broader economic goals to satisfy. So, while this group of macroeconomic policy instruments is too important for land-use scientists and policy analysts to ignore, it is unlikely (and probably even undesirable) that they would be "targeted" to achieve specific land-use objectives. Public expenditures on research and extension and laws and regulations affecting access to and transfer of land and other assets are much more tightly linked to land-use issues and comprise an important set of topics for engagement between researchers and policy makers and other stakeholders. Direct mechanisms to address the "market failures" that underpin many of the environmental problems linked to land-use and land-cover change probably are the most challenging among land-policy research issues because few (if any) workable methods have been developed. However, despite this challenge, as a general rule the "closer" an intervention is linked to the problem it seeks to influence, the better the chances for success without also producing offsetting distortions. A specific example here would be mechanisms to reward poor people for managing landscapes to produce environmental services as well as conventional commodities. The approach to research required for success in developing such policy instruments depends on, but also is very different from, research strategies that are effective in identifying and quantifying basic cause and effect relationships. A further complication is that, for situations in which there are multiple, interacting policy problems (as typically is the case in land-use policy analysis), it is unlikely that any single intervention can address all problems. Moreover, a piecemeal approach easily can make the overall situation worse. An important example here would be deregulation of markets to reduce trade distortions for

forest products without also addressing property rights over forest resources. So, a comprehensive approach to policy analysis and implementation is necessary. Given message 3 above, policies need to reconnect agents and stakeholders of change.

Compared with some other issues, land-use issues may not have high priorities in political agendas. This makes it all the more important that land-use scientists explain the land-use implications of policy options.

7.5   Conclusions

In the last decade, land-change science came into its own. Because of this, we can write this chapter and suggest some messages for policy makers. We also can learn from the scholarship in other areas to suggest a process to improve the links between scientists and policy makers. We now know much more about the rates, causes, pathways and consequences of land-use/cover change, and these are usually specific to different parts of the world (see Chap. 2, 3 and 4). We think that understanding patterns of forces driving rapid land-use change and associated effects (or feedbacks) on the environment and human societies can help policy makers develop more effective strategies and identify specific opportunities for policy intervention. We need to build on the few win-wins, where the goals of sustainability and development are aligned, but also on situations where we make big wins and lose only a little. And policy makers must account for the actions of different actors when crafting new policy.

This chapter argues that land-change science has made some major advances in producing information, which builds on and integrates a long tradition of studying land-systems change in various parts of science. This information is often relevant to land policy. For example, we have learned that human population growth is closely associated with land-use change. However, population growth is not as influential as previously thought, particularly if that growth is in urban rather than rural areas (Mather and Needle 2000). Instead, policy-related and other factors often play more important roles. Further, there are many causes of land-use/cover change, but this complexity is not infinite. Some causes are more important than others and similar forces cause land-cover change in systems as dissimilar as humid forests and drylands. Despite this, we still cannot explain fully what affects the speed and magnitude of land-use/cover change (see Chap. 5). We also know that there are some relatively predictable pathways of land-use/cover change and that these allow some generalization about driving forces and development of scenarios of future change (see Chap. 6). And we know how and why single interventions can change the way people use the land. For example, the introduction of new agricultural technologies can sometimes encourage farmers to rapidly clear tropical forest, but in other places, with differing economic and social contexts, the introduction of a similar technology may discourage farmers from expanding their agricultural land at the expense of the forest. The latter obtains only under quite restrictive conditions and, in reality, the former is much more important. The key insight here, however, it that the outcomes depend as much or more on market access, institutions, and the policy environment – i.e., interacting or mediating factors – than on the specific technology (Angelsen and Kaimowitz 2001b).

But better understanding is just the beginning. The chapters in this collection show how land use involves decisions taken by individuals (e.g., farmers, pastoralists, forest dwellers), but these decisions also are shaped by policy and political economy. The various actors have different knowledge systems, power relations, and interests, which calls for better communication among scientists, policy makers, and society. Effective links between policy makers, local communities and scientists will reduce the risk of unexpected changes in unexpected places, and strengthen the entire process of land management (Lebel 2004).

Influencing policy clearly is not a trivial task. However, land-use problems touch some of the most daunting problems of our times. To rise to this challenge, land-use scientists can develop better and more reliable ways to provide input into decision making, if they take steps to become properly engaged and make the commitment to follow through. Some scientific leaders call not only for increased engagement with policy makers by scientists, but also for the creation of a radical new approach, creating new professions and strong accountability (Lubchenco 1998). Unfortunately, from experience worldwide so far, it is clear that developing, implementing and evaluating effective science-policy links takes time, perhaps a decade or more (Cash et al. 2003). With no time to lose, the best time to start is now.

Chapter 8

Conclusion

Scientific Steering Committee of the Land-Use/Cover Change (LUCC) Project

In this chapter, we first summarize some of the key findings of the LUCC project on its research questions (see Box 1.1), and then outline some of the elements at the frontier in land-use/cover change research.

8.1 Main Findings on Land-Change Science

LUCC Question 1

> How has land cover been changed by human use over the last 300 years?

Human activities have transformed our planet's landscapes for a long time. The pace and intensity of land-cover change increased rapidly over the last three centuries, and accelerated over the last three decades. Since the 1960s and the Green Revolution, an intensification in land-use practices has been observed. The rapid land-cover changes that have been observed (mostly in humid forests) are not randomly or uniformly distributed, but clustered in particular locations; for example, on forest edges and along transportation networks. Spatially diffuse land-cover changes, especially in drylands, are more difficult to observe.

Different processes of land-cover change have taken place in different parts of the world in the last two decades (for example, decreases in cropland in temperate regions and increases in the tropics), and have had different impacts. Land-cover modifications (subtle changes that affect the character of the land cover without changing its overall classification) are as important as land-cover conversions (the replacement of one cover type by another). Reliable data at a global scale is lacking on changes in (sub)tropical dry forests (e.g., Miombo and Chaco forests); forest-cover changes caused by selective logging, fires and insect damage; drainage or other alterations of wetlands; soil degradation in croplands; changes in extent and productive capacity of pastoral lands; dryland degradation; changes related to urban infrastructure; and lifestyle-driven changes. Moreover, many parts of the world are inadequately represented in existing land-cover change data sets.

LUCC Question 2

> What are the major human causes of land-cover change in different geographical and historical contexts?

Land-use change is always caused by multiple interacting factors originating from different levels of organization of coupled human-environment systems. The mix of driving forces of land-use change varies in time and space, according to specific human-environment conditions.

At decadal time scales, land-use changes mostly result from individual and social responses to changing economic conditions, which are mediated by institutional factors. Opportunities and constraints for new land uses are created by markets and policies and are increasingly influenced by global factors. New technologies can lead to rapid shifts in land-use practices. Institutions (political, legal, economic and traditional) at various scales, and their interactions with individual attitudes, values and knowledge systems, have a major impact on land-use change. Globalization can either amplify or attenuate the effect of driving forces of land-use change. Migration is the most important demographic factor causing land-use change at the timescale of a few decades. At a centennial timescale, both increases and decreases of a given population have a large impact on land use. Demographic change is also associated with the development of households and features of their life cycle.

A restricted set of dominant pathways of land-use change can be identified, as certain human-environment conditions repeatedly appear in case studies. For example, development of the forest frontiers by weak state economies, for geopolitical reasons or to promote interest groups; loss of entitlements to land resources (e.g., expropriation for large-scale agriculture, dams, or wildlife conservation) that lead to ecological marginalization of the poor; induced innovation and intensification, especially in peri-urban and market-accessible areas of developing regions; urbanization followed by changes in consumption patterns and income distribution with differential rural impacts.

LUCC Question 3

> How will changes in land use affect land cover in the next 50–100 years?

Improved understanding of the complex dynamic processes underlying land-use change has led to more reliable projections and more realistic scenarios of future change. A wide range of land-use change models, for different scales and research questions and based on a variety of approaches, is now available. Different models of land-use change address different questions, for example, location of change *versus* quantity of change. No model is able to answer all questions. Some models consider an area of land as the unit of analysis, while others are centered on individuals as decision making agents.

Only a few models of land-use change can generate long-term projections of future land-use/cover changes at the global scale, and so a regional approach is usually adopted. Crucial to projections of future land use is understanding the factors that control positive and negative feedback in land-use change. Model reconstructions of past land-use patterns are now better than random patterns or "no-change" assumptions.

Scenarios of land-use change help to explore possible futures under a set of simple conditions by summing up current knowledge in the form of consistent, conditional statements about the future. Scenario building can involve policy makers and stakeholders to define and negotiate relevant scenarios. Existing land-use change scenarios indicate the possibility of long-term and large-scale changes in land use and land cover with implications for many aspects of the Earth System. They indicate that long-term trends may be reversed after some decades. Urbanization and associated changes in lifestyles are likely to become the dominant factor in land-use change in the decades to come.

LUCC Question 4

> How do human and biophysical dynamics affect the coupled human-environment system?

Human-environmental systems are complex adaptive systems in which properties, such as land use, emerge from the interactions amongst the various components of the entire system. These properties themselves feed back to influence the subsequent development of those interactions. Land-use changes have multiple impacts on ecosystem goods and services at a variety of spatial and temporal scales. There are trade-offs between immediate human needs satisfied by land use, and maintaining the capacity of the biosphere to provide goods and services in the long term. Adopting a long-term view of land-use change history in a given region is essential to understanding current changes and to predicting future

ones as legacies of past land-use changes continue to have impacts today. Institutional and technological innovations may lead to negative feedback loops that decrease the rate of land-use change. There are several historical and contemporary examples of land-use transitions associated with other societal and biophysical changes.

LUCC Question 5

> How might changes in climate (variability) and biogeochemistry affect both land use and land cover, and *vice versa*?

Slow and localized land-cover conversion takes place against a background of high temporal frequency regional-scale fluctuations in land-cover conditions caused by climatic variability, and it is often linked through positive feedbacks with land-cover modifications. Abrupt, short-term changes, often caused by the interaction of climatic and land-use factors, have important impacts on ecosystem processes.

Towards a Theory of Land-Use Change

The large complexity of causes, processes and impacts of land change has so far impeded the development of an integrated theory of land-use change. Much progress has been made in understanding under what conditions different theoretical orientations, borrowed from a variety of disciplines, prove useful. However, the need to address land change from the perspective of a coupled human-environment system (or societal-ecological system) is now widely recognized, with the hope that one or more overarching theories of land change may emerge. Such theories must address the behavior of people and society (agency and structure) and the uses to which land units are put, as well as feedbacks from one to the other. Theories must be multi-level with respect to both people and land units, recognizing that they can combine in ways that affects their collective and individual behaviors. They must incorporate the extent to which people and pixels are connected to the broader world in which they exist, and must incorporate both history and the future.

Policy Implications

The use of land is a highly political activity. Misguided or uncoordinated sectoral policies are one of the major causes of land degradation. Lifestyle choices and consumption patterns affect land-use choices, and universal policies for controlling land-use change will not be effective when implemented. Rather, a detailed understanding of the complex set of causes affecting land-use change in a given location is required prior to any policy intervention. Connections between land-use change in one area and impacts elsewhere also deserves full attention. Policy intervention should address the underlying causes as much as the proximate causes of land-use change. To design effective response strategies in the face of rapid land-use change, one needs to understand: *(a)* environmental perception, information processing and transfer by agents; *(b)* determinants of decision making and individual behavior with respect to land management; and *(c)* portfolios of available and feasible responses to land-use change for the different categories of agents. Good and efficient communication of the location of adverse impacts of land-use change to policy makers can allow them to react in a timely manner.

8.2   Frontier in Land-Use/Cover Change Research

At the end of a ten year research project, the list of new issues to be investigated is often longer than the list of research findings. The objective of this article is to highlight some of the important issues at the frontier of land-use/cover change research. The following section provides a sample of topics, which is far from being an exhaustive list. The new Global Land Project (Ojima et al. 2005) will take over a lot of these research issues.

Understanding Land-Use Transitions

Urbanization and migrations are likely to play an ever dominant role in shaping new land uses, further disconnecting spaces of consumption and production worldwide. Migration is generally thought to have a stronger impact on land-use change than mortality and fertility, at least at time scales of a few decades. In future population-environment studies, micro- to macro scale demographic variables should be studied in context rather than as exogenous driving forces. Megacity development tends to dominate discussions on urbanization (e.g., urban lifestyle influences on remote rural areas), but networks of secondary cities and peri-urban areas are also crucial in land-use change as urban-rural linkages are stronger at that level.

Globalization and "export" of land use via international trade also deserves more attention – e.g., in the case of booming economies such as China that pulls products from the entire world with non-negligible land-use impacts in sometimes distant countries. In the same vein, future land-use research needs to better consider constraints such as capital availability, technology, policies, and macro-economic shocks, and the cross-scale interactions between these factors. The expansion of agricultural frontiers remains an important research topic, e.g., in the Amazon, but this expansion is increasingly linked to urbanization and globalization in ways that remain poorly understood.

Managing transitions towards sustainable land use, which is a normative exercise, needs to address these global-local interplays. Transitions are sensitive towards global as well as local and regional constraints and opportunities. Locally, engagement and communication with stakeholders in regions where teams conduct land-use change research need to be more systematic. This will often require, first, establishing interfaces with other disciplines that will be relevant to assess impacts of land-use/cover change, and, second, considering multiple scales of governance structures, institutions, conflicts and interactions between multiple agents.

Vulnerability in the Face of Land-Use Change

There are many research opportunities to understand vulnerability in a multi-dimensional, dynamic way. This research needs to couple social as well as ecological vulnerability and integrate the multiple impacts of land-use change on societies and ecosystems – e.g., on social and economic well-being, food security, health but also water resources, the carbon cycle, and ecosystem functioning. The linkage between land and water use needs to be better understood and incorporated into vulnerability studies. Water impacts on land-use change are an important issue (e.g., irrigation farming in drylands). One of the most important trade-offs facing many societies engaged in intensive agriculture is between water quality and agricultural development. Likewise, new research requires an integration of emerging results from biocomplexity research on patterns of biodiversity at multiple scales, with strong linkages to research on conservation biology and livelihood security.

While land-use change research has tended to focus on so-called "slow variables", a big challenge is to better integrate extreme events of all kinds: climate events (e.g., at ENSO-type time scales, decadal scale, etc.) but also human events (e.g., wars, conflicts, economic shocks). These "fast variables" often determine the resilience and collapse of systems. Surprises happen but integration of surprises in land-use change research has not yet happened to the extent required. The concept of resilience establishes the link between risks from extreme events and social well-being.

Long-Term Social-Ecological Research

The global change scientific community increasingly studies coupled human-environment systems on time scales of hundreds to thousands of years. Land-use change researchers have much to offer to long-term social-ecological research. At these long-time scales, there is a strong footprint of agriculture which needs to be better explored and quantified, including impacts on biogeochemical cycles.

Tools and Methods

Prominent among new tools and methods is integrated modeling. Some of the next steps needed to improve models include better integration of social and biophysical drivers, better modeling of decision-making by agents, an improved ability to model lag times and thresholds in land-use decisions, and multisource data integration (e.g., remote sensing with census and household survey data). Integrated modeling work should rely on global, regional and local scale digital databases, not just on land-cover classes, but also on land management (fertilization, irrigation, etc.), with more participatory open GIS and data sharing.

Future scenarios of land-use change should be formulated in the context of multiple stakeholders. Agent-based models increasingly become a tool of choice for understanding decision making, even though they should not be viewed as a panacea. Spatially-explicit multi-agent simulation models allow simulating surprises and evaluating their potential impacts on the landscape.

Much has been learnt on the causes of land-use change through meta-analyses of large numbers of case studies. A methodological challenge is to move beyond *a posteriori* meta-analyses of results, but rather conduct comparative analyses of case studies by analyzing original data from these case studies. This requires standardized data collection descriptions that allow comparisons, while still recognizing the need to fine tune data collections to the most relevant processes in specific localities. While a standardized land-cover classification system has now been produced, an equivalent scheme for land use is crucially needed.

More generally, land-use change researchers will have to further diversify their portfolio of analytic methods: not just multiple regressions but also narratives, system and agent-based approaches, network analysis, etc.

References

Abbot JIO, Homewood K (1999) A history of change: Causes of miombo woodland decline in a protected area in Malawi. J Appl Ecol 36:422–433

Abrol YP, Sangwan S, Dadhwal VK, Tiwari MK (2002) Land use/ land cover in Indo-Gangetic Plains: History of changes, present concerns and future approaches. In: Abrol YP, Sangwan S, Tiwari MK (eds) Land use: Historical perspectives – Focus on Indo-Gangetic Plains. Allied Publishers, New Delhi, pp 1–28

ACCELERATES (2004) Assessing climate change effects on land use and ecosystems: From regional analysis to the European scale. Section 6: Final report. Louvain-la-Neuve, Belgium, downloaded from: http://www.geo.ucl.ac.be/accelerates/

Achard F (2006) Tropical Ecosystem Environment Observations by Satellite (TREES) project. In: Geist HJ (ed) Our Earth's changing land: An encyclopedia of land-use and land-cover change, vol. 2 (L–Z). Greenwood Press, Westport London, pp 608–610

Achard F, Eva HD, Glinni A, Mayaux P, Richards T, Stibig HJ (1998) Identification of deforestation hot spot areas in the humid tropics. TREES Publications Series B, Research Report No. 4, EUR 18079 EN, European Commission, Luxembourg, 100 pp

Achard F, Eva HD, Stibig H-J, Mayaux P, Galego F, Richards T, Malingreau J-P (2002) Determination of deforestation rates of the world's humid tropical forests. Science 297:999–1002

Achard F, Eva HD, Mayaux P, Stibig H-J, Belward A (2004) Improved estimates of net carbon emissions from land cover change in the tropics for the 1990s. Global Biogeochem Cy 18:GB2008, doi:10.1029/2003GB002142

Agrawal A, Yadama GN (1997) How do local institutions mediate market and population pressures on resources? Forest panchayats in Kumaon, India. Dev Change 28:435–465

Agarwal C, Green GM, Grove JM, Evans TP, Schweik CM (2001) A review and assessment of land use change models: Dynamics of space, time, and human choice. Center for the Study of Institutions, Population, and Environmental Change, Indiana University and USDA Forest Service, Bloomington and South Burlington

Aitkin M, Anderson D, Hinde J (1981) Statistical modelling of data on teaching styles (with discussion). J Roy Stat Soc A 144: 148–161

Alcamo J (2001) Scenarios as tools for international environmental assessment. European Environment Agency, Environmental Issue Report No. 24, 31 pp

Alcamo J, Leemans R, Kreileman E (1998) Global change scenarios of the 21st century: Results from the IMAGE 2.1 model. Elsevier, London, 296 pp

Alcorn J (1990) Indigenous agroforestry strategies meeting farmers' needs. In: Anderson A (ed) Alternatives to deforestation: Steps toward sustainable use of the Amazon rain forest. Columbia University Press, New York, pp 141–148

Alexandratos N (ed) (1995) World agriculture: Towards 2010. An FAO study. FAO and John Wiley & Sons, Ltd., Chichester New York, 488 pp

Allen JC, Barnes DF (1985) The causes of deforestation in developing countries. Ann Assoc Am Geogr 75(2):163–184

Allison GT (1971) Essence of the decision: Explaining the Cuban missile crisis. Little and Brown, Boston

Alroy J (2001) A multi-species overkill simulation of the end-Pleistocene megafaunal extinction. Science 292:1893–1896

Alves DS (2001a) O processo de desmatamento na Amazônia. Parcerias Estratég 12:259–275

Alves DS (2001b) Space-time dynamics of deforestation in Brazilian Amazonia. Int J Remote Sens 23:2903–2908

Alves DS (2002a) An analysis of the geographical patterns of deforestation in Brazilian Amazonia in the 1991–1996 period. In: Wood CH, Porro R (eds) Land use and deforestation in the Amazon. University Press of Florida, Gainesville, pp 95–106

Alves DS (2002b) Space-time dynamics of deforestation in Brazilian Amazonia. Int J Remote Sens 23:2903–2908

Alves DS, Escada MIS, Pereira JLG, Souza CL (2003) Land use intensification and abandonment in Rondônia, Brazilian Amazonia. Int J Remote Sens 24:899–903

Amelung T, Diehl M (1992) Deforestation of tropical rain forests: Economic causes and impact on development. Kieler Studien 241, J.C.B. Mohr, Tübingen

Anderson LE (1996) The causes of deforestation in Brazilian Amazon. J Environ Dev 5:309–328

Andreae MO, Crutzen PJ (1997) Atmospheric aerosols: Biogeochemical sources and role in atmospheric chemistry. Science 276:1052–1058

Angelsen A (1999) Agricultural expansion and deforestation: Modelling the impact of population, market forces and property rights. J Dev Econ 58:185–218

Angelsen A, Kaimowitz D (1999) Rethinking the causes of deforestation: Lessons from economic models. World Bank Res Obser 14(1):73–98

Angelsen A, Kaimowitz D (2001a) Introduction: The role of agricultural technologies in tropical deforestation. In: Angelsen A, Kaimowitz D (eds) Agricultural technologies and tropical deforestation. CAB International, Wallingford New York, pp 1–34

Angelsen A, Kaimowitz D (eds) (2001b) Agricultural technologies and tropical deforestation. CAB International, Wallingford New York

Anken T, Weisskopf P, Zihlmann U, Forrer H, Jansa J, Perhacova K (2004) Long-term tillage system effects under moist cool conditions in Switzerland. Soil Tillage Res 78:171–183

Anselin L (1988) Spatial econometrics: Methods and models. Kluwer Academic, Dordrecht Boston London

Anselin L (2002) Under the hood: Issues in the specification and interpretation of spatial regression models. Agricult Econ 27: 247–267

Antona M, Bousquet F, Le Page C, Weber J, Karsenty A, Guizol P (1998) Economic theory of renewable resource management: A multi-agent system approach. Lect Notes Artif Int 1534:61–78

Archer DR (2003) Scale effects on the hydrological impact of upland afforestation and drainage using indices of flow variability: The River Irthing, England. Hydrol Earth Syst Sc 7(3):325–338

Arentze TA, Timmermans HJP (2000) Albatross: A learning based transportation oriented simulation system. European Institute of Retailing and Service Studies, Eindhoven University, Eindhoven

Arentze TA, Dijst M, Dugundji E, Joh C-H, Kapoen L, Krijgsman S, Maat K, Timmermans HJP, Veldhuisen J (2001) The AMADEUS program: Scope and conceptual development. In: Park CH, Cho JR, Oh J, Hayashi Y, Viegas J (eds) Selected Proceedings of the 9th World Conference on Transport Research, July 22–27, 2001, Seoul, Korea. Elsevier

Arnold JG, Williams JR, Maidment DR (1995) Continuous-time water and sediment-routing model for large basins. J Hydraul Eng-ASCE 121:171–183

Asian Development Bank, Global Environment Facility, United Nations Environment Programme (1999) Asia least cost greenhouse gas abatement strategy. Manila, Philippines: Asian Development Bank, 260 pp

Asner GP (2001) Cloud cover in Landsat observations of the Brazilian Amazon. Int J Remote Sens 22:3855–3862

Asner GP, Archer SA, Hughes RF, Ansley JN, Wessman CA (2003) Net changes in regional woody vegetation cover and carbon storage in North Texas rangelands 1937–1999. Glob Change Biol 9:316–335

Asner GP, DeFries RS, Houghton RA (2004) Typological responses of ecosystems to land use change. In: DeFries RS, Asner GP, Houghton RA (eds) Ecosystems and land use change. Geophysical Monograph 153, American Geophysical Union, Washington D.C., pp 337–344

Asner GP, Knapp DE, Broadbent EN, Oliveira PJC, Keller M, Silva JN (2005) Selective logging in the Brazilian Amazon. Science 310:480–482

Bailey LH (ed) (1909) Cyclopedia of American agriculture: A popular survey of agricultural conditions, practices and ideals in the United States and Canada, vol. 4 – Farm and community. The Macmillan Company, London, 628 pp

Bak P (1998) How nature works. Copernicus/Springer, Berlin Heidelberg

Bakker MM, Govers G, Kosmas C, Vanacker V, Oost K van, Rounsevell M (2005) Soil erosion as a driver of land-use change. Agric Ecosyst Environ 105(3):467–481

Ball JB (2001) Global forest resources: History and dynamics. In: Evans J (ed) The forests handbook, vol. 1. Blackwell, Oxford, pp 3–22

Balmford A, Bennun L, ten Brink B, Cooper D, Côté IM, Crane P, Dobson A, Dudley N, Dutton I, Green RE, Gregory DR, Harrison J, Kennedy ET, Kremen C, Leader-Williams N, Lovejoy TE, Mace G, May R, Mayaux P, Morling P, Phillips J, Redford K, Ricketts TH, Rodríguez JP, Sanjayan M, Schei PJ, van Jaarsveld AS, Walther BA (2005) The Convention on Biological Diversity's 2010 target. Science 307:212–213

Barber CV, Schweithelm J (2000) Trial by fire: Forest fires and forestry policy in Indonesia's era of crisis and reform. World Resources Institute, Washington D.C., 88 pp

Barbier EB (1993) Economic aspects of tropical deforestation in Southeast Asia. Global Ecol Biogeogr 3:215–234

Barbier EB (1996) Impact of market and population pressure on production, incomes, and natural resources in the dry land savannas of West Africa: Bioeconomic modeling at the village level (EPTD Discussion Paper 21). International Food Policy Research Institute, Washington D.C.

Barbier EB (1997) The economic determinants of land degradation in developing countries. Philos Trans R Soc London, Ser B 352:891–899

Barbier EB (2000a) Links between economic liberalization and rural resource degradation in the developing regions. Agricult Econ 23(3):299–310

Barbier EB (2000b) The economic linkages between rural poverty and land degradation: Some evidence from Africa. Agric Ecosyst Environ 82:355–370

Barnosky AD, Koch PL, Feranec RS, Wing SL, Shabel AB (2004) Assessing the causes of Late Pleistocene extinctions on the continents. Science 306:70–75

Barraclough SL, Ghimire KB (1996) Forests and livelihoods: The social dynamics of deforestation in developing countries. Saint Martin's Press, New York

Barrett CB, Barbier EB, Reardon T (2001) Agroindustrialization, globalization, and international development: The environmental implications. Environ Dev Econ 6:419–433

Bartholomé E, Belward AS (2005) GLC2000: A new approach to global land cover mapping from Earth Observation data. Int J Remote Sens 26:1959–1977

Batchelor CH, Sundblad K (1999) Reversibility and the time scales of recovery. In: Falkenmark M, Andersson L, Castensson R, Sundblad K (eds) Water: A reflection of land use. Options for counteracting land and water mismanagement. Stockholm: Swedish Natural Science Research Council, pp 79–91

Batterbury SPJ, Bebbington A (1999) Environmental histories: Access to resources and landscape change. Land Degrad Dev 10:279–289

Bauer E, Claussen M, Brovkin V, Huenerbein A (2003) Assessing climate forcings of the Earth System for the past millennium. Geophys Res Lett 30:1276, doi:10.1029/2002GL016639, 9-1–9-4

Bawa KS, Dayanandan S (1997) Socioeconomic factors and tropical deforestation. Nature 386:562–563

Bebbington A (2000) Reencountering development: Livelihood transitions and place transformation in the Andes. Ann Assoc Am Geogr 90(3):495–520

Becker CD (1999) Protecting a Garua forest in Ecuador: The role of institutions and ecosystem valuation. Ambio 28(2):156–161

Behnke RH, Scoones I (1993) Rethinking range ecology: Implications for rangeland management in Africa. In: Behnke RH, Scoones I, Kerven C (eds) Range ecology at disequilibrium: New models of natural variability and pastoral adaptation in African savannas. Overseas Development Institute, London, pp 1–30

Behrenfeld MJ, Randerson JT, McClain CR, Feldman GC, Los SO, Tucker CJ, Falkowski PG, Field CB, Frouin R, Esaias WE, Kolber DD, Pollack NH (2001) Biospheric primary production during an ENSO transition. Science 291:2594–2597

Bell KP, Bockstael NE (2000) Applying the generalized-moments estimation approach to spatial problems involoving microlevel data. Rev Econ Stat 82:72–82

Benitez JA, Fisher TR (2004) Historical land-cover conversion (1665–1820) in the Choptank watershed, eastern United States. Ecosystems 7(3):219–232

Benjaminsen TA (2001) The population-agriculture-environment nexus in the Malian cotton zone. Global Environ Chang 11:283–295

Beresford Q, Phillips H, Bekle H (2001) The salinity crisis in Western Australia: A case of policy paralysis. Aust J Public Adm 60:30–38

Berger T (2001) Agent-based spatial models applied to agriculture: A simulation tool for technology diffusion, resource use changes and policy analysis. Agricult Econ 25:245–260

Bernard FE (1993) Increasing variability in agricultural production: Meru District, Kenya, in the twentieth century. In: Turner BL II, Hyden G, Kates R (eds) Population growth and agricultural change in Africa. University Press of Florida, Gainesville, pp 80–113

Berrang Ford L (2006a) Malaria. In: Geist HJ (ed) Our Earth's changing land: An encyclopedia of land-use and land-cover change, vol. 2 (L–Z). Greenwood Press, Westport London, pp 400–404

Berrang Ford L (2006b) African trypanosomiasis. In: Geist HJ (ed) Our Earth's changing land: An encyclopedia of land-use and land-cover change, vol. 1 (A–K). Greenwood Press, Westport, London, pp 4–6

Berrang Ford L (2006c) Human Immunodeficiency Virus (HIV)/ Acquired Immunodeficiency Syndrome (AIDS). In: Geist HJ (ed) Our Earth's changing land: An encyclopedia of land-use and land-cover change, vol. 1 (A–K). Greenwood Press, Westport, London, pp 296–298

Betts AK, Ball JH, Beljaars ACM, Miller MJ, Viterbo P (1996) The land surface-atmosphere interaction: A review based on observational and global modeling perspectives. J Geophys Res 101(D3):7209–7225

Bicik I (1995) Possibilities of long-term human-nature interaction analysis: The case of land-use changes in the Czech Republic. In: Simmons IG, Mannion AM (eds) The changing nature of the people-environment relationship: Evidence from a variety of archives. Faculty of Science, University of Prague, Prague, pp 47–60

Biggs R, Scholes RJ (2002) Land-cover changes in South Africa 1911–1993. S Afr J Sci 98:420–424

Billington C, Kapos V, Edwards M, Blyth S, Iremonger S (1996) Estimated original forest cover map: A first attempt. World Conservation Monitoring Center, Cambridge

Bilsborrow RE (1987) Population pressures and agricultural development in developing countries: A conceptual framework and recent evidence. World Dev 15(2):183–203

Bilsborrow R, Geores M (1994) Population, land-use and the environment in developing countries: What can we learn from cross-national data? In: Brown K, Pearce DW (eds) The causes of tropical deforestation: The economic and statistical analysis of factors giving rise to the loss of the tropical forests. University College London Press, London, pp 106–133

Bilsborrow RE, Okoth-Ogendo HWO (1992) Population-driven changes in land use in developing countries. Ambio 21(1):37–45

Bingen J (2004) Pesticides, politics and pest management: Toward a political ecology of cotton in Sub-Saharan Africa. In: Moseley WG, Logan BI (eds) African environment and development: Rhetoric, programs, realities. Ashgate Publishing, Burlington, pp 111–126

Biswas AK, Masakhalia YFO, Odego-Ogwal LA, Palnagyo EP (1987) Land use and farming systems in the Horn of Africa. Land Use Policy 4:419–443

Blench R (2000) 'You can't go home again', extensive pastoral livestock systems: Issues and options for the future. Overseas Development Institute, London, United Nations Food and Agriculture Organization

Board on Agriculture and Natural Resources (2003) Frontiers in agricultural research: Food, health, environment, and communities. National Academy Press, Washington D.C.

Bockstael NE, Irwin EG (2000) Economics and the land use-environment link. In: Folmer H, Tietenberg T (eds) The international yearbook of environmental and resource economics 1999/2000. Edward Elgar Publishing, Cheltenham, pp 1–54

Bonan GB (1999) Frost followed the plow: Impacts of deforestation on the climate of the United States. Ecol Appl 9(4):1305–1315

Bonan GB (2001) Observational evidence for reduction of daily maximum temperatures by croplands in the midwest United States. J Climate 14:2430–2442

Bonta JV, Amerman CR, Harlukowicz TJ, Dick WA (1997) Impact of coal surface mining on three Ohio watersheds' surface water hydrology. J Am Water Resour Asc 33:907–917

Bork H-R, Bork H, Dalchow C, Faust B, Piorr H-P, Schatz Th (1998) Landschaftsentwicklung in Mitteleuropa: Wirkungen des Menschen auf Landschaften. Klett-Perthes, Gotha, 328 pp

Bosch JM, Hewlett JD (1982) A review of catchment experiments to determine the effect of vegetation changes on water yield and evapotranspiration. J Hydrol 55:3–23

Boserup E (1965) The conditions of agricultural growth: The economics of agrarian change under population pressure. Aldine, Chicago, 124 pp

Boserup E (1975) The impact of population growth on agricultural output. Q J Econ 89:257–270

Boserup E (1981) Population and technological change. University of Chicago Press, Chicago

Boserup E (2002) (posthum). Technological society and its relation to global environmental change. In: Munn T, Timmermann P (eds) Encyclopedia of global environmental change, vol. 5: Social and economic dimensions of global environmental change. John Wiley & Sons, Ltd., Chichester New York, pp 86–96

Bounoua L, DeFries RS, Collatz GJ, Sellers P, Khan H (2002) Effects of land cover conversion on surface climate. Climatic Change 52:29–64

Bousquet F, Le Page C (2004) Multi-agent simulations and ecosystem management: A review. Ecol Model 176:313–332

Bousquet F, Bakam I, Proton H, Le Page C (1998) Cormas: Common-pool resources and multiagent systems. Lect Notes Artif Int 1416:826–837

Bousquet F, Barreteau O, Le Page C, Mullon C, Weber J (1999) An environmental modelling approach: The use of multi-agent simulations. In: Blasco F, Weill A (eds) Advances in environmental and ecological modelling. Elsevier, Paris, pp 113–122

Boyd DJ (2001) Life without pigs: Recent subsistence changes among the Irakia Awa, Papua New Guinea. Hum Ecol 29(3):259–282

Braimoh AK, Vlek PLG (2004) Scale-dependent relationships between land-use change and its determinants in the Volta Basin of Ghana. Earth Interact 8:1–23

Braimoh AK, Vlek PLG, Stein A (2004) Land evaluation for maize based on fuzzy set and interpolation. Environ Manage 33(2): 226–238

Braimoh AK, Stein A, Vlek PLG (2005) Identification and mapping of associations among soil variables. Soil Sci 170(2):137–148

Bray DB, Ellis EA, Armijo-Canto N, Beck CT (2004) The institutional drivers of sustainable landscapes: A case study of the 'Mayan Zone' in Quintana Roo, Mexico. Land Use Policy 21:333–346

Brejda JJ, Karlen DL, Smith JL, Allan DL (2000) Identification of regional soil quality factors and indicators: II. Northern Mississippi loess hills and Palouse Prairie. Soil Sci Soc Am J 64: 2125–2135

Briassoulis H (2000) Analysis of land use change: Theoretical and modeling approaches. In: Loveridge S (ed) The web book of regional science at http://www.rri.wvu.edu/regscweb.htm. West Virginia University, Morgantown

Bridge G (2006) Mineral extraction. In: Geist HJ (ed) Our Earth's changing land: An encyclopedia of land-use and land-cover change, vol. 2 (L–Z). Greenwood Press, Westport, London, pp 410–416

Broad R (1994) The poor and the environment: Friend or foes? World Dev 22:811–812

Brookfield HC (1962) Local study and comparative method: An example from central New Guinea. Ann Assoc Am Geogr 52: 242–252

Brookfield H (1999) Environmental damage: Distinguishing human from geophysical causes. Environ Hazards 1:3–11

Brookfield H (2001) Exploring agrodiversity. Columbia University Press, New York

Brookfield H, Padoch C (1994) Appreciating agrodiversity: A look at the dynamism and diversity of indigenous farming practices. Environment 36(5):6–11, 37–43

Brookfield H, Stocking M (1999) Agrodiversity: Definition, description and design. Global Environ Chang 9(1):77–80

Brookfield HC, Lian FJ, Low K-S, Potter L (1990) Borneo and the Malay Peninsula. In: Turner BL II, Clark WC, Kates RW, Richards JF, Mathews JT, Meyer WB (eds) The Earth as transformed by human action: Global and regional changes in the biosphere over the past 300 years. Cambridge University Press, New York, pp 495–512

Brookfield H, Parsons H, Brookfield M (eds) (2003) Agrodiversity: Learning from farmers across the world. United Nations University Press, Tokyo

Brooks TM, Balmford A (1996) Atlantic forest extinctions. Nature 380:380–115

Brooks E, Emel J (1995) The Llano Estacado of the American Southern High Plains. In: Kasperson JX, Kasperson RE, Turner BL II (eds) Regions at risk: Comparisons of threatened environments. United Nations University Press, Tokyo New York Paris, pp 255–303

Brooks TM, Pimm SL, Collar NJ (1997) Deforestation predicts the number of threatened birds in insular Southeast Asia. Conserv Biol 11:382–394

Brooks TM, Pimm SL, Oyugi JO (1999) Time lag between deforestation and bird extinction in tropical forest fragments. Conserv Biol 15:1140–1150

Brouwer F, McCarl BA (eds) (2006) Agriculture and climate beyond 2015: A new perspective on future land use patterns. Environment & Policy, vol. 46, Springer, Dordrecht

Browder JO, Godfrey BJ (1997) Rainforest cities, urbanization, development, and globalization of the Brazilian Amazon. Columbia University Press, New York

Brown LR (1995) Who will feed China? Wake-up call for a small planet. World Watch Institute, Washington D.C.

Brown P, Podolefsky A (1976) Population density, agricultural intensity, land tenure, and group size in the New Guinea highlands. Ethnology XV(3):211–238

Brown LR, Abramovitz J, Bright C, et al. (1996) State of the world 1996: A Worldwatch Institute report on progress toward a sustainable society. World Watch Institute, Washington D.C.

Brown DG, Goovaerts P, Burnicki A, Li MY (2002) Stochastic simulation of land-cover change using geostatistics and generalized additive models. Photogramm Eng Remote Sens 68:1051–1061

Brown DG, Page SE, Riolo R, Rand W (2004) Agent-based and analytical modeling to evaluate the effectiveness of greenbelts. Environ Model Software 19:1097–1109

Brown DG, Page S, Riolo R, Zellner M, Rand W (2005) Path dependence and the validation of agent-based spatial models of land use. Int J Geogr Inf Sci 19:153–174

Bryant RL, Bailey S (1997) Third world political ecology. Routledge, London New York

Bryant RL, Rigg J, Stott P (1993) Forest transformations and political ecology in Southeast Asia. Global Ecol Biogeogr 3:101–111

Bryceson DF (1996) De-agrarianisation and rural employment in Sub-Saharan Africa: A sectoral perspective. World Dev 24: 97–111

Burbridge P, Buddemeier RW, Le Tissier M, Costanza R (2005) Synthesis of main findings and conclusions. In: Crossland CJ, Kremer HH, Lindeboom HJ, Marshall Crossland JI, Tissier MDA (eds) Coastal fluxes in the Anthropocene: The Land-Ocean Interactions in the Coastal Zone Project of the International Geosphere-Biosphere Programme. (The IGBP Series), Springer, Berlin Heidelberg, pp 201–218

Butzer K (1990) The realm of cultural-human ecology: Adaptation and change in historical perspective. In: Turner BL II, Clark WC, Kates RW, Richards JF, Mathews JT, Meyer WB (eds) The Earth as transformed by human action: Global and regional changes in the biosphere over the past 300 years. Cambridge University Press, Cambridge, pp 685–702

Campbell K, Borner M (1995) Population trends and distribution of Serengeti herbivores: Implications for management. In: Sinclair A, Arcese P (eds) Serengeti II: Dynamics, management and conservation of an ecosystem. University of Chicago Press, Chicago, pp 117–145

Canadell J, Pataki D, Pitelka L (2006) Terrestrial ecosystems in a changing world. Cambridge University Press, Cambridge

Cannon T (1991) Hunger and famine: Using a food system's model to analyse vulnerability. In: Bohle H-G, Cannon T, Hugo G, Ibrahim FN (eds) Famine and food security in Africa and Asia: Indigenous response and external intervention to avoid hunger. Bayreuther Geowissenschaftliche Arbeiten 15, Naturwissenschaftliche Gesellschaft, Bayreuth, pp 291–312

Carney J (1993) Converting the wetlands, engendering the environment: The intersection of gender with agrarian change in the Gambia. Econ Geogr 69(4):329–348

Carr DL (2004) Proximate population factors and deforestation in tropical agricultural frontiers. Popul Environ 25(6):585–612

Carr DL (2005) Population, land use and deforestation in the Sierra de Lacandón National Park, Petén, Guatemala. Prof Geogr 57(2): 157–168

Caruso G, Rounsevell MDA, Cojocaru G (2005) Exploring a spatio-dynamic neighbourhood-based model of residential behaviour in the Brussels peri-urban area. Int J Geogr Inf Sci 19:103–124

Carvalho G, Moutinho P, Nepstad D, Mattos L, Santilli M (2004) An Amazon perspective on the forest-climate connection: Opportunity for climate mitigation, conservation and development? Environ Dev Sustainability 6:163–174

Cash DW, Clark WC, Alcock F, Dickson NM, Eckley N, Guston DH, Jager J, Mitchell RB (2003) Knowledge systems for sustainable development. Proc Natl Acad Sci USA 100:8086–8091

Cassman K, Wood S, Choo PS, Dixon J, Gaskell J, Khan S, Lal R, Pretty J, Primavera J, Viglizzo E, Kadungure S, Kanbar N, Porter S, Tharme R (2005) Cultivated systems. In: Scholes R, Rashid H (eds) Millennium Ecosystem Assessment: Working group on conditions and trends. Island Press, Washington D.C.

Castella J-C, Boissau S, Trung TN, Quang DD (2005) Agrarian transition and lowland-upland interactions in mountain areas in northern Vietnam: Application of a multi-agent simulation model. Agricult Sys 86:312–332

Cervigni R (2001) Biodiversity in the balance: Land use, national development and global welfare. Edward Elgar, Cheltenham Northampton, 271 pp

Chameides WL, Kasibhatla PS, Yienger J, Levy H II (1994) Growth in continental-scale metro-agroplexes: Regional ozone pollution and world food production. Science 264:74–77

Chameides WL, Yu H, Liu C, Bergin M, Ziuji Z, Mearns LO, Gao W, Kinag CS, Sylor RD, Chao L, Huang Y, Steiner A, Giorgi F (1999) Case study of the effect of atmospheric aerosols and regional haze on agriculture: An opportunity to enhance crop yields in China through emission controls? Proc Natl Acad Sci USA 96: 13626–13633

Charney J, Stone PH (1975) Drought in the Sahara: A biogeophysical feedback mechanism. Science 187:434–435

Chase TN, Pielke RA, Kittel TGF, Nemani RR, Running SW (1996) Sensitivity of a general circulation model to global changes in leaf area index. J Geophys Res 101:7393–7408

Chase TN, Pielke RA Sr, Kittel TGF, Baron JS, Stohlgren TJ (1999) Potential impacts on Colorado Rocky Mountain weather and climate due to land use changes on the adjacent Great Plains. J Geophys Res 104:16673–16690

Chase TN, Pielke RA, Kittel TGF, Nemani RR, Running SW (2000) Simulated impacts of historical land cover changes on global climate in northern winter. Clim Dynam 16:93–105

Chase TN, Pielke RA, Kittel TGF, Zhao M, Pitman AJ, Running SW, Nemani RR (2001) Relative climatic effects of landcover change and elevated carbon dioxide combined with aerosols: A comparison of model results and observations. J Geophys Res 106: 31685–31691

Chayanov AV (1966) Peasant farm organization. In: Thorner D, Kerblay B, Smith REF (eds) A. V. Chayanov on the theory of peasant economy. Irwin, Homewood, pp 21–57

Checkland P (1981) Systems thinking, systems practice. John Wiley & Sons, Ltd., Chichester New York

Chhabra A, Dadhwal VK (2004) Assessment of major pools and fluxes of carbon in Indian forests. Climatic Change 64(3):341–360

Chhabra A, Palria S, Dadhwal VK (2002) Spatial distribution of phytomass carbon pool in Indian forests. Glob Change Biol 8(12):1230–1239

Chhabra A, Palria S, Dadhwal VK (2003) Soil organic carbon pool in Indian forests. Forest Ecol Manag 173(1–3):187–199

Chomentowski W, Salas B, Skole DL (1994) Landsat Pathfinder project advances deforestation mapping. GIS World 7:34–38

Chomitz KM, Gray DA (1996) Roads, land use, and deforestation: A spatial model applied to Belize. World Bank Econ Rev 10:487–512

Chomitz KM, Kumari K (1996) The domestic benefits of tropical forests: A critical review emphasizing hydrological functions (Policy Research Working Paper 1601). World Bank, Washington D.C., 41 pp

Chomitz KM, Thomas TS (2001) Geographic patterns of land use and land intensity in the Brazilian Amazon. World Bank

Chomitz KM, Thomas TS (2003) Determinants of land use in Amazonia: A fine-scale spatial analysis. Am J Agric Econ 85: 1016–1028

Chong S-K, Becker MA, Moore SM, Weaver GT (1986) Characterization of mined land with and without topsoil. J Environ Qual 15:157–160

Christaller W (1933) Central places of Southern Germany, ed. 1966. Prentice Hall, London

Christensen TR, Prentice IC, Kaplan J, Haxeltine A, Sitch S (1996) Methane flux from northern wetlands and tundra: An ecosystem source modelling approach. Tellus 48B:651–660

Chuluun T, Ojima D (eds) (2002) Fundamental issues affecting sustainability of the Mongolian Steppe: Open symposium on "Change and sustainability of pastoral land use systems in temperate and central Asia", Ulaanbaatar, Mongolia, June 28–July 1, 2001. IISNC, Ulaanbaatar, Mongolia

Cilliers P (1998) Complexity and postmodernism. Routledge, New York

Clarke KC, Gaydos LY (1998) Loose-coupling a cellular automaton model and GIS: Long-term urban growth prediction for San Francisco and Washington/Baltimore. Int J Geogr Inf Sci 12:699–714

Clarke KC, Hoppen S, Gaydos LJ (1996) Methods and techniques for rigorous calibration of a cellular automaton model of urban growth. Proceedings of the Third International Conference/Workshop Integrating GIS and Environmental Modeling, Santa Fe, NM, January 21–25, 1996. National Center for Geographic Information and Analysis, Santa Barbara

Clark JS, Carpenter SR, Barber M, Collins S, Dobson A, Foley JA, Lodge DM, Pascual M, Pielke R Jr, Pizer W, Pringle C, Reid WV, Rose KA, Sala O, Schlesinger WH, Wall DH, Wear D (2001) Ecological forecasts: An emerging imperative. Science 293:657–660

Clark W, Mitchell R, Cash DW, Alcock F (2002) Information as influence: How institutions mediate the impact of scientific assessments on global environmental affairs. John F. Kennedy School of Govt., Harvard Univ., Cambridge, Faculty Working Paper RWP02-044

Claussen M, Brovkin V, Ganopolski A (2002) Africa: Greening of the Sahara. In: Steffen W, Jäger J, Carson D, Bradshaw C (eds) Challenges of a changing Earth: Proceedings of the Global Change Open Science Conference, Amsterdam, The Netherlands, 10–13 July 2001. (The IGBP Series), Springer, Berlin Heidelberg, pp 125–128

Cline-Cole RA, Main HAC, Nichol JE (1990) On fuelwood consumption, population dynamics and deforstation in Africa. World Dev 18(4):513–527

Cochrane MA (2001) Synergistic interactions between habitat fragmentation and fire in evergreen tropical forests. Conserv Biol 15(6):1515–1521

Cochrane MA, Alencar A, Schulze MD, Souza CM, Nepstad DC, Lefèbvre P, Davidson EA (1999) Positive feedbacks in the fire dynamic of closed canopy tropical forests. Science 284(5421):1832–1835

Coe MT, Foley JA (2001) Human and natural impacts on the water resources of the Lake Chad Basin. J Geophys Res 106:3349–3356

Coleman JS (1990) Foundations of social theory. The Belknap Press of Harvard University Press, Cambridge

Conelly WT (1992) Agricultural intensification in a Philippine frontier: Impact on labor efficiency and farm diversity. Hum Ecol 20(2):203–223

Conelly WT, Chaiken MS (2001) Intensive farming, agro-diversity, and food security under conditions of extreme population pressure in Western Kenya. Hum Ecol 28(1):19–51

Conklin H (1957) Hanunoo agriculture. Food and Agriculture Organization of the United Nations, Rome

Contreras-Hermosilla A (2000) The underlying causes of forest decline. CIFOR Occasional Papers 30. Center for International Forestry Research, Bogor, Indonesia

Cook TD, Cooper H, Cordray D, Hartmann H, Hedges L, Light R, Louis T, Mosteller F (1992) Meta-analysis for explanation: A casebook. Russel Sage Foundation, New York

Coomes OT, Grimard F, Burt GJ (2000) Tropical forests and shifting cultivation: Secondary forest fallow dynamics among traditional farmers of the Peruvian Amazon. Ecol Econ 32(1):109–124

Correia FN, Saraiva MD, Da Silva FN (1999) Floodplain management in urban developing areas. Part I. Urban growth scenarios and land-use controls. Water Resour Manag 13(1):1–21

Costanza R (1989) Model goodness of fit: A multiple resolution procedure. Ecol Model 47:199–215

Court J, Young J (2003) Bridging research and policy: Insights from 50 case studies. Overseas Development Institute, London

Cousins SAO (2001) Land-cover transitions based on 17th and 18th century cadastral maps and aerial photographs. Landscape Ecol 16:41–54

Cox PA, Elmqvist T (2000) Pollinator extinction in the Pacific Islands. Conserv Biodivers 14(5):1237–1239

Cramer W, Bondeau A, Schaphoff S, Lucht W, Smith B, Sitch S (2004) Tropical forests and the global carbon cycle: Impacts of atmoshperic carbon dioxide, climate change and rate of deforestation. Philos Trans R Soc Lond B359:331–343

Craswell ET, Lefroy RDB (2001) The role and function of organic matter in tropical soils. Nutr Cycl Agroecosys 61(1):7–18

Craswell ET, Grote U, Henao J, Vlek PLG (2004) Nutrient flows in agricultural production and international trade: Ecology and policy issues. Discussion Paper on Development Issues 72. Center for Development Research, Bonn, 62 pp

Crewe E, Young J (2002) Bridging research and policy: Context, evidence and links. Overseas Development Institute, London

Cronon W (1991) Nature's metropolis: Chicago and the Great West. W. W. Norton & Company, New York, 530 pp

Cropper M, Griffiths C (1994) The interaction of population growth and environmental quality. Am Econ Rev 84:250–254

Cropper M, Griffiths C, Mani M (1999) Roads, population pressures, and deforestation in Thailand 1976–1989. Land Econ 75(1):58–73

Crossland CJ, Baird D, Ducrotoy J-P, Lindeboom HJ (2005a) The coastal zone: A domain of global interactions. In: Crossland CJ, Kremer HH, Lindeboom HJ, Marshall Crossland JI, Le Tissier MDA (eds) Coastal fluxes in the Anthropocene: The Land-Ocean Interactions in the Coastal Zone Project of the International Geosphere-Biosphere Programme. (The IGBP Series), Springer, Berlin Heidelberg, pp 1–38

Crossland CJ, Kremer HH, Lindeboom HJ, Marshall Crossland JI, Le Tissier MDA (eds) (2005b) Coastal fluxes in the Anthropocene: The Land-Ocean Interactions in the Coastal Zone Project of the International Geosphere-Biosphere Programme. (The IGBP Series), Springer, Berlin Heidelberg, 232 pp

Crosson P (1995) Future supplies of land and water for world agriculture. In: Islam N (ed) Population and food in the early 21st century: Meeting future food demand of an increasing world population. IFPRI Occasional Paper. International Food Policy Research Institute, Washington D.C., pp 143–160

Crowley TJ (2000) Causes of climate change over the past 1000 years. Science 289:270–277

Crumley C (2000) From garden to globe: Linking time and space with meaning and memory. In: McIntosh RJ, Tainter JA, McIntosh SK (eds) The way the wind blows: Climate, history, and human action. Columbia University Press, New York, pp 193–208

Crutzen PJ, Andreae MO (1990) Biomass burning in the tropics: Impact on atmospheric chemistry and biogeochemical cycles. Science 250:1669–1678

Csiszar I, Justice CO, McGuire AD, Cochrane MA, Roy DP, Brown F, Conard SG, Frost PGH, Giglio L, Elvidge C, Flannigan MD, Kasischke E, McRae DJ, Rupp TS, Stocks BJ, Verbyla DL (2004) Land use and fires. In: Gutman G, Janetos AC, Justice CO, Moran EF, Mustard JF, Rindfuss RR, Skole D, Turner BL II, Cochrane MA (eds) Land change science: Observing, monitoring and understanding trajectories of change on the Earth's surface. Remote Sensing and Digital Image Processing Series 6. Kluwer Academic, Dordrecht Boston London, pp 329–350

Czaplewski RL (2003) Can a sample of Landsat sensor scenes reliably estimate the global extent of tropical deforestation? Int J Remote Sens 24:1409–1412(4)

Dadhwal VK, Chhabra A (2002) Landuse/landcover change in Indo-Gangetic plains: Cropping pattern and agroecosystem carbon cycle. In: Abrol YP, Sangwan S, Tiwari MK (eds) Land use: Historical perspectives. Focus on Indo-Gangetic Plains. Allied Publishers, New Delhi, pp 249–276

Dahl E (1990) Probable effects of climatic change due to the greenhouse effect on plant productivity and survival in North Europe. In: Holten JI (ed) Effects of climatic change on terrestrial ecosystems. NIA Norsk Institut for Naturforskning, Trondheim, pp 7–17

Dale VH, Pearson SM (1999) Modeling the driving factors and ecological consequences of deforestation in the Brazilian Amazon. In: Mladenoff DJ, Baker WL (eds) Spatial modeling of forest landscape change: Approaches and applications. Cambridge University Press, Cambridge, pp 256–276

Darby C (1956) The clearing of the woodland in Europe. In: Thomas WL Jr. (ed) Man's role in changing the face of the Earth. Chicago University Press, Chicago, pp 183–216

Davis K (1963) The theory of change and response in modern demographic history. Popul Index 29:345–66

De Groot WT (1992) Environmental science theory; Concepts and methods in a one-world problem oriented paradigm. Elsevier Science Publishers, Amsterdam and New York

De Groot RS, Rotmans J (2004) MedAction. Final report. Key action 2: "Global change, climate and biodiversity"; Subaction 2.3.3 "Fighting land degradation and desertification" within the energy, environment and sustainable development. Contract no. EVK2-CT-2000-00085. ICIS, Maastricht

De Koning GHJ, Veldkamp A, Fresco LO (1998) Land use in Ecuador: A statistical analysis at different aggregation levels. Agric Ecosyst Environ 70:321–247

De Nijs TCM, De Niet R, Crommentuijn L (2004) Constructing land-use maps of the Netherlands in 2030. J Environ Manage 72:35–42

Deacon RT (1994) Deforestation and the rule of law in a cross-section of countries. Land Econ 70(4):414–430

Deacon RT (1995) Assessing the relationship between government policy and deforestation. J Environ Econ Manag 28:1–18

Deacon RT (1999) Deforestation and ownership: Evidence from historical accounts and contemporary data. Land Econ 75(3):341–359

Deadman P, Robinson D, Moran E, Brondizio E (2004) Colonist household decisionmaking and land-use change in the Amazon Rainforest: An agent-based simulation. Environ Plann B31:693–709

DeFries RS, Achard F (2002) New estimates of tropical deforestation and terrestrial carbon fluxes: Results of two complementary studies. LUCC Newsletter 8:7–9

DeFries RS, Eshleman KN (2004) Land-use change and hydrologic processes: A major focus for the future. Hydrol Process 18: 2183–2186

DeFries RS, Townshend JRG (1994) NDVI-derived land cover classifications at a global scale. Int J Remote Sens 15:3567–3586

DeFries R, Field C, Fung I, Justice C, Los S, Matson P, Matthews E, Mooney H, Potter C, Prentice K, Sellers P, Townshend J, Tucker C, Ustin S, Vitousek P (1995a) Mapping the land-surface for global atmosphere-biosphere models: Toward continuous distributions of vegetations functional-properties. J Geophys Res-Atmos 100:20867–20882

DeFries RS, Hansen M, Townshend J (1995b) Global discrimination of land cover types from metrics derived from AVHRR pathfinder data. Remote Sens Environ 54:209–222

DeFries RS, Hansen M, Townshend JRG, Sohlberg R (1998) Global land cover classifications at 8 km spatial resolution: The use of training data derived from Landsat imagery in decision tree classifiers. Int J Remote Sens 19:3141–3168

DeFries RS, Bounoua L, Collatz GJ (2002a) Human modification of the landscape and surface climate in the next fifty years. Glob Change Biol 8:438–458

DeFries RS, Houghton RA, Hansen MC, Field CB, Skole D, Townshend J (2002b) Carbon emissions from tropical deforestation and regrowth based on satellite observations for the 1980s and 1990s. Proc Natl Acad Sci USA 99(22):14256–14261

DeFries RS, Asner GP, Houghton RA (2004a) Trade-offs in land-use decisions: Towards a framework for assessing multiple ecosystem responses to land-use change. In: DeFries RS, Asner GP, Houghton RA (eds) Ecosystems and land use change. Geophysical Monograph 153. American Geophysical Union, Washington D.C., pp 1–9

DeFries RS, Foley JA, Asner GP (2004b) Land-use choices: Balancing human needs and ecosystem function. Front Ecol Environ 2:249–257

DeFries RS, Asner GP, Houghton RA (eds) (2004c) Ecosystems and land use change. Geophysical Monograph 153. American Geophysical Union, Washington D.C.

DeFries RS, Pagiola S, Akcakaya HR, Arcenas A, Babu S, Balk D, Confalonieri U, Cramer W, Fritz S, Green R, Gutierrez-Espeleta E, Kane R, Latham J, Matthews E, Ricketts T, Yue TX (2005) Analytical approaches for asssessing ecosystem conditions and human well-being. In: Scholes R, Rashid H (eds) Millennium Ecosystem Assessment working group on conditions and trends. Island Press, Washington D.C., pp 15–47

Deichmann U, Balk D, Yetman G (2001) Transforming population data for interdisciplinary usages: From census to grid. CIESIN, New York, 20 pp

Deininger K, Binswanger HP (1995) Rent seeking and the development of large-scale agriculture in Kenya, South Africa, and Zimbabwe. Econ Dev Cult Change 43(1):493–522

Deininger KW, Minten B (1999) Poverty, policies, and deforestation: The case of Mexico. Econ Dev Cult Change 47(2):313–344

DeKoninck R, Dery S (1997) Agricultural expansion as a tool of popula-tion redistribution in Southeast Asia. J Southe Asian Stud 28(1):1–26

Dendoncker N, Bogaert P, Rounsevell MDA (2006) A statistical methodology to downscale aggregated land use data and scenarios. J Land Use Sci, submitted.

DES (Directorate of Economics and Statistics) (2004) Agricultural statistics at a glance. Directorate of Economics and Statistics, Department of Agriculture and Cooperation, Ministry of Agriculture, Government of India, New Delhi, 221 pp

Desert Encroachment Control, Rehabilitation Programme (1976) General administration for natural resources. Ministry of Agriculture, Food and Natural Resources and the Agricultural Research Council, National Council for Research (Republic of Sudan) in collaboration with UNEP, UNDP and FAO

Di Gregorio A, Jansen LJM (2000a) Land cover classification system. Food and Agriculture Organization, Rome, 179 pp, http://www.glcn-lccs.org/

Di Gregorio A, Jansen LJM (2000b) Land cover classification system (LCCS): Classification concepts and user manual. Food and Agriculture Organization, http://www.fao.org/documents/show_cdr.asp?url_file=/DOCREP/003/X0596E/X0596e00.htm

Diamond JM (1995) Easter's end. Discovery: The World of Science 16:62–69

Diamond J (1997) Guns, germs, and steel: The fates of human societies. W. W. Norton & Company, New York London

Diamond J (2005) Collapse: How societies choose to fail or succeed. Viking Penguin, New York, 575 pp

Dicken P (2003) Global shift: Reshaping the global economic map in the 21st century, 4th ed. The Guilford Press, New York London, 632 pp

Dietz T, Ostrom E, Stern P (2003) The struggle to govern the commons. Science 302:1907–1912

Dirzo R (2001) Tropical forests. In: Chapin FS, Sala OE, Huber-Sann-wald E (eds) Global biodiversity in a changing environment: Scenarios for the 21st century. Springer, Berlin Heidelberg, pp 251–276

Dirzo R, Raven PH (2003) Global state of biodiversity and loss. Annu Rev Env Resour 28:137–167

Dixon RK, Brown S, Houghton RA, Solomon AM, Trexler MC, Wisniewski J (1994) Carbon pools and flux of global forest ecosystems. Science 263:185–190

Dockerty T, Lovett A, Sünnenberg G, Appleton K, Parry M (2005) Visualising the potential impacts of climate change on rural landscapes. Comput Environ Urban 29(3):297–320

Döll P, Siebert S (2000) A digital global map of irrigated areas. ICID J 49:55–66

Domínguez P, González-Asensio A (1995) Situación de los acuíferos del Campo de Dalías (Almería) en relación con su declaración de sobreexplotación. Hidrogeología y Recursos Hidraúlicos, AEHS, Madrid XXI:443–467

Döös BR (2002) Population growth and loss of arable land. Global Environ Chang 12(4):303–311

Döös BR, Shaw R (1999) Can we predict the future food production? A sensitivity analysis. Global Environ Chang 9:261–283

Doran JW, Parkin TB (1996) Quantitative indicators of soil quality: A minimum data set. In: Doran JW, Jones AJ (eds) Methods for assessing soil quality. SSSA Spec. Pub. 49. Soil Science Society of America, Madison, pp 25–37

Dovers SR (1995) A framework for scaling and framing policy problems in sustainability. Ecol Econ 12:93–106

Dregne HE (1977) Generalized map of the status of desertification of arid lands. A/CONF 74/31, United Nations Conference on Desertification, Nairobi, Kenya

Dregne HE (1983) Desertification of arid lands. Harwood Academic Publishers, New York

Dregne HE, Chou N (1992) Global desertification and cost. In: Dregne HE (ed) Degradation and restoration of arid lands. Texas Tech University, Lubbock, pp 249–282

Dregne HE (2002) Land degradation in the drylands. Arid Land Res Manage 16:99–132

Drescher AW (1996) Urban microfarming in central southern Africa: A case study of Lusaka, Zambia. Afr Urban Q 11(2/3):210–216

Ducrot R, Le Page C, Bommel P, Kuper M (2004) Articulating land and water dynamics with urbanization: An attempt to model natural resources management at the urban edge. Comput Environ Urban 28:85–106

Dudley N, Stolton S (2003) Running pure: The importance of forest protected areas to drinking water. World Bank, WWF Alliance for Forest Conservation and Sustainable Use, Washington D.C.

Dumanski J, Pieri C (2000) Land quality indicators: Research plan. Agric Ecosyst Environ 81:92–102

Dumanski J, Pettapiece WW, McGregor RJ (1998) Relevance of scale dependent approaches for integrating biophysical and socioeconomic information and development of agroecological indicators. Nutr Cycl Agroecosys 50:13–22

Duncan BN, Martin RV, Staudt AC, Yevich R, Logan JA (2003) Inter-annual and seasonal variability of biomass burning emissions constrained by satellite observations. J Geophys Res-Atmos 108, DOI 10.1029/2002JD002378

Dwyer E, Pinnock S, Grégoire J-M, Pereira JMC (2000) Global spatial and temporal distribution of vegetation fire as determined from satellite observations. Int J Remote Sens 21:1289–1302

Dynesius M, Nilsson C (1994) Fragmentation and flow regulation of river systems in the northern third of the world. Science 266: 753–762

Easterling WE (1997) Why regional studies are needed in the development of full-scale integrated assessment modelling of global change processes. Global Environ Chang A 7:337–356

Eastman JR, Fulk M (1993) Long sequence time series evaluation using standardized principal components. Photogramm Eng Remote Sens 59(8):991–996

Eastman JL, Coughenour MB, Pielke RA (2001) The effects of CO_2 and landscape change using a coupled plant and meteorological model. Glob Change Biol 7(7):797–815

Eder JF (1991) Agricultural intensification and labor productivity in a Philippine vegetable gardening community: A longitudinal study. Hum Organ 50(3):245–255

Ehrhardt-Martinez K (1998) Social determinants of deforestation in developing countries: A cross-national study. Soc Forces 77(2):567–586

Ehrlich P, Holdren J (1971) The impact of population growth. Science 171:1212–1217

Eickhout B, Meijl Hv, Tabeau A, Zeijts Hv (2004) Between liberalization and protection: Four long-term scenarios for trade, poverty and the enviroonment. GTAP Conference Paper, Presented at the 7th Annual Conference on Global Economic Analysis, Washington D.C., www.gtap.agecon.purdue.edu

Eickhout B, van Meijl H, Tabeau A, van Rheenen T (2005) Economic and ecological consequences of four European land-use scenarios. Land Use Policy

Ellis J, Swift DM (1988) Stability of African pastoral ecosystems: Alternative paradigms and implications for development. J Range Manage 41:450–459

Eltahir EAB (1996) Role of vegetation in sustaining large-scale atmospheric circulations in the tropics. J Geophys Res 101(D2):4255–4268

Eltahir EAB, Bras RL (1996) Precipitation recycling. Rev Geophys 34:367–378

Elvidge CD, Baugh KE, Kihn EA, Kroehl HW, Davis ER (1997) Mapping of city lights using DMSP operational linescan system data. Photogramm Eng Remote Sens 63:727–734

Elvidge CD, Baugh KE, Safran J, Tuttle BT, Howard AT, Hayes PJ, Jantzen J, Erwin EH (2001) Nighttime lights of the world: 1994–95. Photogramm Eng Remote Sens 56:81–99

Elvidge CD, Sutton PC, Wagner TW, Ryzner R, Vogelmann JE, Goetz SJ, Smith AJ, Jantz C, Seto KC, Imhoff ML, Wang YG, Milesi C, Nemani R (2004) Urbanization. In: Gutman G, Janetos AC, Justice CO, Moran EF, Mustard JF, Rindfuss RR, Skole D, Turner BL II, Cochrane MA (eds) Land change science: Observing, monitoring and understanding trajectories of change on the Earth's surface. Remote Sensing and Digital Image Processing Series No. 6, Kluwer Academic, Dordrecht Boston London, pp 315–328

Entwisle B, Stern PC (eds) (2005) Population, land use, and environment: Research directions. National Academy Press, Washington D.C.

Entwisle B, Walsh SJ, Rindfuss RR, Chamratrithirong A (1998) Land-use/land-cover and population dynamics. In: Liverman D, Moran EF, Rindfuss RR, Stern PC (eds) People and pixels: Linking remote sensing and social science. National Academy Press, Washington D.C., pp 121–144

Eshleman KN (2004) Hydrological consequences of land use change: A review of the state-of-the-science. In: DeFries RS, Asner G, Houghton RA (eds) Ecosystems and land use change. Geophysical Monograph 153, American Geophysical Union, Washington D.C., pp 13–29

Eswaran H, Reich P (1998) Desertification: A global assessment and risks to sustainability. In: Proceedings of the 16th International Congress of Soil Sciences, Montpellier, France

Eswaran H, Reich P, Beinroth F (2003) Global desertification tension zones. USDA, Natural Resources Conservation Service, www.nrcs.usda.gov/technical/worldsoils/landdeg/papers/tzpaper.html

European Commission (2000) Addressing desertification and land degradation. Office of the Official Publications of the European Commission, Luxembourg

Eva H, Lambin EF (1998a) Remote sensing of biomass burning in tropical regions: Sampling issues and multisensor approach. Remote Sens Environ 64(3):292–315

Eva H, Lambin EF (1998b) Burnt areas mapping in Central Africa from ATSR data. Int J Remote Sens 19(18):3471–3473

Eva H, Lambin EF (2000) Fires and land-cover change in the tropics: A remote sensing analysis at the landscape scale. J Biogeogr 27:765–776

Eva HD, Grégoire J-M, Mayaux P (2004) Support for fire management in Africa's protected areas: The contribution of the European Commission's Joint Research Centre. EUR 21296 EN, Luxembourg: Office for Official Publications of the European Communities, 63 pp, www-gem.jrc.it/tem/PDF_publis/publications.htm

Evans TP, Manire A, de Castro F, Brondizio E, McCracken S (2001) A dynamic model of household decision-making and parcel level landcover change in the Eastern Amazon. Ecol Model 143:95–113

Evrard O, Persoons E, Vandaele K, Wesemael B van (2006) Effectiveness of erosion mitigation measures to prevent muddy floods: A case study in the Belgian loam belt. Agric Ecosyst Environ, submitted

Ewell PT, Merrill-Sands D (1987) Milpa in Yucatán: A long-fallow maize system and its alternatives in the Maya peasant economy. In: Turner BL II, Brush SB (eds) Comparative farming systems. Guildford, New York, pp 95–129

Ewert F (2006) Green revolution. In: Geist HJ (ed) Our Earth's changing land: An encyclopedia of land-use and land-cover change, vol. 1 (A–K). Greenwood Press, Westport, London, pp 276–278

Fairhead J, Leach M (1996) Misreading the African landscape: Society and ecology in a forest-savanna mosaic. Cambridge University Press, Cambridge

Fairhead J, Leach M (1998) Reframing deforestation: Global analyses and local realities. Studies in West Africa. Routledge, London

Falkenmark M (1999) A land-use decision is also a water decision. In: Falkenmark M, Andersson L, Castensson R, Sundblad K (eds) Water: A reflection of land use. Options for counteracting land and water mismanagement. Swedish Natural Science Research Council, Stockholm, pp 58–78

Faminow MD (1997) Spatial economics of local demand for cattle products in Amazon development. Agric Ecosyst Environ 62:1–11

Fang J, Xie Z (1994) Deforestation in preindustrial China: The Loess Plateau region as an example. Chemosphere 29:983–999

FAO (Food and Agriculture Organization of the United Nations) (1992) FAO/UNESCO soil map of the world: Digital version. FAO, Rome, www.grid.unep.ch/data/grid/gnv6.php

FAO (Food and Agriculture Organization of the United Nations) (1996) Our land, our future. FAO, Rome, 36 pp

FAO (Food and Agriculture Organization of the United Nations) (1999) State of food and agriculture 1999. FAO, Rome

FAO (Food and Agriculture Organization of the United Nations) (2000a) Land resource potential and constraints at regional and country levels. FAO, Rome, 122 pp

FAO (Food and Agriculture Organization of the United Nations) (2000b) Agriculture: Towards 2015/30. Technical interim report, April 2000. FAO, Rome

FAO (Food and Agriculture Organization of the United Nations) (2001a) FRA 2000 Main Report. FAO Forestry Paper 140, FAO, Rome, 479 pp

FAO (Food and Agriculture Organization of the United Nations) (2001b) FRA 2000: Pan-tropical survey of forest cover changes 1980–2000. FRA Working Paper No. 49, FAO, Rome, 15 pp

FAO (Food and Agriculture Organization of the United Nations) (2001c) FAO statistical databases. http://apps.fao.org

FAO (Food and Agriculture Organization of the United Nations) (2003) State of the world forests 2003. FAO, Rome

FAO (Food and Agriculture Organization of the United Nations) (2004a) FAOSTAT data. FAO, Rome, http://apps.fao.org

FAO (Food and Agriculture Organization of the United Nations) (2004b) Forests and floods: Drowning in fiction or thriving on facts. Draft, FAO, Rome

Farahani HJ, Peterson GA, Westfall DG, Sherrod LA, Ahuja LR (1998) Soil water storage in dryland cropping systems: The significance of cropping intensification. Soil Sci Soc Am J 62:984–991

Fearnside PM (1997) Transmigration in Indonesia: Lessons from its environmental and social impacts. Environ Manage 21(4):553–570

Fearnside PM (2000) Global warming and tropical land-use change: Greenhouse gas emissions from biomass burning, decomposition and soils in forest conversion, shifting cultivation and secondary vegetation. Climatic Change 46(1/2):115–158

Fearnside PM (2003) Deforestation control in Mato Grosso: A new model for slowing the loss of Brazil's Amazon forest. Ambio 32:343–45

Federal Environment Agency (2004) 7th report on the state of the environment in Austria. Federal Environment Agency, Vienna, Austria, www.umweltbundesamt.at/umweltkontrolle/ukb2004

Feng Q, Endo KN, Cheng GD (2001) Towards sustainable development of the environmentally degraded arid rivers of China: A case study from Tarim River. Environ Geol 41:229–238

Feng Z, Yang Y, Zhang Y, Zhang P, Li Y (2006) Grain-for-green policy and its impacts on grain supply in West China. Land Use Policy, forthcoming

Field C, Raupach M (eds) (2004) The global carbon cycle: Integrating humans, climate and the natural world. Island Press, Washington D.C.

Fischer G, Schrattenholzer L (2001) Global bioenergy potentials through 2050. Biomass Bioenerg 20:151–159

Fischer G, Sun LX (2001) Model based analysis of future land use development in China. Agric Ecosyst Environ 85(1–3):163–176

Fischer G, Velthuizen H, Shah M, Nachtergaele F (2002) Global agro-ecological assessment for agriculture in the 21st century: Methodology and results. International Institute for Applied Systems Analysis and Food and Agriculture Organization, Laxenburg, Austria

Fisher MJ, Rao IM, Ayarza MA, Lascano CE, Sanz JI, Thomas RJ, Vera RR (1994) Carbon storage by introduced deep-rooted grasses in the South American savannas. Nature 371:236–238

Fishwick P, Luker P (eds) (1991) Qualitative simulation modeling and analysis. Springer, Berlin Heidelberg, 341 pp

Flint EP, Richards JF (1991) Historical-analysis of changes in land-use and carbon stock of vegetation in South and Southeast Asia. Can J Forest Res 21:91–110

Fold N, Pritchard B (eds) (2005) Cross-continental agro-food chains. Routledge, London

Foley JA, Costa MH, Delire C, Ramankutty N, Snyder P (2003) Green surprise? How terrestrial ecosystems could affect Earth's climate. Front Ecol Environ 1(1):38–44

Foley JA, Kucharik CJ, Twine TE, Coe MT, Donner SD (2004) Land use, land cover, and climate change across the Mississippi Basin: Impacts on selected land and water resouces. In: DeFries RS, Asner GP, Houghton RA (eds) Ecosystems and land use change. Geophysical Monograph 153, American Geophysical Union, Washington D.C., pp 249–261

Foley JA, DeFries R, Asner GP, Barford C, Bonan G, Carpenter SR, Chapin FS, Coe MT, Daily GC, Gibbs HK, Helkowski JH, Holloway T, Howard EA, Kucharik CJ, Monfreda C, Patz JA, Prentice IC, Ramankutty N, Snyder PK (2005) Global consequences of land use. Science 309:570–574

Folke C, Jansson Å, Larsson J, Costanza R (1997) Ecosystem appropriation by cities. Ambio 26(3):167–172

Folland CK, Karl TR, Christy JR, Clarke RA, Gruza GV, Jouzel J, Mann ME, Oerlemans J, Salinger MJ, Wang S-W (2001) Observed climate variability and change. In: Houghton JT, Ding Y, Griggs DJ, Noguer M, van der Linden PJ, Dai X, Maskell K, Johnson CA (eds) Climate change 2001: The scientific basis. Contributions of Working Group I to the third assessment report of the Intergovernmental Panel on Climate Change (IPCC). Cambridge University Press, Cambridge New York, pp 99–181

Ford RE (1993) Marginal coping in extreme land pressures: Ruhengeri, Rwanda. In: Turner BL II, Hyden G, Kates RW (eds) Population growth and agricultural change in Africa. University Press of Florida, Gainesville, pp 145–186

Forman R, Godron M (1986) Landscape ecology. John Wiley & Sons, Ltd., Chichester New York

Fox J, Krummel J, Yarnasarn S, Ekasingh M, Podger N (1995) Land use and landscape dynamics in northern Thailand: Assessing change in three upland watersheds. Ambio 24(6):328–334

Fox J, Truong DM, Rambo AT, Tuyen NP, Cuc LT, Leisz S (2000) Shifting cultivation: A new old paradigm for managing tropical forests. BioScience 50:521–528

Fox J, Rindfuss RR, Walsh SJ, Mishra V (eds) (2003) People and the environment: Approaches for linking household and community surveys to remote sensing and GIS. Kluwer Academic, Dordrecht Boston London, 319 pp

Franchito SH, Rao VB (1992) Climatic change due to land surface alterations. Climatic Change 22:1–34

Frederick WH, Worden RL (eds) (1992) A country study: Indonesia. The Library of Congress, http://lcweb2.loc.gov/frd/cs/cshome.html

Freitas H (2006) Eutrophication. In: Geist HJ (ed) Our Earth's changing land: An encyclopedia of land-use and land-cover change, vol. 1 (A–K). Greenwood Press, Westport, London, pp 218–219

Friedl MA, McIver DK, Hodges JCF, Zhang XY, Muchoney D, Strahler AH, Woodcock CE, Gopal S, Schneider A, Cooper A, Baccini A, Gao F, Schaaf C (2002) Global land cover mapping from MODIS: Algorithms and early results. Remote Sens Environ 83:287–302

Fu C (2002) Can human-induced land-cover change modify the monsoon system? In: Steffen W, Jäger J, Carson D, Bradshaw C (eds) Challenges of a changing Earth: Proceedings of the Global Change Open Science Conference, Amsterdam, The Netherlands, 10–13 July 2001. (The IGBP Series), Springer, Berlin Heidelberg, pp 133–136

Fujita M, Krugman P, Mori T (1999) On the evolution of hierarchical urban systems. Eur Econ Rev 43:209–251

Gallopín G, Raskin P (2002) Global sustainability: Bending the curve. Routledge, London

Gallopín G, Hammond A, Raskin P, Swart R (1997) Branch points: Global scenarios and human choice. PoleStar Series Report No. 7, Stockholm Environment Institute, Boston

Gardner RH (1998) Pattern, process, and the analysis of spatial scales. In: Peterson DL, Parker VT (eds) Ecological scale: Theory and applications. Columbia University Press, New York, pp 17–34

Garrett J (1998) Research that matters: The impact of IFPRI's policy research. International Food Policy Research Institute, Washington D.C.

Gashumba JK, Mwambu PM (1981) Sleeping sickness epidemic in Busoga, Uganda. Trop Doct 11:175–178

Gedney N, Valdes PJ (2000) The effect of Amazonian deforestation on the Northern Hemisphere circulation and climate. Geophys Res Lett 27:3053–3056

Geertman S, Stillwell J (2004) Planning support systems: An inventory of current practice. Comput Environ Urban 28:291–310

Geist HJ (1999a) Exploring the entry points for political ecology in the international research agenda on global environmental change. Z Wirtsch.geogr 43:158–168

Geist HJ (1999b) Global assessment of deforestation related to tobacco farming. Tob Control 8:18–28

Geist HJ (2000) Transforming the fringe: Tobacco-related wood usage and its environmental implications. In: Majoral R, Delgado-Cravidão F, Jussila H (eds) Marginality, landscape, and environment. Ashgate Publishing, Aldershot, pp 87–118

Geist HJ (2003a) The role of population as an underlying driving force of deforestation and desertification: Insights from two meta-analytical studies. Popul Geogr 25(1/2):29–40

Geist HJ (2005) The causes and progression of desertification. Ashgate Publishing, Aldershot, 258 pp

Geist HJ (2006a) Meta-analysis. In: Geist HJ (ed) Our Earth's changing land: An encyclopedia of land-use and land-cover change, vol. 2 (L–Z). Greenwood Press, Westport, London, pp 406–407

Geist HJ (2006b) Tragedy of enclosure. In: Geist HJ (ed) Our Earth's changing land: An encyclopedia of land-use and land-cover change, vol. 2 (L–Z). Greenwood Press, Westport, London, pp 601–603

Geist HJ, Lambin EF (2001) What drives tropical deforestation? A meta-analysis of proximate and underlying causes of deforestation based on subnational case study evidence. LUCC Report Series No. 4, LUCC International Project Office, Louvain-la-Neuve, 116 pp

Geist HJ, Lambin EF (2002) Proximate causes and underlying driving forces of tropical deforestation. BioScience 52(2):143–150

Geist HJ, Lambin EF (2003) Is poverty the cause of tropical deforestation? Int Forest Rev 5(1):64–67

Geist HJ, Lambin EF (2004) Dynamic causal patterns of desertification. BioScience 54(9):817–829

Geist H, Lambin E, Palm C, Tomich T (2006) Agricultural transitions at dryland and tropical forest margins: Actors, scales and trade-offs. In: Brouwer F, McCarl BA (eds) Agriculture and climate beyond 2015: A new perspective on future land use patterns (Environment & Policy Vol. 46). Springer, Dortrecht, pp 53–73

Gell-Mann M (1994) The quark and the jaguar. Freeman, New York

Genxu W, Guodong C (1999) Water resource development and its influence on the environment in arid areas of China: The case of the Hei River Basin. J Arid Environ 43:121–131

Geoghegan J (2006a) Socializing the pixel. In: Geist HJ (ed) Our Earth's changing land: An encyclopedia of land-use and land-cover change, vol. 2 (L–Z). Greenwood Press, Westport, London, pp 536–538

Geoghegan J (2006b) Pixelizing the social. In: Geist HJ (ed) Our Earth's changing land: An encyclopedia of land-use and land-cover change, vol. 2 (L–Z). Greenwood Press, Westport, London, pp 472–475

Geoghegan J, Wainger LA, Bockstael NE (1997) Spatial landscape indices in a hedonic framework: An ecological economics analysis using GIS. Ecol Econ 23:251–264

Geoghegan J, Pritchard JL, Ogneva-Himmelberger Y, Chowdhury RR, Sanderson S, Turner BL II (1998) 'Socializing the Pixel' and 'Pixelizing the Social' in land-use and land-cover change. In: Liverman D, Moran EF, Rindfuss RR, Stern PC (eds) People and pixels: Linking remote sensing and social science. National Academy Press, Washington D.C., pp 51–69

Geoghegan J, Villar SC, Klepeis P, Mendoza PM, Ogneva-Himmelberger Y, Chowdhury RR, Turner II BL, Vance C (2001) Modeling tropical deforestation in the southern Yucatán Peninsular region: Comparing survey and satellite data. Agric Ecosyst Environ 85:25–46

George PS, Chattopadhyah S (2001) Population and land use in Kerala. In: Indian National Science Academy, Chinese Academy of Sciences, U.S. National Academy of Sciences (eds) Growing populations, changing landscapes: Studies from India, China, and the United States. National Academy Press, Washington D.C., pp 79–105

Geoscience Australia (2004a) Vegetation: Post-European settlement (1988). Australian Government, http://www.ga.gov.au/nmd/products/thematic/vegmap.jsp

Geoscience Australia (2004b) Vegetation: Pre-European settlement (1788). Australian Government, http://www.ga.gov.au/nmd/products/thematic/vegmap.jsp

Giambelluca TW (2002) Hydrology of altered tropical forest. Hydrol Process 16:1665–1669

Giannini A, Saravanan R, Chang P (2003) Oceanic forcing of Sahel rainfall on interannual to interdecadal time scales. Science 302:1027–1030

Gibson CC, Ostrom E, Anh TK (2000) The concept of scale and the human dimensions of global change: A survey. Ecol Econ 32:217–239

Gibson CA, McKean MA, Ostrom E (2003) People and forests: Communities, institutions, and governance. MIT Press, Cambridge

Gillison AN (2005) The potential role of aboveground biodiversity indicators in assessing best-bet alternatives to slash and burn. In: Palm C A, Vosti SA, Sanchez PA, Ericksen PJ (eds) Slash-and-burn agriculture: The search for alternatives. Columbia University Press, New York

Gilmour DA, Bonell M, Cassells DS (1987) The effects of forestation on soil hydraulic properties in the middle hills of Nepal: A preliminary assessment. Mt Res Dev 7:239–249

Glantz MH (1998) Creeping environmental problems in the Aral Sea Basin. In: Kobori I, Glantz MH (eds) Central Eurasian water crisis: Caspian, Aral, and Dead Seas. United Nations University Press, Tokyo New York Paris, pp 25–52

Glazovsky NF (1995) The Aral Sea Basin. In: Kasperson JX, Kasperson RE, Turner BL II (eds) Regions at risk: Comparisons of threatened environments. United Nations University Press, Tokyo New York Paris, pp 92–139

Gleick PH (2003) Water use. Annu Rev Environ Resour 28:275–314

Gockowski J, Nkamleu B, Wendt J (2001) Implications of resource use intensification for the environment and sustainable technology systems in the central African rainforest. In: Lee D, Barrett D (eds) Agricultural intensification, economic development and the environment: Tradeoffs or synergies? CAB International, Wallingford New York, pp 197–219

Godoy R, Wilkie D, Franks J (1997) The effects of markets on neotropical deforestation: A comparative study of four Amerindian societies. Curr Anthropol 38(5):875–878

Goetz SJ, Shortle JS, Bergstrom JC (2004) Land use problems and conflicts: Causes, consequences and solutions. Routledge, London

GOFC-GOLD (2005) GLOBCOVER: Post processing kick-off and project development for a new global land cover product. GOFC-GOLD Newsletter No. 6:1–3

Goldberg ED (1994) Coastal zone space: Prelude to conflict? United Nations Education, Science and Culture Organization, Paris

Goldman A (1993) Agricultural innovation in three areas of Kenya: Neo-Boserupian theories and regional characterization. Econ Geogr 69(1):44–71

Goldstein H (1995) Multilevel statistical models. Halstaed, New York

Goldstein NC, Candau JT, Clarke KC (2004) Approaches to simulating the "March of Bricks and Mortar". Comput Environ Urban 28:125–147

Gonzalez P (2001) Desertification and a shift of forest species in the West African Sahel. Climate Res 17:217–228

Goodale CL, Apps MJ, Birdsey RA, Field CB, Heath LS, Houghton RA, Jenkins JC, Kohlmaier GH, Kurz W, Liu S, Nabuurs G-J, Nilsson S, Shvidenko AZ (2002) Forest carbon sinks in the Northern Hemisphere. Ecol Appl 12:891–899

Goodman D, Watts MJ (eds) (1997) Globalising food: Agrarian questions and global restructuring. Routledge, London

Goodwin BJ, Fahrig L (1998) Spatial scaling and animal population dynamics. In: Peterson DL, Parker VT (eds) Ecological scale: Theory and applications. Columbia University Press, New York, pp 193–206

Goudsblom J, De Vries B (eds) (2004) Mappae mundi: Humans and their habitats in a long-term socio-ecological perspective. Myths, maps, and models. Amsterdam University Press, Amsterdam, 448 pp

Gower ST (2003) Patterns and mechanisms of the forest carbon cycle. Annu Rev Env Resour 28:169–204

Gray LC, Kevane M (2001) Evolving tenure rights and agricultural intensification in southwestern Burkina Faso. World Dev 29(4):573–587

Gray LC, Moseley WG (2005) A geographical perspective on poverty-environment interactions. Geogr J/RGS 171(1):doi:10.1111/j.1475-4959

Green GM, Schweik CM, Randolph JC (2005) Linking disciplines across space and time: Useful concepts and approaches for land-cover change studies. In: Moran EF, Ostrom E (eds) Seeing the forest and the trees: Human-environment interactions in forest ecosystems. MIT Press, Cambridge London, pp 61–80

Grigg DB (1974) The agricultural systems of the world: An evolutionary approach. Cambridge University Press, New York, 358 pp

Grigg DB (1987) The Industrial Revolution and land transformation. In: Wolman MGF, Fournier FGA (eds) Land transformation in agriculture. pp 79–109

Grindle MS (1980) Politics and policy implementation in the third world. Princeton University Press, Princeton

Groombridge B, Jenkins MD (2002) Global biodiversity: Earth's living resources in the 21st century. World Conservation, Cambridge

Grover K, Quegan S, da Costa Freitas C (1999) Quantitative estimation of tropical forest cover by SAR. IEEE T Geosci Remote 37:479–490

Grübler A (1994) Technology. In: Meyer WB, Turner BL (eds) Changes in land use and land cover: A global perspective. Cambridge University Press, Cambridge, pp 287–328

Gumbo DJ, Ndiripo TW (1996) Open space cultivation in Zimbabwe: A case study of Greater Harare, Zimbabwe. Afr Urban Q 11(2/3):210–216

Guo LB, Gifford RM (2002) Soil carbon stocks and land use change: A meta analysis. Glob Change Biol 8(4):345–360

Guston DH (2001) Boundary organizations in environmental policy and science: An introduction. Sci Technol Hum Val 26:399–408

Gutman G, Janetos AC, Justice CO, Moran EF, Mustard JF, Rindfuss RR, Skole D, Turner BL II, Cochrane MA (eds) (2004) Land change science: Observing, monitoring and understanding trajectories of change on the Earth's surface. Remote Sensing and Digital Image Processing Series 6, Kluwer Academic, Dordrecht Boston London

Guyer JI, Lambin EF (1993) Land use in an urban hinterland: Ethnography and remote sensing in the study of African intensification. Am Anthropol 95:839–859

Haberl H, Erb K-H, Krausmann F, Loibl W, Schulz NB, Weisz H (2001) Changes in ecosystem processes induced by land use: Human appropriation of net primary production and its influence on standing crop in Austria. Global Biogeochem Cy 15:929–942

Haberl H, Schulz NB, Plutzar C, Erb K-H, Krausmann F, Loibl W, Moser D, Sauberer N, Weisz H, Zechmeister H, Zulka P (2004a) Human appropriation of net primary production and species diversity in agricultural landscapes. Agric Ecosyst Environ 102(2):213–218

Haberl H, Wackernagel M, Krausmann F, Erb K-H, Monfreda C (2004b) Ecological footprints and human appropriation of net primary production: A comparison. Land Use Policy 21:279–288

Haberl H, Plutzar C, Erb K-H, Gaube V, Pollheimer M, Schulz NB (2005) Human appropriation of net primary production as determinant of avifauna diversity in Austria. Agric Ecosyst Environ 110(3/4):119–131

Hagen A (2002) Fuzzy set approach to assessing similarity of categorical maps. Int J Geogr Inf Sci 17:235–249

Hails RS (2002) Assessing the risks associated with new agricultural practices. Nature 418:685–88

Hairiah K, van Noordwijk M, Weise S (2005) Sustainability of tropical land-use systems following forest conversion. In: Palm CA, Vosti SA, Sanchez PA, Ericksen PJ (eds) Slash-and-burn agriculture: The search for alternatives. Columbia University Press, New York

Hall DO, Rosillo-Calle F, Williams RH, Woods J (1993) Biomass for energy: Supply prospects. In: Johansson TB, Kelly H, Reddy AKN, Williams RH (eds) Renewable energy, sources for fuels and electricity. Earthscan, Island Press, London Washington D.C. Covelo, pp 653–698

Hamilton LS (1987) What are the impacts of deforestation in the Himalayas on the Ganges-Brahmaputra lowlands and Delta? Relations between assumptions and facts. Mt Res Dev 7:256–263

Hammond PM (1995) Magnitude and distribution of biodiversity. In: Heywood VH (ed) Global biodiversity assessment. Cambridge University Press, Cambridge, pp 113–138

Hansen MC, DeFries RS, Townshend JRG, Sohlberg R (2000) Global land cover classification at 1 km spatial resolution using a decision tree classifier. Int J Remote Sens 21:1331–1365

Hansen AJ, Rasker R, Maxwell B, Rotella JJ, Johnson JD, Parmenter AW, Langner U, Cohen WB, Lawrence RL, Kraska MPV (2002) Ecological causes and consequences of demographic change in the New West. BioScience 52:151–162

Hansen MC, DeFries RS, Townshend JRG, Carroll M, Dimiceli C, Sohlberg RA (2003) Global percent tree cover at a spatial resolution of 500 meters: First results of the MODIS vegetation continuous fields algorithm. Earth Interact 7:1–15

Hansen MC, DeFries RS (2004) Detecting long-term global forest change using continuous fields of tree-cover maps from 8-km advanced very high resolution radiometer (AVHRR) data for the years 1982–1999. Ecosystems 7:695–717

Hao WM, M-H Liu (1994) Spatial and temporal distribution of tropical biomass burning. Global Biogeochem Cy 8(4):495–503

Hecht SB (1993) The logic of livestock and deforestation in Amazonia: Considering land markets, value of ancillaries, the larger macroeconomic context, and individual economic strategies. BioScience 43(10):687–695

Hecht SB (2005) Soybeans, development and conservation on the Amazon frontier. Dev Change 36(2):375–404

Heilig GK (1994) Neglected dimensions of global land-use change: Reflections and data. Popul Dev Rev 20(4):831–859

Heilig GK (1999) Can China feed itself? – A system for evaluation of policy options. International Institute for Applied Systems Analysis, http://www.iiasa.ac.at/Research/LUC/ChinaFood/index.htm

Helfman ES (1962) Land, people, and history. David McKay Company, New York, 271 pp

Henry S, Boyle P, Lambin EF (2003) Modelling interprovincial migration in Burkina Faso, West Africa: The role of socio-demographic and environmental factors. Appl Geogr 23:115–136

Henry S, Piche V, Ouedraogo D, Lambin EF (2004) Environmental influence on migration decisions in Burkina Faso. Popul Environ 25(5):397–422

Herold M, Schmullius C (2004) Report on the harmonization of global and regional land cover products. Workshop report at FAO, Rome, Italy, 14–16 July 2004, GOFC-GOLD report series 20, http://www.fao.org/gtos/gofc-gold/series.html

Herold M, Goldstein NC, Clarke KC (2003) The spatiotemporal form of urban growth: Measurement, analysis and modeling. Remote Sens Environ 86:286–302

Herold M, Woodcock C, Di Gregorio A, Mayaux PAB, Latham J, Schmullius CC (2006) A joint initiative for harmonization and validation of land cover datasets. IEEE T Geosci Remote, forthcoming

Hewlett JD, Bosch JM (1984) The dependence of storm flows on rainfall intensity and vegetal cover: South Africa. J Hydrol 75:365–381

Hewlett JD, Helvey JD (1970) Effects of forest clearfelling on the storm hydrograph. Water Resour Res 6(3):768–782

Heywood VH (ed) (1995) Global biodiversity assessment. Cambridge University Press, Cambridge

Hijmans RJ, Van Ittersum MK (1996) Aggregation of spatial units in linear programming models to explore land use options. Neth J Agr Sci 44:145–162

Hill RD (1994) Upland development policy in the People's Republic of China. Land Use Policy 11:8–16

Hilton-Taylor C (2000) IUCN red list of threatened species. World Conservation Union, Gland

Himiyama Y (ed) (1992) Atlas of land use change in modern Japan. Hokkaido University of Education, Asahikawa

Himiyama Y (1998) Land use/cover changes in Japan: From the past to the future. Hydrol Process 12(13/14):1995–2001

Himiyama Y, Mather A, Bicik I, Milanova EV (eds) (2001, 2002, 2005) Land use/cover changes in selected regions in the world, vol. I–IV. International Geographical Union Commisssion on Land Use and Land Cover Change, Asahikawa

Ho SPS, Lin GCS (2004) Non-agricultural land use in post-reform China. China Quart (179):758–781

Hole F, Smith R (2004) Arid land agriculture in northeastern Syria: Will this be a tragedy of the commons? In: Gutman G, Janetos AC, Justice CO, Moran EF, Mustard JF, Rindfuss RR, Skole D, Turner BL II, Cochrane MA (eds) Land change science: Observing, monitoring and understanding trajectories of change on the Earth's surface. Remote Sensing and Digital Image Processing Series 6, Kluwer Academic, Dordrecht Boston London, pp 209–222

Homewood KM (2004) Policy, environment and development in African rangelands. Environ Sci Policy 7:125–143

Homewood K, Lambin EF, Coast E, Kariuki A, Kikulai I, Kiveliai J, Said M, Serneels S, Thompson M (2001) Long-term changes in Serengeti-Mara wildebeest and land cover: Pastoralism, population, or policies? Proc Natl Acad Sci USA 98(22):12544–12549

Hoogwijk M, Faaij A, van den Broek R, Berndes G, Gielen D, Turkenburg W (2003) Exploration of the ranges of the global potential of biomass for energy. Biomass Bioenerg 25:119–133

Hopkins NS (1987) Mechanized irrigation in Upper Egypt: The role of technology and the state in agriculture. In: Turner BL II, Brush SB (eds) Comparative farming systems. Guilford, New York, pp 223–247

Hoshino S (2001) Multilevel modeling on farmland distribution in Japan. Land Use Policy 18:75–90

Houghton RA (2003) Revised estimates of the annual net flux of carbon to the atmosphere from changes in land use and land management 1850–2000. Tellus 55B:378–390

Houghton RA, Goodale CL (2004) Effects of land-use change on the carbon balance of terrestrial ecosystems. In: DeFries RS, Asner GP, Houghton RA (eds) Ecosystems and land use change. American Geophysical Union, Washington D.C., pp 85–98

Houghton RA, Hackler JL (2001) Carbon flux to the atmosphere from land-use changes: 1850 to 1990. NDP-050/R1, Carbon Dioxide Information Analysis Center, Oak Ridge National Laboratory, Oak Ridge

Houghton RA, Hackler JL (2003) Sources and sinks of carbon from land-use change in China. Global Biogeochem Cy 17:1034, 10.1029/2002GB001970

Houghton RA, Hackler JL (2006) Emissions of carbon from land-use change in Sub-Saharan Africa. J Geophys Res-Biogeo-sciences, forthcoming

Houghton RA, Hobbie JE, Melillo JM, Moore B, Peterson BJ, Shaver GR, Woodwell GM (1983) Changes in the carbon content of terrestrial biota and soils between 1860 and 1980: A net release of CO_2 to the atmosphere. Ecol Monogr 53:235–262

Houghton RA, Boone RD, Melillo JM, Palm CA, Woodwell GM, Myers N, Moore B, Skole DL (1985) Net flux of carbon dioxide from tropical forest in 1980. Nature 316:617–620

Houghton RA, Lefkowitz DS, Skole DL (1991) Changes in the landscape of Latin America between 1850 and 1985. I. Progressive loss of forests. Forest Ecol Manag 38:143–172

Houghton RA, Hackler JL, Lawrence KT (1999) The U.S. carbon budget: Contributions from land-use change. Science 285:574–578

Houghton RA, Skole DL, Nobre CA, Hackler JL, Lawrence KT, Chomentowski WH (2000) Annual fluxes or carbon from deforestation and regrowth in the Brazilian Amazon. Nature 403: 301–304

Houghton JT, Ding Y, Griggs DJ, Noguer M, van der Linden PJ, Dai X, Maskell K, Johnson CA (eds) (2001) Climate change 2001: The scientific basis. Contributions of working group I to the third assessment report of the Intergovernmental Panel on Climate Change (IPCC). Cambridge University Press: Cambridge, New York, 944 pp

Hsin H, van Tongeren F, Dewbre J, van Meijl H (2004) A new representation of agricultural production technology in GTAP. GTAP Resource No. 1504, http://www.gtap.agecon.purdue.edu

Huigen MGA (2004) First principles of the MameLuke multi-actor modelling framework for land use change, illustrated with a Philippine case study. J Environ Manage 72:5–21

Humphries S (1998) Milk cows, migrants, and land markets: Unraveling the complexities of forest-to-pasture conversion in Northern Honduras. Econ Dev Cult Change 47(1):95–124

Humphries S, Gonzales J, Jimenez J, Sierra F (2000) Searching for sustainable land use practices in Honduras: Lessons from a programme of participatory research with hillside farmers. Network Paper No. 16, Overseas Development Institute, London, 104 pp

Hwang M (2006) Coastal zone. In: Geist HJ (ed) Our Earth's changing land: An encyclopedia of land-use and land-cover change, vol. 1 (A–K). Greenwood Press, Westport, London, pp 124–130

Ijumba JN, Lindsay SW (2001) Impact of irrigation on malaria in Africa: Paddies paradox. Med Vet Entomol 15:1–11

IMAGE-Team (2001) The IMAGE 2.2 implementation of the SRES scenarios: A comprehensive analysis of emissions, climate change and impacts in the 21st century. RIVM CD-ROM publication 481508018, National Institute for Public Health and the Environment, Bilthoven

Imbernon J (1999a) Changes in agricultural practice and landscape over a 60-year period in North Lampung, Sumatra. Agric Ecosyst Environ 76(1):61–66

Imbernon J (1999b) A comparison of the driving forces behind deforestation in the Peruvian and the Brazilian Amazon. Ambio 28(6):509–513

Imhoff ML, Bounoua L, Ricketts T, Loucks C, Harriss R, Lawrence WT (2004) Global patterns in human consumption of net primary production. Nature 429:870–873

Indrabudi H, Gier A de, Fresco LO (1998) Deforestation and its driving forces. A case study of Riam Kanan watershed, Indonesia. Land Degrad Dev 9(4):311–322

Innis MQ (1935) An economic history of Canada. Ryerson Press, Toronto, pp 302

INPE (Instituto Nacional de Pesquisas Especiais) (2000) Monitoring of the Brazilian Amazonian Forest. INPE, Sao Paulo

IPCC (Intergovernmental Panel on Climate Change) (1997) Special report on the regional impacts of climate change: An assessment of vulnerability. Cambridge University Press, Cambridge

IPCC (Intergovernmental Panel on Climate Change) (2000a) Special report on emissions scenarios. Cambridge University Press, Cambridge

IPCC (Intergovernmental Panel on Climate Change) (2000b) Land use, land-use change, and forestry. A Special Report of the IPCC. Cambridge University Press, Cambridge

Irwin EG, Bockstael NE (2002) Interacting agents, spatial externalities and the endogenous evolution of land use patters. J Econ Geogr 2:31–54

Irwin EG, Bockstael NE (2004) Land use externalities, open space preservation, and urban sprawl. Reg Sci Urban Econ 34:705–725

Irwin EG, Geoghegan J (2001) Theory, data, methods: Developing spatially-explicit economic models of land use change. Agric Ecosyst Environ 85:7–24

Ives JD (1989) Deforestation in the Himalayas. The cause of increased flooding in Bangladesh and northern India? Land Use Policy 6:187–193

Jacobsen T, Adams RM (1958) Salt and silt in ancient Mesopotamian agriculture. Science 128:1251–1258

Jansen H, Stoorvogel JJ (1998) Quantification of aggregation bias in regional agricultural land use models: Application to Guacimo County, Costa Rica. Agricult Sys 58:417–439

Janzen DH (1988) Tropical dry forests: The most endangered major tropical ecosystem. In: Wilson EO, Peter FM (eds) Biodiversity. National Academy Press, Washington D.C., pp 130–137

Jeleček L (1995) Changes in the production and techniques in the agriculture of Bohemia 1870–1945. In: Havinden MA, Collins EJT (eds) Agriculture in the industrial state. University of Reading, Rural History Centre, Reading, pp 126–145

Jeleček L (2006) Agricultural revolution. In: Geist HJ (ed) Our Earth's changing land: An encyclopedia of land-use and land-cover change, vol. 1 (A–K). Greenwood Press, Westport, London, pp 25–27

Jenerette GD, Wu J (2001) Analysis and simulation of land-use change in the central Arizona-Phoenix region, USA. Landscape Ecol 16:611–626

Jepson W (2005) A disappearing biome? Reconsidering land-cover change in the Brazilian Savanna. Geogr J/RGS 171:99; doi:10.1111/ j.1475-4959.2005.00153.x

Jepson W (2006) Brazilian Cerrado. In: Geist HJ (ed) Our Earth's changing land: An encyclopedia of land-use and land-cover change, vol. 1 (A–K). Greenwood Press, Westport, London, pp 76–78

Jepson P, Jarvie JK, MacKinnon K, Monk KA (2001) The end of Indonesia's lowland forests? Science 292:859–861

Jiang H (2002) Culture, ecology, and nature's changing balance: Sandification on Mu Us Sandy Land, Inner Mongolia, China. In: Reynolds JF, Stafford-Smith DM (eds) Global desertification: Do humans cause deserts? Dahlem Workshop Report 88, Dahlem University Press, Berlin, pp 181–196

Jiang H, Zhang P, Zheng D, Wang F (1995) The Ordos Plateau of China. In: Kasperson JX, Kasperson RE, Turner BL II (eds) Regions at risk: Comparisons of threatened environments. United Nations University Press, Tokyo New York Paris, pp 420–459

Johnson SH III (1986) Agricultural intensification in Thailand: Complementary role of infrastructure and agricultural policy. In: Easter WK (ed) Irrigation investment, technology, and management strategies for development. Westview, Boulder, pp 111–127

Jolly CL, Torrey BB (eds) (1993) Population and land use in developing countries: Report of a workshop. National Academy Press, Washington D.C.

Jones DW (1983) Location, agricultural risk, and farm income diversification. Geogr Anal 15:231–246

Jones PG, Thornton PK (2002) Spatial modeling of risk in natural resource management. Conserv Ecol 5(2), www.consecolorg/vol5/ iss2/art27

Joos F, Gerber S, Prentice IC, Otto-Bliesner BL, Valdes P (2004) Transient simulations of Holocene atmospheric carbon dioxide and terrestrial carbon since the Last Glacial Maximum. Global Biogeochem Cy 18:1–18, doi:10.1029/2003GB002156

Jordan AM (1979) Trypanosomiasis control and land use in Africa. Outlook Agr 10:123–129

Jordan AM (1986) Trypanosomiasis control and African rural development. Longman, Harlow

Joshi SR, Ahmad F, Gurung MB (2004) Status of *Apis laboriosa* populations in Kaski District, western Nepal. J Apic Res 43(4):176–180

Kabat P, Claussen M, Dirmeyer PA, Gash JHC, Bravo de Guenni L, Meybeck M, Pielke RA Sr, Vörösmarty CJ, Hutjes RWA, Lütkemeier S (eds) (2004) Vegetation, water, humans and the climate: A new perspective on an interactive system. Springer, Berlin Heidelberg

Kaimowitz D, Angelsen A (1998) Economic models of tropical deforestation: A review. Centre for International Forestry Research, Jakarta, 139 pp

Kaimowitz D, Thiele G, Pacheco P (1999) The effects of structural adjustment on deforestation and forest degradation in lowland Bolivia. World Dev 27(3):505–520

Kalnay E, Cai M (2003) Impact of urbanization and land-use change on climate. Nature 423:528–531

Karlen D (2004) Soil quality as an indicator of sustainable tillage practices. Soil Tillage Res 78(2):129–130

Karlen DL, Mausbach MJ, Doran JW, Cline RG, Harris RF, Schuman GE (1997) Soil quality: A concept, definition, and framework for evaluation. Soil Sci Soc Am J 61:4–10

Kasfir N (1993) Agricultural transformation in the robusta coffee/banana zone of Bushenyi, Uganda. In: Turner BL II, Hyden G, Kates R (eds) Population growth and agricultural change in Africa. University Press of Florida, Gainesville, pp 41–79

Kasischke ES, Williams D, Barry D (2002) Analysis of the patterns of large fires in the boreal forest region of Alaska. Int J Wildland Fire 11(2):131–144

Kasperson JX, Kasperson RE, Turner BL II (eds) (1995) Regions at risk: Comparisons of threatened environments. United Nations University Press, Tokyo

Kasperson JX, Kasperson RE, Turner BL II (1999) Risk and criticality: Trajectories of regional environmental degradation. Ambio 28:562–568

Kasperson RE, Archer E, Caceres D, Dow K, Downing T, Elmqvist T, Folke C, Han G, Iyengar K, Vogel C, Wilson K, Ziervogel G (2005) Vulnerable people and places. In: Scholes R, Rashid H (eds) Millenium Ecosystem Assessment. Working group on conditions and trends, Island Press, Washington D.C.

Kates RW (2000) Population and consumption: What we know, what we need to know. Environment 42(3):10–19

Kates RW, Haarman V (1991) Poor people and threatened environments: Global overviews, country comparisons, and local studies. Research Report 91-20, The Alan Shawn Feinstein World Hunger Program, Brown University, Providence

Kates RW, Haarman V (1992) Where the poor live: Are the assumptions correct? Environment 34:4–11, 25–28

Kauffman JB (2004) Death rides the forest: Perceptions of fire, land use, and ecological restoration of western forests. Conserv Biol 18(4):878–882

Kemp-Benedict E, Heaps C, Raskin P (2002) Global scenario group futures: Technical notes. SEI PoleStar Series Report No. 9 (revised and expanded), Stockholm Environment Institute, Boston

Keynes JM (1936) Essays in persuasion: Economic possibilities for our grandchildren. In: Keynes JM (ed) The collected writings of J. M. Keynes, vol. IX. MacMillan Press, London, pp321–332

Keys E (2004) Jalapeño pepper cultivation: Emergent commercial land use. In: Turner BL II, Geoghegan J, Foster D (eds) Integrated land-change science and tropical deforestation in the southern Yucatán: Final frontiers. Oxford University Press, Oxford, pp 207–219

Keys E, McConnell WJ (2005) Global change and the intensification of agriculture in the tropics. Global Environ Chang 15:320–337

Kiersch B (2001) Land use impacts on water resources: A literature review. Land-water linkages in rural watersheds. Electronic Workshop (Discussion Paper 1). Food and Agriculture Organization of the United Nations, Rome

Kimble GHT (1962) Tropical Africa, vol. 1. (Land and livelihood). Anchor Books, Doubleday & Company, Garden City, New York

Klein Goldewijk K (2001) Estimating global land use change over the past 300 years: The HYDE Database. Global Biogeochem Cy 15:417–433

Klein Goldewijk K (2004) Footprints from the past: Blueprint for the future? In: DeFries RS, Asner GP, Houghton RA (eds) Ecosystems and land use change. Geophysical Monograph 153. American Geophysical Union, Washington D.C., pp 203–215

Klein Goldewijk K, Ramankutty N (2004) Land cover change over the last three centuries due to human activities: The availability of new global data sets. GeoJournal 61:335–344

Klepeis P, Chowdhury RR (2004) Institutions, organizations, and policy affecting land change: Complexity within and beyond the *ejido*. In: Turner BL II, Geoghegan J, Foster D (eds) Integrated land-change science and tropical deforestation in the southern Yucatán: Final frontiers. Oxford University Press, Oxford, pp 145–169

Klepeis P, Turner BL II (2001) Integrated land history and global change science: The example of the Southern Yucatán Peninsular region project. Land Use Policy 18:27–39

Klijn JA, Vullings LAE, van de Berg M, van Meijl H, van Lammeren R, van Rheenen T, Eickhout B, Veldkamp A, Verburg PH, Westhoek H (2005) EURURALIS 1.0: A scenario study on Europe's rural areas to support policy discussion. Background document. Wageningen University and Research Centre/Environmental Assessment Agency (RIVM)

Kohler W (2004) Eastern enlargement of the EU: A comprehensive welfare assessment. J Policy Model 26:865–888

Kok K (2004) The role of population in understanding Honduran land use patterns. J Environ Manage 72:73–89

Kok K, Patel M (eds) (2003) Target area scenarios: First sketch. MedAction Deliverable 7. Report No. I03-E003, ICIS, Maastricht

Kok K, Rothman DS (2003) Mediterranean scenarios: First draft. MedAction Deliverable 3. Report No. I03-E001, ICIS, Maastricht

Kok K, Veldkamp A (2000) Using the CLUE framework to model changes in land use on multiple scales. In: Bouman BAM, Jansen HGP, Schipper RA, Hengsdijk H, Nieuwenhuyse A (eds) Tools for land use analysis on different scales, with case studies for Costa Rica. Kluwer Academic, Dordrecht Boston London, pp 35–63

Kok K, Veldkamp A (2001) Evaluating impact of spatial scales on land use pattern analysis in Central America. Agric Ecosyst Environ 85:205–222

Kok K, Winograd M (2002) Modeling land-use change for Central America, with special reference to the impact of hurricane Mitch. Ecol Model 149:53–69

Kok K, Farrow A, Veldkamp A, Verburg PH (2001) A method and application of multi-scale validation in spatial land use models. Agric Ecosyst Environ 85:223–238

Kok K, Rothman DS, Greeuw S, Patel M (2003) European scenarios: From VISIONS to MedAction. MedAction Deliverable 2. Report No. I03-E004, ICIS, Maastricht

Kok K, Verburg PH, Veldkamp A (2004) International workshop "Integrated Assessment of the Land System: The future of land use", Amsterdam, the Netherlands, 28–30 October 2004. LUCC Newsletter 10:10–11

Kondrashov LG (2001) Russian Far East forest disturbances and socio-economic problems of restoration. Forest Ecol Manag 201:65–74

Krausmann F (2006) Industrialization. In: Geist HJ (ed) Our Earth's changing land: An encyclopedia of land-use and land-cover change, vol. 1 (A–K). Greenwood Press, Westport, London, pp 304–308

Krugman P (1999) The role of geography in development. Int Regional Sci Rev 22:142–161

Krumm K, Kharas H (eds) (2004) East Asia integrates: A trade policy agenda for shared growth. World Bank, Washington D.C.

Kuhlman T, Koomen E, Groen J, Bouwman A (2006) Simulating agricultural land use change in The Netherlands. In: Brouwer F, McCarl BA (eds) Agriculture and climate beyond 2015: A new perspective on future land use patterns (Environment & Policy Vol. 46). Springer, Dortrecht

Kuhn KG, Campbell-Lendrum DH, Davies CR (2002) A continental risk map for malaria mosquito (Diptera: *Culicidae*) vectors in Europe. J Med Entomol 39(4):621–630

Kull CA (1998) Leimavo revisited: Agrarian land-use change in the highlands of Madagascar. Prof Geogr 50(2):163–176

Lacerda LD, Souza M de, Ribeiro MG (2004) The effects of land use change on mercury distribtuion in soils of Alta Forest, southern Amazon. Environ Pollut 129:247–255

Lal R (1997) Degradation and resilience of soils. Philos Trans R Soc Lond B352:997–1010

Lal R (2000) Physical management of soils of the tropics: Priorities for the 21st century. Soil Sci 165(3):191–207

Lal R (2002a) Carbon sequestration in dryland ecosystems of West Asia and North Africa. Land Degrad Dev 13:45–59

Lal R (2002b) Soil carbon dynamics in cropland and rangeland. Environ Pollut 116:353–362

Lambin EF (1997) Modelling and monitoring land-cover change processes in tropical regions. Prog Phys Geog 21:375–393

Lambin EF (2003) Linking socioeconomic and remote sensing data at the community or at the household level: Two case studies from Africa. In: Fox J, Rindfuss RR, Walsh SJ, Mishra V (eds) People and the environment: Approaches for linking household and community surveys to remote sensing and GIS. Kluwer Academic, Dordrecht Boston London, pp 223–240

Lambin EF (2005) Conditions for sustainability of human-environment systems: Information, motivation, and capacity. Global Environ Chang 15(3):177–180

Lambin EF, Ehrlich D (1997a) The identification of tropical deforestation fronts at broad spatial scales. Int J Remote Sens 18(17):3551–3568

Lambin EF, Ehrlich D (1997b) Land-cover changes in Sub-Saharan Africa (1982–1991): Application of a change index based on remotely sensed surface temperature and vegetation indices at a continental scale. Remote Sens Environ 61(2):181–200

Lambin EF, Geist HJ (2001) Global land-use/land-cover changes: What have we learned so far? IGBP Global Change Newsletter No. 46, June 2001, pp 27–30

Lambin EF, Geist HJ (2003a) The land managers who have lost control of their land use: Implications for sustainability. Trop Ecol 44:15–24

Lambin EF, Geist HJ (2003b) Regional differences in tropical deforestation. Environment 45(6):22–36

Lambin EF, Geist HJ (guest eds) (2005) Focus: Land-use and land-cover change. IHDP Update – Newsletter of the International Human Dimensions Programme on Global Environmental Change No. 3, August 2005

Lambin EF, Baulies X, Bockstael N, Fischer G, Krug T, Leemans R, Moran EF, Rindfuss RR, Sato Y, Skole D, Turner BL II, Vogel C (1999) Land-use and land-cover change (LUCC): Implementation strategy. IGBP Report 48, IHDP Report 10, International Geosphere-Biosphere Programme, International Human Dimensions on Global Environmental Change Programme, Stockholm Bonn, 125 pp

Lambin EF, Rounsevell MDA, Geist HJ (2000) Are agricultural land-use models able to predict changes in land-use intensity? Agric Ecosyst Environ 82:321–331

Lambin EF, Turner BL II, Geist HJ, Agbola SB, Angelsen A, Bruce JW, Coomes O, Dirzo R, Fischer G, Folke C, George PS, Homewood K, Imbernon J, Leemans R, Li X, Moran EF, Mortimore M, Ramakrishnan PS, Richards JF, Skånes H, Stone GD, Svedin U, Veldkamp TA, Vogel C, Xu J (2001) The causes of land-use and land-cover change: Moving beyond the myths. Global Environ Chang 11(4):261–269

Lambin EF, Chasek PS, Downing TE, Kerven C, Kleidon A, Leemans R, Lüdeke M, Prince SD, Xue Y (2002) The interplay between international and local processes affecting desertification. In: Reynolds JF, Stafford-Smith DM (eds) Global desertificaton: Do humans cause deserts? Dahlem Workshop Report No. 88, Dahlem University Press, Berlin, pp 387–401

Lambin EF, Geist HJ, Lepers E (2003) Dynamics of land use and cover change in tropical and subtropical regions. Annu Rev Env Resour 28:205–241

Lamprey HF (1975) Report on the desert encroachment reconnaissance in northern Sudan 21 Oct. to 10 Nov. 1975. UNESCO/UNEP

Lamprey HF (1988) Report on the desert encroachment reconnaissance in northern Sudan 21 Oct. to 10 Nov. 1975. Desertific Control Bull 17:1–7

Lamprey R, Reid RS (2004) Expansion of human settlement in Kenya's Maasai Mara: What future for pastoralism and wildlife? J Biogeogr 31:997–1032

Laney RM (2002) Disaggregating induced intensification for land-change analysis: A case study from Madagascar. Ann Assoc Am Geogr 92(4):702–726

Laney RM (2004) A process-led approach to modeling land change in agricultural landscapes: A case study from Madagascar. Agric Ecosyst Environ 101:135–153

Larson WE, Pierce FJ (1991) Conservation and enhancement of soil quality. In: International Board for Soil Research and Management (ed) Evaluation for sustainable land management in the developing world, vol 2, (Technical papers) (IBSRAM Proceedings 12-2). International Board for Soil Research and Management, Bangkok, pp 175–203

Larsson L-I, Frisk M (2000) Bringing the past to life. GEOEurope 9:40–41

Laurance WF, Laurance SG, Ferreira LV, Merona JMR, Gascon C, Lovejoy TE (1997) Biomass collapse in Amazonian forest fragments. Science 278:1117–1118

Laurance WF, Cochrane MA, Bergen S, Fearnside PM, Delamônica P, Barber C, D'Angelo S, Fernandes T (2001) The future of the Brazilian Amazon. Science 291:438–439

Laurance WF, Albernaz AKM, Fearnside PM, Vasconcelos HL, Ferreira LV (2004) Deforestation in Amazonia. Science 304:1109–1109

Lavorel S, Flannigan MD, Lambin EF, Scholes MC (2005) Vulnerability of land systems to fire: Interactions between humans, climate, the atmosphere and ecosystems. Mitigation and Adaptation Strategies for Global Change, vol. 10

Lawton RO, Nair US, Pielke, RA, Welch RM (2001) Climate impact of tropical lowland deforestation on nearby montane cloud forests. Science 294:584–587

Leach M, Fairhead J (2000) Challenging neo-Malthusian deforestation analyses in West Africa's dynamic forest landscapes. Popul Dev Rev 26(1):17–43

Leaf MJ (1987) Intensification in peasant farming: Punjab in the green revolution. In: Turner BL II, Brush SB (eds) Comparative farming systems. Guilford, New York, pp 248–275

Leak SGA (1999) Tsetse biology and ecology: Their role in the epidmiology and control of trypanosomosis. International Livestock Research Institute, Nairobi, Kenya, CAB International, Wallingford New York

Lebel L (2004) Nobody knows best. Polit Law Econ 10:111–127

Ledec G (1985) The political economy of tropical deforestation. In: Leonhard HJ (ed) Diverting nature's capital: The political economy of environmental abuse in the Third World. Holmes & Meier, New York London, pp 179–226

Leemans R (1999) Modeling for species and habitats: New opportunities for problem solving. Sci Total Environ 240:51–73

Leemans R, van Amstel A, Battjes C, Kreileman E, Toet S (1996) The land cover and carbon cycle consequences of large-scale utilizations of biomass as an energy source. Global Environ Chang 6(4):335–357

Leemans R, Lambin EF, McCalla A, Nelson JPP, Watson B (2003) Drivers of change in ecosystems and their services. In: Millennium Ecosystem Assessment (ed) Ecosystems and human well-being: A framework for assessment. Island Press, Washington D.C., pp 85–105

Leichenko R, Solecki W (2005) Exporting the American dream: The globalization of suburban consumption landscapes. Reg Stud 39(2):241–253

Lele U, Viana V, Veríssimo A, Vosti S, Perkins K, Husain SA (2000) Brazil, forests in the balance: Challenges of conservation with development. World Bank, Operation Evaluation Department, Washington D.C.

Lepers E, Lambin EF, Janetos AC, DeFries RS, Achard F, Ramankutty N, Scholes RJ (2005) A synthesis of information on rapid land-cover change for the period 1981–2000. BioScience 55(2):115–124

Li X, Yeh AG (2001) Calibration of cellular automata by using neural networks for the simulation of complex urban systems. Environ Plann A 33:1445–1462

Li X, Yeh AG (2002) Neural-network-based cellular automata for simulating multiple land use changes using GIS. Int J Geogr Inf Sci 16:323–343

Li X, Yeh AG (2004) Data mining of cellular automata's transition rules. Int J Geogr Inf Sci 18:723–744

Li C, Qiu J, Frolking S, Xiao X, Salas W, Moore B III, Boles S, Huang Y, Sass R (2002) Reduced methane emissions from large-scale changes in water management of China's rice paddies during 1980–2000. Geophys Res Lett 29(20):1972ff

Liebig MA, Tanaka DL, Wienhold BJ (2004) Tillage and cropping effects on soil quality indicators in the northern Great Plains. Soil Tillage Res 78:131–141

Lin GCS, Ho SPS (2005) The state, land system, and land development processes in contemporary China. Ann Assoc Am Geogr 95(2):411–436

Lin NF, Tang J, Han FX (2001) Eco-environmental problems and effective utilization of water resources in the Kashi Plain, western Terim Basin, China. Hydrogeol J 9:202–207

Lines J (1995) The effects of climatic and land-use changes on the insect vectors of human disease. In: Harrington R, Stork NE (eds) Insects in a changing environment. Academic Press, London, pp 158–177

Liu JG (2001) Integrating ecology with human demography, behavior, and socioeconomics: Needs and approaches. Ecol Model 140(1/2):1–8

Liu JG, Linderman M, Ouyang Z, An L, Yang J, Zhang H (2001) Ecological degradation in protected areas: The case of Wolong Nature Reserve for giant pandas. Science 292:98–101

Liu J, Daily GC, Ehrlich PR, Luck GW (2003a) Effects of household dynamics on resource consumption and biodiversity. Nature 421:530–533

Liu JG, Ouyang Z, Pimm S, Raven P, Wang X, Miao H, Han N (2003b) Protecting China's biodiversity. Science 300:1240–1241

Liverman D, Moran EF, Rindfuss RR, Stern PC (eds) (1998) People and pixels: Linking remote sensing and social science. National Academy Press, Washington D.C., 244 pp

Lofgren BM (1995) Sensitivity of land-ocean circulations, precipitation, and soil moisture to perturbed land surface albedo. J Climate 8:2521–2542

Lomborg B (2001) The skeptical environmentalist: Measuring the real state of the world. Cambridge University Press, Cambridge

Loveland TR, Reed BC, Brown JF, Ohlen DO, Zhu Z, Yang L, Merchant JW (2000) Development of a global land cover characteristics database and IGBP DISCover from 1 km AVHRR data. Int J Remote Sens 21:1303–1330

Lubchenco J (1998) Entering the century of the environment: A new social contract for science. Science 279:491–497

LUCC Scientific Steering Committee (2005) Key findings of LUCC on its research questions. IGBP Global Change Newsletter No. 63, September 2005, pp 12–14

Luhman N (1985) A sociological theory of law. Routledge and Kegan Paul, London

Lupo F, Reginster I, Lambin EF (2001) Monitoring land-cover changes in West Africa with SPOT vegetation: Impact of natural disasters in 1998–1999. Int J Remote Sens 22(13):2633–2639

Lutz W, Sanderson WC, Scherbov S (2004) The end of world population growth. In: Lutz W, Sanderson WC, Scherbov S (eds) The end of world population growth in the 21st century: New challenges for human capital formation and sustainable development. Sterling, Earthscan, London, pp 17–83

Mabbutt JA (1984) A new global assessment of the status and trends of desertification. Environ Conserv 11:103–113

Magurran AE, May RM (1999) Evolution of biological diversity. Oxford University Press, Oxford

Mahar D (2002) Agro-ecological zoning in Rondônia, Brazil: What are the lessons? In: Hall A (ed) Amazonia at the crossroads. Institute of Latin American Studies, London, pp 115–128

Maizel M, White RD, Root R, Gage S, Stitt S, Osborne L, Muehlbach G (1988) Historical interrelationships between population settlement and farmland in the conterminous United States 1790 to 1990. In: Sisk TD (ed) Perspectives on the land use history of North America: A context for understanding our changing environment. U.S. Geological Survey, Biological Resources Division, Biological Science Report, USGS/BRD/BSR 1998-0003

Malingreau JP, Stephens G, Fellows L (1985) Remote sensing of forest fires: Kalimantan and North Borneo in 1982–1983. Ambio 14:314–321

Mallee H (1996) Reform of the Houkou system: Introduction. Chinese Soc Anthrop 29:3–14

Malthus TR (1798) An essay on the principle of population, as it affects the future improvement of society with remarks on the speculations of Mr. Godwin, M. Condorcet, and other writers. Printed for J. Johnson, London

Manel S, Williams HC, Ormerod SJ (2001) Evaluating presence-absence models in ecology: The need to account for prevalence. J Appl Ecol 38:921–931

Manies KL, Mladenoff DJ (2000) Testing methods to produce landscape-scale presettlement vegetation maps from the U.S. public land survey records. Landscape Ecol 15:741–754

Mann ME, Bradley RS, Hughes MK (1999) Northern Hemisphere temperatures during the past millenium: Inferences, uncertainties, and limitations. Geophys Res Lett 26:759–762

Manne LL, Brooks TM, Pimm SL (1999) Relative risk of extinction of passerine birds on continents and islands. Nature 399:258–261

Margules CR, Pressey RL (2000) Systematic conservation planning. Nature 405:243–253

Marks RB (1998) Tigers, rice, silk, and silt: Environment and economy in Late Imperial South China. Cambridge University Press, Cambridge, 407 pp

Marquette CM (1998) Land use patterns among small farmer settlers in the Northeastern Ecuadorian Amazon. Hum Ecol 26(4):573–598

Marsh GP (1864) Man and nature, or physical geography as modified by human action. Harvard University Press, Cambridge (reprint edition; edited by David Lowenthal 1965)

Martens P, Moser SC (2001) Health impacts of climate change. Science 292(5519):1065–1066

Martens P, Rotmans J (eds) (2002) Transitions in a globalising world. Swets & Zeitlinger, Lisse

Martin PS, Klein RG (eds) (1984) Quaternary extinctions: A prehistoric revolution. University of Arizona Press, Tucson

Maslow H (1943) A theory of human motivation. Psychol Rev 50:370–396

Master LL, Stein BA, Kutner LS, Hammerson GA (2000) Vanishing assets: Conservation status of the US species. In: Stein BA, Kutner LS, Adams JS (eds) Precious heritage: The status of biodiversity in the United States. Oxford University Press, New York, pp 93–118

Matarazzo B, Nijkamp P (1997) Meta-analysis for comparative environmental case studies: Methodological issues. Int J Soc Econ 24(7–9):799–811

Mather AS (1992) The forest transition. Area 24(4):367–379

Mather AS (2001) The transition from deforestation to reforestation in Europe. In: Angelsen A, Kaimowitz (eds) Agricultural technologies and tropical deforestation. CAB International, Wallingford New York, pp 35–52

Mather AS (2004) Forest transition theory and the reforesting of Scotland. Scot Geogr J 120:83–98

Mather AS (2006a) Proximate causes. In: Geist HJ (ed) Our Earth's changing land: An encyclopedia of land-use and land-cover change, vol. 2 (L–Z). Greenwood Press, Westport, London, pp 490–495

Mather AS (2006b) Driving forces. In: Geist HJ (ed) Our Earth's changing land: An encyclopedia of land-use and land-cover change, vol. 1 (A–K). Greenwood Press, Westport, London, pp 179–185

Mather AS (2006c) Land-use policies. In: Geist HJ (ed) Our Earth's changing land: An encyclopedia of land-use and land-cover change, vol. 2 (L–Z). Greenwood Press, Westport, London, pp 375–379

Mather AS (2006d) Forest transition. In: Geist HJ (ed) Our Earth's changing land: An encyclopedia of land-use and land-cover change, vol. 1 (A–K). Greenwood Press, Westport, London, pp 241–246

Mather AS, Needle CL (1998) The forest transition: A theoretical basis. Area 30(2):117–124

Mather AS, Needle CL (2000) The relationships of population and forest trends. Geogr J 166(1):2–13

Mather AS, Needle CL, Coull JR (1998) From resource crisis to sustainability: The forest transition in Denmark. Int J Sust Dev World 5(3):182–193

Mather AS, Fairbairn J, Needle CL (1999) The course and drivers of the forest transition: The case of France. J Rural Stud 15(1):65–90

Mathevet R, Bousquet F, Le Page C, Antona M (2003) Agent-based simulations of interactions between duck population, farming decisions and leasing of hunting rights in the Camargue (Southern France). Ecol Model 165:107–126

Matson PA, Parton WJ, Power AG, Swift MJ (1997) Agricultural intensification and ecosystem properties. Science 277:504–509

May RM (2000) The dimensions of life on Earth. In: Raven PH, Williams T (eds) Nature and human society: The quest for a sustainable world. National Academy of Sciences, Washington D.C., pp 30–45

May RM, Lawton JH, Stork NE (1995) Assessing extinction rates. In: Lawton JH, May RM (eds) Extinction rates. Oxford University Press, Oxford, pp 1–24

Mayaux P, Bartholomé E, Fritz S, Belward A (2004) A new land-cover map of Africa for the year 2000. J Biogeogr 31:861–877

Mayaux P, Holmgren P, Achard F, Eva H, Stibig H-J, Branthomme A (2005) Tropical forest cover change in the 1990s and options for future monitoring. Philos Trans R Soc Lond B Biol Sci 360: 373–384

McCarthy JJ, Canziani OF, Leary NA, Dokken DJ, White KS (eds) (2001) Climate change 2001: Impacts, adaptation and vulnerability. Contribution of working group II to the third assessment report of the Intergovernmental Panel on Climate Change. Cambridge University Press, Cambridge New York

McConnell W (2002) Misconstrued land use in Vohibazaha: Participatory planning in the periphery of Madagascar's Mantadia National Park. Land Use Policy 19(3):217–230

McConnell WJ, Keys E (2005) Meta-analysis of agricultural change. In: Moran EF, Ostrom E (eds) Seeing the forest and the trees: Human-environment interactions in forest ecosystems. MIT Press, Cambridge London, pp 325–353

McConnell W, Moran EF (eds) (2001) Meeting in the middle: The challenge of meso-level integration. LUCC Report Series No. 5, LUCC Focus 1 Office, Indiana University, Bloomington, 69 pp

McCracken SD, Brondizio ES, Nelson D, Moran EF, Siqueira AD, Rodriguez-Pedraza C (1999) Remote sensing and GIS at farm property level: Demopgraphy and deforestation in the Brazilian Amazon. Photogramm Eng Remote Sens 65(11):1311–1320

McCracken SD, Boucek B, Moran EF (2002) Deforestation trajectories in a frontier region of the Brazilian Amazon. In: Walsh S, Crews-Meyer K (eds) Linking people, place, and policy: A GIScience approach. Kluwer Academic, Dordrecht Boston London, pp 215–234

McCully P (1996) Silenced rivers: The ecology and politics of large dams. Zed Books, London

McGuffie K, Henderson-Sellers A, Zhang H, Durbidge TB, Pitman AJ (1995) Global climate sensitivity to tropical deforestation. Global Planet Chang 10:97–128

McGuire AD, Sitch S, Clein JS, Dargaville R, Esser G, Foley J, Heimann M, Joos F, Kaplan J, Kicklighter DW, Meier RA, Melillo JM, Moore B III, Prentice IC, Ramankutty N, Reichenau T, Schloss A, Tian H, Williams LJ, Wittenberg U (2001) Carbon balance of the terrestrial biosphere in the twentieth century: Analyses of CO_2, climate and land-use effects with four process-based ecosystem models. Global Biogeochem Cy 15(1):183–206

McKellar FL, Lutz W, Prinz C, Goujon A (1995) Population, households, and CO_2 emissions. Popul Dev Rev 21(4):849–865

McKibben B (1989) The end of Nature. Random House, New York, 226 pp

McMichael AJ (2001) Human frontiers, environment and disease: Past patterns, uncertain futures. Cambridge University Press, Cambridge

McMichael AJ, Martens WJM (1995) The health impacts of global climate change: Grappling with scenarios, predictive models, and multiple uncertainties. Ecosyst Health 1(1):23–33

McNeely JA (1994) Lessons from the past: Forests and biodiversity. Biodivers Conserv 3:3–20

McNeely JA, Scherr SJ (2003) Ecoagriculture: Strategies to feed the world and save biodiversity. Island Press, Washington D.C.

McNeill JR (2000) An environmenthal history of the twentieth-century world. W. W. Norton & Company, New York, 421 pp

Mduma SRA, Sinclair ARE, Hilborn R (1999) Food regulates the Serengti wildebeest: A 40 year record. J Anim Ecol 68:1101–1122

Meadows DH, Meadows DL, Randers J, Behrens WW III (1972) The limits to growth: A report for The Club of Rome's project on the predicament of mankind. Universe Books, New York

Meentemeyer V (1989) Geographical perspectives of space, time, and scale. Landscape Ecol 3:163–173

Mendelsohn R (1994) Property rights and tropical deforestation. Oxford Econ Pap 46(5):750–756

Mendelsohn R, Balick M (1995) Private property and rainforest conservation. Conserv Biol 9(5):1322–1323

Menzies MW (1973) Grain marketing methods in Cananda: The theory, assumptions, and approach. Am J Agric Econ 91: 791–799

Mertens B, Lambin EF (2000) Land-cover-change trajectories in southern Cameroon. Ann Assoc Am Geogr 90:467–494

Mertens B, Sunderlin WD, Ndoye O, Lambin EF (2000) Impact of macro-economic changes on deforestation in South Cameroon: Integration of household survey and remotely-sensed data. World Dev 28(6):983–999

Merz J, Nakarmi G, Shrestha SK, Dahal BM, Dongol BS, Schaffner M, Shakya S, Sharma S, Weingartner R (2004) Public water sources in rural catchments of Nepal's Middle Mountains: Issues and constraints. Environ Manage 34(1):26–37

Messina JP, Walsh SJ (2001) 2.5D morphogenesis: Modeling landuse and landcover dynamics in the Ecuadorian Amazon. Plant Ecol 156:75–88

Meybeck M, Vörösmarty C (2004) Human-driven changes to continental aquatic systems. In: Steffen W, Sanderson A, Tyson PD, Jäger J, Matson PA, Moore B III, Oldfield F, Richardson K, Schellnhuber HJ, Turner BL II, Wasson RJ (eds) Global change and the Earth System: A planet under pressure. (The IGBP Series), Springer, Berlin Heidelberg, pp 112–113

Meyer WB, Turner BL II (2002) The Earth transformed: Trends, trajectories and patterns. In: Johnston RJ, Taylor PJ, Watts MJ (eds) Geographies of global change: Remapping the world, 2nd ed. Blackwell, Oxford, pp 364–376

Meyer WB, Adger WN, Brown K, Graetz D, Gleick P, Richards JF, Maghalães A (1998) Land and water use. In: Rayner S, Malone E (eds) Human choice and climate change, vol. 2, (Resources and technology). Battelle Press, Columbus, pp 79–143

Middleton N, Thomas D (eds) (1992) World atlas of desertification. United Nations Environment Programme, Edward Arnold Publishers, London, 69 pp

Middleton N, Thomas D (eds) (1997) World atlas of desertification, 2nd ed. United Nations Environment Programme, Edward Arnold Publishers, London, 192 pp

Milchunas DG, Sala OE, Lauenroth WK (1988) A generalized model of the effects of grazing by large herbivores on grassland community structure. Amer Nat 132:87–106

Millennium Ecosystem Assessment (2003) Ecosystems and human well-being: A framework for assessment. Island Press, Washington D.C., 245 pp

Millennium Ecosystem Assessment (2005) Ecosystems and human well-being: Synthesis. Island Press, Washington D.C., 137 pp

Miller EJ, Kriger DS, Hunt JD (1999) TCRP web document 9: Integrated urban models for simulation of transit and land-use policies. Final report. University of Toronto Joint Program in Transportation and DELCAN Corporation, Toronto

Miller EJ, Douglas Hunt J, Abraham JE, Salvini PA (2004) Microsimulating urban systems. Comput Environ Urban 28:9–44

Millington AC, Velez-Liendo XM, Bradley AV (2003) Scale dependence in multitemporal mapping of forest fragmentation in Bolivia: Implications for explaining temporal trends in landscape ecology and applications to biodiversity conservation. ISPRS J Photogramm 57:289–299

Milly PCD, Wetherald RT, Dunne KA, Delworth TL (2002) Increasing risk of great floods in a changing climate. Nature 415: 514–517

Misselhorn AA (2005) What drives food insecurity in southern Africa? A meta-analysis of household economy studies. Global Environ Chang 15(1):33–43

Mittermeier R, Mittermeier CG, Gil PR, Pilgrim J, Fonseca G, Brooks T, Konstant WR (2003) Wilderness: Earth's last wild places. University of Chicago Press, Chicago, 576 pp

Mölders N (1999) On the effects of different flooding stages of the Oder and different land-use types on the distributions of evapotranspiration, cloudiness and rainfall in the Brandenburg-Polish border area. Contrib Atmos Phys 72:1–25

Mölders N (2000) Similarity of microclimate as simulated in response to landscapes of the 1930s and 1980s. J Hydrometeorol 1:330–352

Mollicone D, Achard F, Eva HD, Belward AS, Federici S, Lumicisi A, Rizzo VC, Stibig H-J, Valentini R (2003) Land use change monitoring in the framework of the UNFCCC and its Kyoto protocol: Report on current capabilities of satellite remote sensing technology. EUR 20867 EN, European Communities, Luxembourg, 48 pp

Molyneux DH (1997) Patterns of change in vector-borne diseases. Ann Trop Med Parasitol 91(7):827–839

Molyneux DH (1998) Vector-borne parasitic diseases: An overview of recent changes. Int J Parasitol 28:927–934

Monson RK, Holland E (2001) Biospheric trace gas fluxes and their control over tropospheric chemistry. Annu Rev Ecol Syst 32: 547–576

Mooney HA, Canadell J, Chapin FS III, Ehleringer JR, Körner C, McMurtrie RE, Parton WJ, Pitelka LF, Schulze E-D (1999) Ecosystem physiology responses to global change. In: Walker B, Steffen W, Canadell J, Ingram J (eds) The terrestrial biosphere and global change: Implications for natural and managed ecosystems. Cambridge University Press, Cambridge, pp 141–189

Moore N, Rojstaczer S (2002) Irrigation's influence on precipitation: Texas High Plains, U.S.A. Geophys Res Lett 29(16):21–24

Moorman TB, Cambardella CA, James DE, Karlen DL, Kramer LA (2004) Quantification of tillage and landscape effects on soil carbon in small Iowa watersheds. Soil Tillage Res 78:225–236

Moran EF (1981) Developing the Amazon. Indiana University Press, Bloomington

Moran EF (ed) (1995) The comparative analysis of human societies: Toward common standards for data collection and reporting. Rienner, Boulder

Moran EF (2005) Human-environment interactions in forest ecosystems: An introduction. In: Moran EF, Ostrom E (eds) Seeing the forest and the trees: Human-environment interactions in forest ecosystems. MIT Press, Cambridge London, pp 3–21

Moran EF, Brondizio ES (1998) Land-use change after deforestation in Amazonia. In: Liverman D, Moran EF, Rindfuss RR, Stern PC (eds) People and pixels: Linking remote sensing and social science. National Academy Press, Washington D.C., pp 94–120

Moran EF, Ostrom E (eds) (2005) Seeing the forest and the trees: Human-environment interactions in forest ecosystems. MIT Press, Cambridge London

Moran EF, Brondizio ES, McCracken SD (2002) Trajectories of land use: Soils, succession, and crop choice. In: Wood CH, Porro R (eds) Deforestation and land use in the Amazon. University of Florida Press, Gainesville, pp 193–217

Moran EF, Siqueira A, Brondizio E (2003) Household demographic structure and its relationship to deforestation in the Amazon Basin. In: Fox J, Rindfuss RR, Walsh SJ, Mishra V (eds) People and the environment: Approaches for linking household and community surveys to remote sensing and GIS. Kluwer Academic, Dordrecht Boston London, pp 61–89

Moran EF, Skole DL, Turner BL II (2004) The development of the international Land Use and Land Cover Change (LUCC) research program and its links to NASA's Land Cover and Land Use Change (LCLUC) initiative. In: Gutman G, Janetos AC, Justice CO, Moran EF, Mustard JF, Rindfuss RR, Skole D, Turner BL II, Cochrane MA 2004 (eds) Land change science: Observing, monitoring and understanding trajectories of change on the Earth's surface. Remote Sensing and Digital Image Processing Series 6. Kluwer Academic, Dordrecht Boston London, pp 1–15

Mortimore M (1993a) Population growth and land degradation. GeoJournal 31(1):15–21

Mortimore M (1993b) The intensification of peri-urban agriculture: The Kano close-settled zone 1964–1986. In: Turner BL II, Hyden G, Kates RW (eds) Population growth and agricultural change in Africa. University Press of Florida, Gainesville, pp 358–400

Mortimore M, Adams WM (1999) Working the Sahel: Environment and society in northern Nigeria. Routledge, London

Mortimore M, Adams WM (2001) Farmer adaption, change and "crisis" in the Sahel. Global Environ Chang 11:49–57

Mortimore M, Tiffen M (1994) Population growth and a sustainable environment. Environment 36(8):10ff

Moseley WG (2001) African evidence on the relation of poverty, time preference and the environment. Ecol Econ 38(3):317–326

Moseley WG (2006) Poverty. In: Geist HJ (ed) Our Earth's changing land: An encyclopedia of land-use and land-cover change, vol. 2 (L–Z). Greenwood Press, Westport, London, pp 478–481

Müller D, Zeller M (2002) Land use dynamics in the Central Highlands of Vietnam: A spatial model combining village survey data and satellite imagery interpretation. Agricult Econ 27:333–354

Munday PL (2004) Habitat loss, resource specialization, and extinction on coral reefs. Glob Change Biol 10(10):1642–1647

Munroe D, Southworth J, Tucker C (2002) The dynamics of land-cover change in Western Honduras: Exploring spatial and temporal complexity. Agricult Econ 27(3):355–369

Muriuki GW, Njoka TJ, Reid RS, Nyariki DM (2005) Tsetse control and land-use change in Lambwe valley, south-western Kenya. Agric Ecosyst Environ 106(1):99–107

Mustard JF, Fisher TR (2004) Land use and hydrology. In: Gutman G, Janetos AC, Justice CO, Moran EF, Mustard JF, Rindfuss RR, Skole D, Turner BL II, Cochrane MA (eds) Land change science: Observing, monitoring and understanding trajectories of change on the Earth's surface. Remote Sensing and Digital Image Processing 6. Kluwer Academic, Dordrecht Boston London, pp 257–276

Mustard JF, DeFries RS, Fisher T, Moran E (2004) Land-use and land-cover change pathways and impacts. In: Gutman G, Janetos AC, Justice CO, Moran EF, Mustard JF, Rindfuss RR, Skole D, Turner BL II, Cochrane MA (eds) Land change science: Observing, monitoring and understanding trajectories of change on the Earth's surface. Remote Sensing and Digital Image Processing No. 6. Kluwer Academic, Dordrecht Boston London, pp 411–429

Myers N (1997) Consumption: Challenge to sustainable development. Science 276(5309):53–55

Myers N, Kent J (2001) Perverse subsidies: How tax dollars can undercut the environment and the economy. Island Press, Washington D.C., 277 pp

Myneni RB, Keeling CD, Tucker CJ, Asrar G, Nemani RR (1997) Increased plant growth in the northern high latitudes from 1981 to 1991. Nature 386:698–702

Nagata H (1996) The effect of forest disturbance on avian community structure at two lowland forests in peninsular Malaysia. In: Abidin A, Hasan A, Akbar Z (eds) Conservation and faunal biodiversity in Malaysia. University Malaysia, Bangi

Nagendra H (2006) Community involvement. In: Geist HJ (ed) Our Earth's changing land: An encyclopedia of land-use and land-cover change, vol. 1 (A–K). Greenwood Press, Westport, London, p 137

Nagendra H, Southworth J, Tucker C (2003) Accessibility as a determinant of landscape transformation in Western Honduras: Linking pattern and process. Landscape Ecol 18:141–158

Nagendra H, Munroe DK, Southworth J (2004) From pattern to process: Landscape fragmentation and the analysis of land use/land cover change. Agric Ecosyst Environ 101:111–115

Nakicenovic N, Swart R (2000) Special report on emission scenarios. Intergovernmental Panel on Climate Change, Cambridge University Press, Cambridge

Nakicenovic N, Grübler A, McDonald A (1998) Global energy perspectives. Cambridge University Press, Cambridge

Nayak SR, Bahuguna A (2001) Application of RS data to monitor mangroves and other coastal vegetation of India. Indian J Mar Sci 30(4):195–213

Nayak SR, Bahuguna A, Chauhan P, Chauhan HB, Rao RS (1997) Remote sensing applications for coastal environmental management in India. MAEER'S MIT Pune Journal 4(15/16):113–125

Naylor R (2000) Agriculture and global change. In: Ernst WG (ed) Earth systems: Processes and issues. Cambridge University Press, Cambridge, pp 462–475

Naylor RL, Goldburg RJ, Mooney H, Beveridge M, Clay J, Folke C, Kautsky N, Lubchenco J, Primavera J, Williams M (1998) Nature's subsidies to shrimp and salmon farming. Science 282:883–884

Naylor RL, Bonine KM, Ewel KC, Waguk E (2002) Migration, markets, and mangrove resource use on Kosrae, Fedrated State of Micronesia. Ambio 31:340–350

Neilson S (2001) Knowledge utilization and public policy processes: A literature review. Evaluation Unit, International Development Research Council, Ottawa

Nelson GC, Hellerstein D (1997) Do roads cause deforestation? Using satellite images in econometric analysis of land use. Am J Agric Econ 79:80–88

Nelson GC, Harris V, Stone SW (2001) Deforestation, land use, and property rights: Empirical evidence from Darien, Panama. Land Econ 77:187–205

Nelson GC, De Pinto A, Harris V, Stone S (2002) Land use and road improvements: A spatial perspective. Int Regional Sci Rev 27: 297–325

Nelson GC, Bennett E, Berhe AA, Cassman KG, DeFries R, Dietz T, Dobson A, Dobermann A, Janetos A, Levy M, Marco D, Nakićenović N, O'Neill B, Norgaard R, Petschel-Held G, Ojima D, Pingali P, Watson R, Zurek M (2005) Drivers of change in ecosystem condition and services. In: Millennium Ecosystem Assessment (ed) Ecosystems and human well-being: Scenarios. Island Press, Washington D.C., pp 173–222

Nepstad DC, Veríssimo A, Alencar A, Nobre C, Lima E, Lefèbvre P, Schlesinger P, Potter C, Moutinho P, Mendoza E, Cochrane M, Brooks V (1999) Large-scale impoverishment of Amazonian forests by logging and fire. Nature 398:505–508

Nepstad D, Lefèbvre P, Da Silva UL, Tomasella J, Schlesinger P, Solorzano L, Moutinho P, Ray D, Benito JG (2004) Amazon drought and its implications for forest flammability and tree growth: A basin-wide analysis. Glob Change Biol 10(5):704–717

Netting RMcC, Stone GD, Stone MP (1993) Agricultural expansion, intensification, and market participation among the Kofyar, Jos Plateau, Nigeria. In: Turner BL II, Hyden G, Kates RW (eds) Population growth and agricultural change in Africa. University Press of Florida, Gainesville, pp 206–249

Newson MD, Calder IR (1989) Forests and water resources: Problems of prediction on a regional scale. Philos Trans R Soc London, Ser B 324:283–198

Niamir-Fuller M (1999) International aid for rangeland development: Trends and challenges. In: Freudenberger D (ed) International rangelands congress, Townsville, Australia. pp 147–152

Niasse M (2002) Equity dimensions of dam-based water resources development: Winners and losers. In: Steffen W, Jäger J, Carson D, Bradshaw C (eds) Challenges of a changing Earth: Proceedings of the Global Change Open Science Conference, Amsterdam, The Netherlands, 10–13 July 2001. (The IGBP Series), Springer, Berlin Heidelberg, pp 39–43

Nicholson SE (2002) What are the key components of climate as a driver of desertification? In: Reynolds JF, Stafford-Smith DM (eds) Global desertification: Do humans cause deserts? Dahlem Workshop Report No. 88, Dahlem University Press, Berlin, pp 41–57

Nicholson SE, Tucker CJ, Ba MB (1998) Desertification, drought and surface vegetation: An example from the West African Sahel. B Am Meteorol Soc 79(5):815–830

Nielsen TL, Zöbisch MA (2001) Multi-factorial causes of land-use change: Land-use dynamics in the agropastoral village of Im Mial, northwestern Syria. Land Degrad Dev 12:143–161

Nijkamp P, Rossi E, Vindigni G (2004) Ecological footprints in plural: A meta-analytic comparison of empirical results. Reg Stud 38(7):747–765

Norris DE (2004) Mosquito-borne diseases as a consequence of land use change. Ecohealth 1:19–24

NRCS (Natural Resources Conservation Service) (2001) State of the land, natural resources conservation service. U.S. Department of Agriculture, http://www.nrcs.usda.gov/technical/land/

O'Dowd CD, Aalto PP, Hämeri K, Kulmala M, Hoffman T (2002) Atmospheric particles from organic vapours. Nature 416: 497–499

O'Neill RV, King AW (1998) Homage to St. Michael; or, Why are ther so many books on scale? In: Peterson DL, Parker VT (eds) Ecological scale: Theory and applications. Columbia University Press, New York, pp 3–16

Oaks SC, Mitchell VS, Pearson GW, Carpenter CCJ (eds) (1991) Malaria: Obstacles and opportunities. National Academy Press, Washington D.C.

Odgaard BV, Rasmussen P (2000) Origin and temporal development of macro-scale vegetation patterns in the cultural landscape of Denmark. J Ecol 88:733–748

Odum EP (1989) Ecology and our endangered life-support systems. Sinauer Associates, Sunderland, 282 pp

OECD (Organization for Economic Cooperation and Development) (2001) OECD Environmental Outlook. OECD, Paris

Oedekoven KH (1963) Forest history of the Near East. Unasylva 68, http://www.fao.org/documents/show_cdr.asp?url_file=/docrep/e3200e/e3200e03.htm

Ojima DS, Galvin KA, Turner BL II (1994) The global impact of land-use change. BioScience 44(5):300–304

Ojima D, Chullun T, Bolortsetseg B, Tucker CJ, Hicke J (2004) Eurasian land use impacts on rangeland productivity. In: DeFries RS, Asner GP, Houghton RA (eds) Ecosystems and land use change. Geophysical Monograph 153, American Geophysical Union, Washington D.C., pp 293–301

Ojima D, Moran E, McConnell W, Stafford Smith M, Laumann G, Morais J, Young B (2005) Science plan and implementation strategy. IGBP Report No. 53/IHDP Report No. 19, IGBP Secretariat, Stockholm

Okin GS (2002) Toward a unified view of biophysical land degradation processes in arid and semi-arid lands. In: Reynolds JF, Stafford-Smith DM (eds) Global desertification: Do humans cause deserts? Dahlem Workshop Report No. 88, Dahlem University Press, Berlin, pp 95–109

Okin GS, Murray B, Schlesinger WH (2001) Degradation of sandy arid shrubland environments: Observations, process modelling, and management implications. J Arid Environ 47:123–144

Okoth-Ogendo HWO, Oucho JO (1993) Population growth and agricultural change in Kisii District, Kenya: A sustained symbiosis? In: Turner BL II, Hyden G, Kates RW (eds) Population growth and agricultural change in Africa. University Press of Florida, Gainesville, pp 187–205

Oldeman LR (1998) Soil degradation: A threat to food security? ISRIC Report 98/01, International Soil Reference and Information Centre, Wageningen

Oldeman LR, Hakkeling RTA, Sombroek WG (1991) World map of the status of human-induced soil degradation: An explanatory note, revised ed. United Nations Environment Proramme, Nairobi; International Soil Reference and Information Centre, Wageningen, 34 pp (with maps)

Oreskes N, Shrader-Frechette K, Belitz K (1994) Verification, validation, and confirmation of numerical models in the Earth sciences. Science 263:641–646

Orgaz F, Fernández MD, Bonachela S, Gallardo M, Fereres E (2005) Evapotranspiration of horticultural crops in an unheated plastic greenhouse. Agricult Water Manag 72:81–96

Ostrom E, Burger J, Field CB, Norgaard RB, Policansky D (1999) Sustainability: Revisiting the commons. Local lessons, global challenges. Science 284:278–282

Otsamo A, Goran A, Djers A, Hadi T, Kuusipalo J, Vuokko R (1997) Evaluation of reforestation potential of 83 tree species planted on *Imperata cylindrica* dominated grassland. A case study from South Kalimantan, Indonesia. New Forest 14:127–143

Otterman J (1974) Baring high-albedo soils by overgrazing: A hypothesised desertification mechanism. Science 86:531–533

Ottichilo WK, De Leeuw J, Skidmore AK, Prins HHT, Said MY (2000) Population trends of large non-migratory wild herbivores and livestock in the Masai Mara ecosystem, Kenya, between 1977 and 1997. Afr J Ecol 38(3):202–216

Overmars KP, Verburg PH (2005a) Analysis of land use drivers at the watershed and household level: Linking two paradigms at the Philippine forest fringe. Int J Geogr Inf Sci 19:125–152

Overmars KP, Verburg PH (2006) Multi-level modelling of land use from field to village level in the Philippines. Agricult Sys, doi:10.1016/j.agsy.2005.10.006

Overmars KP, de Koning GHJ, Veldkamp A (2003) Spatial autocorrelation in multi-scale land use models. Ecol Model 164:257–270

Ozdogan M, Salvucci GD (2004) Irrigation induced changes in potential evapotranspiration in southeastern Turkey: Test and application of Bouchet's complementary hypothesis. Water Resour Res 40: doi: 1029/2003WR002822

Pacala SW, Hurtt GC, Baker D, Peylin D, Houghton RA, Birdsey RA, Heath L, Sundquist ET, Stallard RF, Ciais P, Moorcroft P, Caspersen JP, Shevliakova E, Moore B, Kohlmaier G, Holland E, Gloor M, Harmon ME, Fan S-M, Sarmiento JL, Goodale CL, Schimel D, Field CB (2001) Consistent land- and atmosphere-based U.S. carbon sink estimates. Science 292:2316–2320

Pacheco P (2006a) Agricultural frontier. In: Geist HJ (ed) Our Earth's changing land: An encyclopedia of land-use and land-cover change, vol. 1 (A–K). Greenwood Press, Westport, London, pp 13–19

Pacheco P (2006b) Amazonia. In: Geist HJ (ed) Our Earth's changing land: An encyclopedia of land-use and land-cover change, vol. 1 (A–K). Greenwood Press, Westport, London, pp 42–47

Pacheco P (2006c) Arc of deforestation. In: Geist HJ (ed) Our Earth's changing land: An encyclopedia of land-use and land-cover change, vol. 1 (A–K). Greenwood Press, Westport, London, pp 51–54

Page SE, Siegert F, Rieley JO, Boehm HDV, Jayal A, Limin S (2002) The amount of carbon released from peat and forest fires in Indonesia during 1997. Nature 420:61–65

Palm CA, van Noordwijk M, Woomer PL, Alegre J, Arevalo L, Castilla C, Cordeiro DG, Hairiah K, Kotto-Same J, Moukam A, Parton WJ, Ricse A, Rodrigues V, Sitompul SM (2005) Carbon losses and sequestration following land use change in the humid tropics. In: Palm CA, Vosti SA, Sanchez PA, Ericksen PJ (eds) Slash-and-burn agriculture: The search for alternatives. Columbia University Press, New York

Pan WKY, Bilsborrow RE (2005) The use of a multi-level statistical model to analyze factors influencing land use: A study of the Ecuadorian Amazon. Global Biogeochem Cy 47(2–4):232–252

Pan WKY, Walsh SJ, Bilsborrow RE, Frizzelle BG, Erlien CM, Baquero F (2004) Farm-level models of spatial patterns of land use and land cover dynamics in the Ecuadorian Amazon. Agric Ecosyst Environ 101:117–134

Panayotou T (1993) Green markets: The economics of sustainable development. International Center for Economic Growth, San Francisco

Parker DC, Meretsky V (2004) Measuring pattern outcomes in an agent-based model of edge-effect externalities using spatial metrics. Agric Ecosyst Environ 101:233–250

Parker DC, Manson SM, Janssen MA, Hoffman M, Deadman P (2003) Multi-agent systems for the simulation of land-use and land-cover change: A review. Ann Assoc Am Geogr 93:314–337

Parmesan C, Yohe G (2003) A globally coherent fingerprint of climate change impacts across natural systems. Nature 421:37–42

Parry ML, Rosenzweig C, Iglesias A, Lovermore M, Fischer G (2004) Effects of climate change on global food production under SRES emissions and socio-economic scenarios. Global Environ Chang 14:53–67

Pathfinder Humid Tropical Deforestation Project (1998) Amazon and Central Africa forest change products. Geography Department, University of Maryland, College Park

Patz JA, Norris DE (2004) Land use change and human health. In: DeFries RS, Asner G, Houghton RA (eds) Ecosystems and land use change. Geophysical Monograph 153, American Geophysical Union, Washington D.C., pp 159–167

Patz JA, Wolfe ND (2002) Global ecological change and human health. In: Aguirre AA, Ostfeld RS, House C, Pearl MC (eds) Conservation medicine: Ecology and health in practice. Oxford University Press, New York

Patz JA, Strzepek K, Lele S, Hedden M, Greene S, Noden B, Hay SI, Kalkstein L, Beier JC (1998) Predicting key malaria transmission factors, biting and entomologic inoculation rates, using modeled soil moisture in Kenya. Trop Med Int Health 3(10):818–827

Patz JA, Graczyk TK, Geller N, Vittor AY (2000) Effects of environmental change on emerging parasitic diseases. Int J Parasitol. 30:1395–1405

Patz JA, Daszak P, Tabor GM, Aguirre AA, Pearl M, Epstein J, Wolfe ND, Kilpatrick AM, Foufopoulos J, Molyneux D, Bradley DJ (2004) Unhealthy landscapes: Policy recommendations on land use change and infectious disease emergence. Environ Health Perspect 101:1092–1098

Patz JA, Confalonieri UEC, Amerasinghe FP, Chua KB, Daszak P, Hyatt AD, Molyneux D, Thomson M, Yameogo L, Mwelecele-Malecela-Lazaro, Vasconcelos P, Rubio-Palis Y, Campbell-Lendrum D, Jaenisch T, Mahamat H, Mutero C, Waltner-Toews D, Whiteman C (2005) Human health: Ecosystem regulation of infectious diseases. In: Millennium Ecosystem Assessment (ed) Conditions and trends assessment. Island Press, Washington D.C., pp 391–416

Pei SJ, Sajise PE (eds) (1993) Regional study on biodiversity concepts, frameworks and methods. Proceedings of the Southeast Asian Universities Agroecosystem Network (SUAN) and Program on Environment (ENV), East-West Center Workshop, Xishuangbanna, Yunnan Province, October 24–30, 1993

Pender JL, Scherr SJ, Durón G (2001) Pathways of development in the hillsides of Honduras: Causes and implications for agricultural production, poverty, and sustainable resource use. In: Lee DR, Barrett CB (eds) Tradeoffs or synergies? Agricultural Intensification, Economic Development and the Environment, CAB International, Wallingford New York

Pender J, Jagger P, Nkonya E, Sserunkuuma D (2004) Development pathways and land management in Uganda. World Dev 32:767–792

Peng S, Huang J, Sheehy JE, Laza RC, Visperas RM, Zhong X, Centeno GS, Khush GS, Cassman KG (2004) Rice yields decline with higher night temperature from global warming. Proc Natl Acad Sci USA 101(27): 9971–9975

Penner JE, Andreae M, Annegarn H, Barrie L, Feichter J, Hegg, D, Jayaraman A, Leaitch R, Murphy D, Nganga J, Pitari G (2001) Aerosols: Their direct and indirect effects. In: Houghton JT, Ding Y, Griggs DJ, Noguer M, van der Linden PJ, Dai X, Maskell K, Johnson CA (eds) Climate change 2001: The scientific basis – Contribution of working group I to the third assessment report of the Intergovernmental Panel on Climate Change. Cambridge University Press, Cambridge, pp 291–348

Pereira JMC, Pereira BS, Barbosa P, Stroppiana D, Vasconcelos MJP, Grégoire JM (1999) Satellite monitoring of fire in the EXPRESSO study area during the 1996 dry season experiment: Active fires, burnt area, and atmospheric emissions. J Geophys Res-Atmos 104(D23):30701–30712

Perkins DM (1969) Agricultural development in China: 1368–1968. Aldine Publishing Company, Chicago, 395 pp

Perry BD, Randolph TF, McDermott JJ, Thornton PK (2002) Investing in animal health research to alleviate poverty. International Livestock Research Institute, Nairobi

Perz SG (2002) The changing social contexts of deforestation in the Brazilian Amazon. Soc Sci Quart 83(1):35–52

Perz SG, Skole DL (2003) Social determinants of secondary forests in the Brazilian Amazon. Soc Sci Res 32:25–60

Peters NE, Meybeck M (2000) Water quality degradation effects on freshwater. Water Int 25(2):185–193

Peterson GD, Cumming GS, Carpenter SR (2003) Scenario planning: A tool for conservation in an uncertain world. Conserv Biol 17:358–366

Petit CC, Lambin EF (2001) Long-term land-cover changes in the Belgian Ardennes (1775–1929): Model-based reconstruction versus historical maps. Glob Change Biol 8(7):616–631

Petit C, Scudder T, Lambin EF (2001) Quantifying processes of land-cover change by remote sensing: Resettlement and rapid land-cover changes in south-eastern Zambia. Int J Remote Sens 22(17):3435–3456

Petschel-Held G (2004) The syndromes approach to place-based assessment. In: Steffen W, Sanderson A, Tyson PD, Jäger J, Matson PA, Moore B III, Oldfield F, Richardson K, Schellnhuber HJ, Turner BL II, Wasson RJ 2004. Global change and the Earth System: A planet under pressure. (The IGBP Series), Springer, Berlin Heidelberg, pp 92–93

Petschel-Held G, Lüdeke MKB, Reusswig F (1999) Actors, structures and environments. A comparative and transdisciplinary view on regional case studies of global environmental change. In: Lohnert B, Geist HJ (eds) Coping with changing environments: Social dimensions of endangered ecosystems in the developing world. Ashgate, Aldershot, pp 255–293

Pfaff ASP (1999) What drives deforestation in the Brazilian Amazon? Evidence from satellite and socioeconomic data. J Environ Econ Manag 37(1):26–43

Phillips JD (1993) Biophysical feedbacks and the risks of desertification. Ann Assoc Am Geogr 83:630–40

Pichón FJ (1997a) Colonist land-allocation decisions, land use, and deforestation in the Ecuadorian Amazon frontier. Econ Dev Cult Change 45(4):707–744

Pichón FJ (1997b) Settler households and land-use patterns in the Amazon frontier: Farm-level evidence from Ecuador. World Dev 25(1):67–91

Pielke RA, Avissar R (1990) Influence of landscape structure on local and regional climate. Landscape Ecol 4(2/3):133–155

Pielke RA, Avissar R, Raupach MR, Dolman AJ, Zeng X, Denning AS (1998) Interactions between the atmosphere and terrestrial ecosystems: Influence on weather and climate. Glob Change Biol 4:461–475

Pieri C, Dumanski J, Hamblin A, Young A (1995) Land quality indicators. World Bank Discussion Paper No. 315, World Bank, Washington D.C.

Pijanowski BC, Brown DG, Shellito BA, Manik GA (2002a) Using neural networks and GIS to forecast land use changes: A land transformation model. Comput Environ Urban 26:553–575

Pijanowski BC, Shellito B, Pithadia S, Alexandridis K (2002b) Forecasting and assessing the impact of urban sprawl in coastal watersheds along eastern Lake Michigan. Lakes Reservoirs: Res Manage 7:271–285

Pijanowski BC, Pithadia S, Shellito BA, Alexandridis K (2005) Calibrating a neural network-based urban change model for two metropolitan areas of the upper midwest of the United States. Int J Geogr Inf Sci 19:197–216

Pimentel D, Berger B, Filiberto D, Newton M, Wolfe B, Karabinakis E, Clark S, Poon E, Abbett E, Nandagopal S (2004) Water resources: Agricultural and environmental issues. BioScience 54(10):909–918

Pimm S, Askins RA (1995) Forest losses predict bird extinctions in eastern North America. Proc Natl Acad Sci USA 92:9343–9347

Pimm SL, Brooks TM (2000) The sixth extinction: How large, where and when? In: Raven PH, Williams T (eds) Nature and human society: The quest for a sustainable world. National Academy Press, Washington D.C., pp 46–62

Pimm SL, Russell GJ, Gittleman JL, Brooks TM (1995) The future of biodiversity. Science 269:347–350

Pitman A, Zhao M (2000) The relative impact of observed changes in land cover and carbon dioxide as simulated by a climate model. Geophys Res Lett 27:1267–1273

Platt RH (2004) Land use and society: Geography, law, and public policy. Island Press, Washington D.C.

Plisnier PD, Serneels S, Lambin EF (2000) Impact of ENSO on East African ecosystems: A multivariate analysis based on climate and remote sensing data. Global Ecol Biogeogr 9:481–497

Polcher J, Laval K (1994) A statistical study of the regional impacts of deforestation on climate in the LMD GCM. Clim Dynam 10:205–219

Polsky C (2004) Putting space and time in Ricardian climate change impact studies: The case of agriculture in the U.S. Great Plains. Ann Assoc Am Geogr 94:549–564

Polsky C, Easterling WE III (2001) Ricardian climate sensitivities: Accounting for adaptation across scales. Agric Ecosyst Environ 85:133–144

Pontius RG (2000) Quantification error *versus* location error in comparison of categorical maps. Photogramm Eng Remote Sens 66:1011–1016

Pontius RG (2002) Statistical methods to partition effects of quantity and location during comparison of categorical maps at multiple resolutions. Photogramm Eng Remote Sens 68:1041–1049

Pontius RG, Malanson J (2005) Comparison of the accuracy of land change models: Cellular automata Markov *versus* Geomod. Int J Geogr Inf Sci 19:243–265

Pontius RG, Pacheco P (2004) Calibration and validation of a model of forest disturbance in the Western Ghats, India 1920–1990. GeoJournal 61(4):325–334

Pontius RG, Cornell JD, Hall CAS (2001) Modeling the spatial pattern of land-use change with GEOMOD2: Application and validation for Costa Rica. Agric Ecosyst Environ 85:191–203

Pontius RG, Huffaker D, Denman K (2004b) Useful techniques of validation for spatially explicit land-change models. Ecol Model 179:445–461

Postel SL, Daily GC, Ehrlich PR (1996) Human appropriation of renewable fresh water. Science 271:785–787

Poteete A, Ostrom E (2004) An institutional approach to the study of forest resources. In: Poulsen J (ed) Human impacts on tropical forest biodiversity and genetic resources. CAB International, Wallingford New York

Prather M, Ehhalt D, Dentener FJ, Derwent R, Dlugokencky EJ, Holland E, Isaksen I, Katima J, Kirchhoff V, Matson P, Midgley WM (2001) Atmospheric chemistry and greenhouse gases. In: JT Houghton, Y Ding, DJ Griggs, M Noguer, PJ van der Linden, X Dai, K Maskell, and CA Johnson (eds) Climate change 2001: The scientific basis. Contribution of working group I to the third assessment report of the Intergovernmental Panel on Climate Change. Cambridge University Press, Cambridge, pp 238–287

Prentice IC, Farquhar GD, Fasham MJR, Goulden ML, Heimann M, Jaramillo VJ, Kheshgi HS, Le Quéré C, Scholes RJ, Wallace DWR (2001) The carbon cycle and atmospheric carbon dioxide. In: Houghton JT, Ding Y, Griggs DJ, Noguer M, van der Linden PJ, Dai X, Maskell K, Johnson CA (eds) Climate change 2001: The scientific basis. Contributions of working group I to the third assessment report of the Intergovernmental Panel on Climate Change (IPCC). Cambridge University Press, Cambridge New York, pp 183–237

Prince SD (2004) Mapping desertification in southern Africa. In: Gutman G, Janetos AC, Justice CO, Moran EF, Mustard JF, Rindfuss RR, Skole D, Turner BL II, Cochrane MA (eds) Land change science: Observing, monitoring and understanding trajectories of change on the Earth's surface. Remote Sensing and Digital Image Processing 6. Kluwer Academic, Dordrecht Boston London, pp 163–184

Prince SD, De Colstoun EB, Kravitz LL (1998) Evidence from rain-use efficiencies does not indicate extensive Sahelian desertification. Glob Change Biol 4:359–374

Puigdefábregas J (1995) Desertification: Stress beyond resilience, exploring a unifying process structure. Ambio 24:311–313

Puigdefábregas J (1998) Ecological impacts of global change on drylands and their implications for desertification. Land Degrad Dev 9(5):393–406

Pulido-Bosch A, Bensi S, Molina L, Vallejos A, Calaforra JM, Pulido-Leboeuf P (2000a) Nitrates as indicators of aquifer interconnection. Applications to the Campo de Dalías (SE Spain). Environ Geol 39(7):791–799

Pulido-Bosch A, Pulido-Leboeuf P, Molina L, Vallejos A, Martin-Rosales (2000b) Intensive agriculture, wetlands, quarries and water management. A case study (Campo de Dalías, SE Spain). Environ Geol 40(1–2):163–168

Purseglove J (1988) Taming the flood: A history of rivers and wetlands. Oxford University Press, Oxford, 307 pp

Putterman L (1997) On the past and future of China's township and village-owned enterprises. World Dev 25:1639–1655

Pyne S (1991) Burning bush: A fire history of Australia. Henry Holt & Co., New York, 520 pp

Rabin M (1998) Psychology and economics. J Econ Lit 36:11–46

Ragin C (1987) The comparative method: Moving beyond qualitative and quantitative strategies. University of California Press, Berkeley

Rajasuriya A, Zahir H, Muley EV, Subramanian BR, Venkataraman K, Wafar MVM, Khan SMMH, Whittington E (2000) Status of coral reefs in South Asia: Bangladesh, India, Maldives and Sri Lanka. In: Wilkinson C (ed) Status of coral reefs of the world. Australian Institute of Marine Sciences, Townsville, pp 95–116

Ramankutty N, Foley JA (1998) Characterizing patterns of global land use: An analysis of global croplands data. Global Biogeochem Cy 12:667–685

Ramankutty N, Foley JA (1999) Estimating historical changes in global land cover: Croplands from 1700 to 1992. Global Biogeochem Cy 13(4):997–1027

Ramankutty N, Foley JA, Olejniczak NJ (2002) People on the land: Changes in global population and croplands during the 20th century. Ambio 31(3):251–257

Raskin P, Banuri T, Gallopín G, Gutman P, Hammond A, Kates R, Swart R (2002) Great transition: The promise and lure of the times ahead. PoleStar Series Report No. 10, Stockholm Environment Institute, Boston

Raynaut C (1997) Societies and nature in the Sahel: Rethinking environmental degradation. Routledge, London

Reale O, Dirmeyer PA (1998) Modeling the effects of vegetation on Mediterranean climate during the Roman classical period. Part I: Climate history and model sensitivity. Center for Ocean-Land-Atmosphere Studies, Calverton

Reardon T, Vosti SA (1995) Links between rural poverty and the environment in developing countries: Asset categories and investment poverty. World Dev 23:1495–1506

Redford KH (1992) The empty forest. BioScience 42:412–426

Redman CL (1999) Human impact on ancient environments. University of Arizona Press, Tucson, 288 pp

Redwood III J (2002) World Bank approaches to the Brazilian Amazon: The bumpy road toward sustainable development. The World Bank, Washington D.C.

Reenberg A (1994) Land use dynamics in the Sahelian Zone in Eastern Niger: Monitoring change in cultivation strategies in drought prone areas. J Arid Environ 27:179–92

Reenberg A (2001) Agricultural land use pattern dynamics in the Sudan-Sahel: Towards an event-driven framework. Land Use Policy 18:309–19

Reenberg A, Paarup-Laursen B (1997) Determinants for land use strategies in a Sahelian agro-ecosystem: Anthropological and ecological geographical aspects of natural resource management. Agricult Sys 53:209–29

Reenberg A, Nielsen TL, Rasmussen K (1998) Field expansions and reallocations in a desert margin region: Land use pattern dynamics in a fluctuating biophysical and socio-economic environment. Global Environ Chang 8(4):309–27

Reid RS, Wilson CJ, Kruska RL, Mulatu W (1997) Impacts of tsetse control and land-use on vegetative structure and tree species composition in southwestern Ethiopia. J Appl Ecol 34:731–747

Reid RS, Kruska RL, Deichmann U, Thornton PK, Leak SGA (2000) Human population growth and the extinction of the tsetse fly. Agric Ecosyst Environ 77:227–236

Reid RS, Thornton PK, McCrabb GJ, Kruska RL, Atieno F, Jones PG (2004) Is it possible to mitigate greenhouse gas emissions in pastoral ecosystems of the tropics? Dev Sustainability 6:91–109

Reid RS, Nkedianye D, Kshatriya M, Said MY, Ogutu J, Kristjanson PK, Kifugo S (2005) Fragmentation of an urban savanna, Kitengela, Kenya. In: Galvin KA, Reid RS, Behnke RH, Hobbs NT (eds) Fragmentation in semi-arid and arid landscapes: Consequences for human and natural systems. Kluwer Academic, Dordrecht Boston London

Reis EJ, Margullis S, Laurance WF, Cochrane MA, Bergen S, Fearnside PM, Delamônica P, Barber C, D'Angelo S, Fernandes T (2001) The future of the Brazilian Amazon. Science 291:438

Remigio AA (1993) Philippine forest resource policy in the Marcos and Aquino governments: A comparative assessment. Global Ecol Biogeogr 3:192–212

Repetto R, Gillis M (eds) (1988) Public policies and the misuse of forest resources. Cambridge University Press, Cambridge

Reynolds JF (2001) Desertification. In: Levin S (ed) Encyclopedia of biodiversity, vol. 2. Academic Press, San Diego, pp 61–78

Reynolds JF, Stafford-Smith DM (eds) (2002) Global desertification: Do humans cause deserts? Dahlem University Press, Berlin, 437 pp

Reynolds JF, Maestre FT, Kemp PR, Stafford-Smith DM, Lambin EF (2006) Natural and human dimensions of land degradation in drylands: Causes and consequences. In: Canadell J, Pataki D, Pitelka L (eds) Terrestrial ecosystems in a changing world. Cambridge University Press, Cambridge, forthcoming

Rice M, Stevenson LA, Mastura SA (eds) (2004) Initial synthesis of land-use and land-cover change research in Asia and the Pacific. Asia-Pacific Network for Global Change Research, Tokyo

Richards JF (1990) Land transformation. In: Turner BL II, Clark WC, Kates RW, Richards JF, Mathews JT, Meyer WB (eds) The Earth as transformed by human action: Global and regional changes in the biosphere over the past 300 years. Cambridge University Press, Cambridge, pp 163–178

Richards JF, Flint EP (1994) Historic land use and carbon estimates for south and southeast Asia: 1880–1980. Num. Data Package-046, Carbon Dioxide Infor. Anal. Cent., Oak Ridge Natl. Lab., Oak Ridge

Richards TS, Gallego J, Achard F (2000) Sampling for forest cover change assessment at the pan-tropical scale. Int J Remote Sens 21:1473–1490

Riebsame WE (1990) The United States Great Plains. In: Turner BL II, Clark WC, Kates RW, Richards JF, Mathews JT, Meyer WB (eds) The Earth as transformed by human action: Global and regional changes in the biosphere over the past 300 years. Cambridge University Press, Cambridge, pp 561–575

Riebsame WE, Parton WJ (1994) Integrated modeling of land use and cover change. BioScience 44:350–357

Riley WJ, Ortiz-Monasterio I, Matson PA (2001) Nitrogen leaching and soil nitrate, nitrite, and ammonium levels under irrigated wheat in northern Mexico. Nutr Cycl Agroecosys 61(3):223–236

Rindfuss RR, Walsh SJ, Mishra V, Fox J, Dolcemascolo GP (2002) Linking household and remotely sensed data, methodological and practical problems. In: Fox J, Rindfuss RR, Walsh SJ, Mishra V (eds) People and the environment. Approaches for linking houshold and community surveys to remote sensing and GIS. Kluwer Academic, Dordrecht Boston London, pp 1–31

Rindfuss RR, Prasartkul P, Walsh SJ, Entwisle B, Sawangdee Y, Vogler JB (2003) Household-parcel linkages in Nang Rong, Thailand: Challenges of large samples. In: Fox J, Rindfuss RR, Walsh SJ, Mishra V (eds) People and the environment: Approaches for linking household and community surveys to remote sensing and GIS. Kluwer Academic, Dordrecht Boston London, pp 131–172

Rindfuss RR, Turner BL II, Entwisle B, Walsh SJ (2004a) Land cover/use and population. In: Gutman G, Janetos AC, Justice CO, Moran EF, Mustard JF, Rindfuss RR, Skole D, Turner BL II, Cochrane MA (eds) Land change science: Observing, monitoring and understanding trajectories of change on the Earth's surface. Remote Sensing and Digital Image Processing Series 6, Kluwer Academic, Dordrecht Boston London, pp 351–366

Rindfuss RR, Walsh SJ, Turner BL II, Fox J, Mishra V (2004b) Developing a science of land change: Challenges and methodological issues. Proc Natl Acad Sci USA 101(39):13976–13981

Ringrose S, Matheson W (1992) The use of Landsat MSS imagery to determine the aerial extent of woody vegetation cover change in the west-central Sahel. Global Ecol Biogeogr 2:16–25

Ritters KH, O'Neill RV, Hunsaker CT, Wickham JD, Yankee DH, Timmins SP, Jones KB, Jackson BL (1995) A factor analysis of landscape pattern and structure metrics. Landscape Ecol 10:23–39

Ritzer G (1998) The McDonaldization of society: Explorations and extensions. Sage, London

Roetter RP, Hoanh CT, Laborte AG, Van Keulen H, Van Ittersum MK, Dreiser C, Van Diepen CA, De Ridder N, Van Laar HH (2005) Integration of systems network (SysNet) tools for regional land use scenario analysis in Asia. Environ Model Software 20:291–307

Rojstaczer S, Sterling SM, Moore NJ (2001) Human appropriation of photosynthesis products. Science 294:2549–2552

Rollefson GO, Kohler-Rollefson I (1992) Early neolithic exploitation patterns in the Levant: Cultural impact on the environment. Popul Environ 13:243–54

Root TL, Price JT, Hall KR, Schneider SH, Rosenzweig C, Pounds JA (2003) Fingerprints of global warming on wild animals and plants. Nature 421:57–60

Rosegrant MW, Paisner MS, Meijer S, Witcover J (2001) 2020 global food outlook: Trends, alternatives and choices. International Food Policy Research Institute, Washington D.C.

Rosegrant MW, Meijer S, Cline SA (2002) International model for policy analysis of agricultural commodities and trade (IMPACT): Model description. International Food Policy Research Institute, Washington D.C., 28 pp

Rosenberg R, Andre RG, Somchit L (1990) Highly efficient dry season transmission of malaria in Thailand. T Roy Soc Trop Med H 84:22–28

Rotmans J, Van Asselt MBA, Anastasi C, Greeuw SCH, Mellors J, Peters S, Rothman DS, Rijkens-Klomp N (2000) Visions for a sustainable Europe. Futures 32:809–831

Roulet M, Lucotte M, Farella N, Serique G, Coelho H, Sousa Passos CJ, Jesus da Silva E de, Scavone de Andreade P, Mergler D, Guimarães J-R, Amorim M (1999) Effects of recent human colonization on the presence of mercury in Amazonian ecosystems. Water, Air, Soil Pollut 112:297–313

Rounsevell MDA, Ewert F, Reginster I, Leemans R, Carter TR (2005) Future scenarios of European agricultural land use: II. Projecting changes in cropland and grassland. Agric Ecosyst Environ 107:101–116

Rounsevell MDA, Reginster I, Araujo MB, Carter TR, Dendoncker N, Ewert F, House JI, Kankanpää S, Leemans R, Metzger MJ, Schmit C, Smith P, Tuck G (2006) A coherent set of future land use change scenarios for Europe. Agric Ecosyst Environ, forthcoming

Ruben R, Moll H, Kuyvenhoven A (1998) Integrating agricultural research and policy analysis: Analytical framework and policy applications for bio-economic modelling. Agricult Sys 58:331–349

Rubio JL, Bochet E (1998) Desertification indicators as diagnostic criteria for desertification risk assessment in Europe. J Arid Environ 39:113–120

Ruddiman WF (2003) The anthropogenic greenhouse era began thousands of years ago. Climatic Change 61:261–293

Rudel TK (1993) Tropical deforestation, small farmers and land clearing in the Ecudorian Amazon. Columbia University Press, New York

Rudel TK (1998) Is there a forest transition? Deforestation, reforestation, and development. Rural Sociol 63(4):533–552

Rudel TK (2001) Did a green revolution restore the forests of the American South? In: Angelsen A, Kaimowitz D (eds) Agricultural technologies and tropical deforestation. CAB International, Wallingford New York, pp 53–68

Rudel TK (2005) Tropical forests: Regional paths of destruction and regeneration in the late 20th century. Columbia University Press, New York

Rudel TK, Roper J (1996) Regional patterns and historical trends in tropical deforestation 1976–1990: A qualitative comparative analysis. Ambio 25(3):160–166

Rudel TK, Roper J (1997) The paths to rain forest destruction: Crossnational patterns of tropical deforestation. World Dev 25: 53–65

Rudel TK, Perez-Lugo M, Zichal H (2000) When fields revert to forest: Development and spontaneous reforestation in post-war Puerto Rico. Prof Geogr 52(3):386–397

Rudel TK, Bates D, Machinguiashi R (2002a) Ecologically noble Amerindians? Cattle ranching and cash cropping among Shuar and colonists in Ecuador. Lat Am Res Rev 37(1):144–159

Rudel TK, Bates D, Machinguiashi R (2002b) A tropical forest transition? Agricultural change, out-migration, and secondary forests in the Ecuadorian Amazon. Ann Assoc Am Geogr 92(1):87–102

Rudel TK, Coomes OT, Moran E, Achard F, Angelsen A, Xu J, Lambin EF (2005) Forest transitions: Towards a global understanding of land use change. Global Environ Chang 15:23–31

Ruiz-Pérez M, Belcher B, Achdiawan R, Alexiades M, Aubertin C, Caballero J, Campbell B, Clement C, Cunningham T, Fantini A, de Foresta H, García Fernández C, Gautam KH, Hersch Martínez P, de Jong W, Kusters K, Kutty MG, López C, Fu M, Martínez Alfaro MA, Nair TR, Ndoye O, Ocampo R, Rai N, Ricker M, Schreckenberg K, Shackleton S, Shanley P, Sunderland T, Youn Y (2004) Markets drive the specialization strategies of forest peoples. Ecol Soc 9(2):4, http://www.ecologyandsociety.org/vol9/iss2/art4

Rutten MMEM (1992) Selling wealth to buy poverty: The process of individualisation of land ownership among the Maasai pastoralists of Kajiado District, Kenya 1890–1990. Breitenbach Publishers, Saarbrücken

Rykiel EJ Jr (1996) Testing ecological models: The meaning of validation. Ecol Model 90:229–244

Saatchi SS, Nelson B, Podest E, Holt J (2000) Mapping land cover types in the Amazon Basin using 1 km JERS-1 mosaic. Int J Remote Sens 21:1201–1234

Sack RD (1990) The realm of meaning: The inadequacy of human-nature theory and the view of mass consumption. In: Turner BL II, Clark WC, Kates RW, Richards JF, Mathews JT, Meyer WB (eds) The Earth as transformed by human action: Global and regional changes in the biosphere over the past 300 years. Cambridge University Press, Cambridge, pp 659–672

Sack RD (1992) Place, modernity and the consumer's world. Johns Hopkins Press, Baltimore

Sader SA, Chowdhury RR, Schneider LC, Turner BL II (2004) Forest change and human driving forces in Central America. In: Gutman G, Janetos AC, Justice CO, Moran EF, Mustard JF, Rindfuss RR, Skole D, Turner BL II, Cochrane MA (eds) Land change science: Observing, monitoring and understanding trajectories of change on the Earth's surface. Remote Sensing and Digital Image Processing Series 6, Kluwer Academic, Dordrecht Boston London, pp 57–76

Sagan C, Toon OB, Pollack JB (1979) Anthropogenic albedo changes and the Earth's climate. Science 206:1363–1368

Said MY (2003) Multiscale perspectives of species richness in East Africa. Ph.D. ITC, Enschede, Netherlands

Saiko TA, Zonn IS (2000) Irrigation expansion and dynamics of desertification in the Circum-Aral region of Central Asia. Appl Geogr 20(4):349–367

Sala OE, Chapin FS III, Armesto JJ, Berlow E, Bloomfield J, Dirzo R, Huber-Sanwald E, Huenneke LF, Jackson RB, Kinzig A, Leemans R, Lodge DM, Mooney HA, Oesterheld M, Poff NL, Sykes MT, Walker BH, Walker M, Wall DH (2000) Biodiversity: Global biodiversity scenarios for the year 2100. Science 287:1770–1774

Salati E, Dourojeanni MJ, Novaes FC, Oliveira AE, Perritt RW, Schubart HOR, Umana JC (1990) Amazonia. In: Turner BL II, Clark WC, Kates RW, Richards JF, Mathews JT, Meyer WB (eds) The Earth as Transformed by Human Action: Global and regional changes in the biosphere over the past 300 years. Cambridge University Press, New York, pp 479–493

Salomons W, Kremer HH, Turner KR (2005) The catchment to coast continuum. In: Crossland CJ, Kremer HH, Lindeboom HJ, Marshall Crossland JI, Le Tissier MDA (eds) Coastal fluxes in the Anthropocene: The Land-Ocean Interactions in the Coastal Zone Project of the International Geosphere-Biosphere Programme. (The IGBP Series), Springer, Berlin Heidelberg, pp 145–200

Sampaio-Nunes D (1995) The role of biomass in the European energy policy. In: Chartier P, Beenackers AACM, Grassi G (eds) Biomass for energy, environment, agriculture and industry, vol. 1. Pergamon Press, Elsevier, Oxford New York Tokyo, pp 20–30

Sanchez PA, Shepherd KD, Soule MJ, Place FM, Mokwunye AU, Buresh RJ, Kwesiga FR, Izac AN, Ndiritu CG, Woomer PL (1997) Soil fertility replenishment in Africa: An investment in natural resource capital. In: Buresh RJ, Sanchez PA (eds) Replenishing soil fertility in Africa. SSSA Special Publication, Soil Science Society of America, American Society of Agronomy, Madison

Sánchez-Azofeifa GA, Castro K, Rivard B, Kalascka M, Harriss RC (2003) Remote sensing research priorities in tropical dry forest environments. Biotropica 35:134–142

Sanchez-Maranon M, Soriano M, Delgado G, Delgado R (2002) Soil quality in Mediterranean mountain environments: Effects of land use change. Soil Sci Soc Am J 66:948–958

Sanderson EW, Jaiteh M, Levy MA, Redford KH, Wannebo AV, Woolmer G (2002) The human footprint and the last of the wild. BioScience 52(10):891–904

Sayago D, Machado L (2004) O pulo do grilo: O Incra e a questão fundiária na Amazônia. In: Sayago D, Tourrand J-F, Bursztyn M (eds) Amazônia: Cenas e cenários. Editora UnB, Brasilia, pp 217–235

Schelhas J (1996) Land use choice and change: Intensification and diversification in the lowland tropics of Costa Rica. Hum Organ 55(3):298–306

Scherr S (2000) A downward spiral? Research evidence on the relationship between poverty and natural resource degradation. Food Policy 25:479–498

Schimel DS, Enting IG, Heinmann M, Wigley TML, Raynaud D, Alves D, Siegenthaler U (1995) CO2 and the carbon cycle. In: Houghton JT, Meira Filho LG, Bruce J, Lee H, Callander BA, Haites E, Harris N, Makell K (eds) Climate Change 1994. Cambridge University Press, Cambridge

Schlebecker JT (1973) The use of the land. Coronado Press, Kansas, 218 pp

Schlebecker JT (1975) Whereby we thrive: A history of American farming 1607-1972. The Iowa State University Press, Ames, 342 pp

Schlesinger WH, Gramenopoulos N (1996) Archival photographs show no climate-induced changes in woody vegetation in the Sudan. Glob Change Biol 2:137–141

Schlesinger WH, Reynolds JF, Cunnigham GL, Huenneke LF, Jarrell WM, Virginia RA, Whitford WG (1990) Biological feedbacks in global desertification. Science 247:1043–1048

Schmidt KA, Ostfeld RS (2001) Biodiversity and the dilution effect in disease ecology. Ecology 82:609–619

Schneider LC, Pontius RG (2001) Modeling land-use change in the Ipswich Watershed, Massachusetts, USA. Agric Ecosyst Environ 85:83–94

Schneider RR (1995) Government and the economy on the Amazon frontier. Environmental Paper No. 11, World Bank, Washington D.C.

Schneider N, Eugster W, Schichler B (2004) The impact of historical land-use changes on the near-surface atmospheric conditions on the Swiss Plateau. Earth Interact 8(12):1–27

Scholes RJ (2002) The past, present and future of carbon on land. In: Steffen W, Jäger J, Carson D, Bradshaw C (eds) Challenges of a changing Earth: Proceedings of the Global Change Open Science Conference, Amsterdam, The Netherlands, 10–13 July 2001. (The IGBP Series), Springer, Berlin Heidelberg, pp 81–85

Scholes MC, Matrai PA, Andreae MO, Smith KA, Manning MR (2003) Biosphere-atmosphere interactions. In: Brasseur GP, Prinn RG, Pszenny AAP (eds) Atmospheric chemistry in a changing World: An integration of a decade of tropospheric chemistry research. (The IGBP Series), Springer, Berlin Heidelberg, pp 19–71

Schueler TR (1994) The importance of imperviousness. Watershed Prot Techniques 1(3):100–111

Schultz MG (2002) On the use of ATSR fire count data to estimate the seasonal and interannual variability of vegetation fire emissions. Atmos Chem Phys Discuss 2:1159–1179

Schweik CM, Adhikari K, Pandit KN (1997) Land-cover change and forest institutions: A comparison of two sub-basins in the southern Siwalik hills of Nepal. Mt Res Dev 17:99–116

Scurlock JMO, Hall DO (1990) The contribution of biomass to global energy use. Biomass 21:75–81

Seixas J (2000) Assessing heterogeneity from remote sensing images: The case of desertification in southern Portugal. Int J Remote Sens 21(13/14):2645–2663

Sellers PJ (1992) Biophysical models of land surface processes. In: Trenberth K (ed) Climate system modeling. Cambridge University Press, London, pp 451–490

Selvam V (2003) Environmental classification of mangrove wetlands of India. Curr Sci 84:757–765

Sen A (1981) Poverty and famines. Clarendon, Oxford

Serneels S, Lambin EF (2001a) Impact of land-use changes on the wildebeest (*Connochaetes taurinus*) in the northern part of the Serengeti-Mara ecosystem. J Biogeogr 28(3):391–408

Serneels S, Lambin EF (2001b) Proximate causes of land use change in Narok district Kenya: A spatial statistical model. Agric Ecosyst Environ 85:65–81

Serneels S, Said MY, Lambin EF (2001) Land-cover changes around a major East African wildlife reserve: The Mara Ecosystem (Kenya). Int J Remote Sens 22(17):3397–3420

Seto KC, Kaufmann RK, Woodcock CE (2000) Landsat reveals China's farmland reserves, but they're vanishing fast. Nature 406:121

Seto KC, Woodcock CE, Song C, Huang X, Lu J, Kaufmann RK (2002) Monitoring land-use change in the Pearl River Delta using Landsat TM. Int J Remote Sens 23:1985–2004

Seto KC, Woodcock CE, Kaufmann RK (2004) Changes in land cover and land use in the Pearl River Delta, China. In: Gutman G, Janetos AC, Justice CO, Moran EF, Mustard JF, Rindfuss RR, Skole D, Turner BL II, Cochrane MA (eds) Land change science: Observing, monitoring and understanding trajectories of change on the Earth's surface. Remote Sensing and Digital Image Processing Series 6, Kluwer Academic, Dordrecht Boston London, pp 223–236

Sever TL (1998) Validating prehistoric and current social phenomena upon the landscape of the Petén, Guatemala. In: Liverman D, Moran EF, Rindfuss RR, Stern PC (eds) People and pixels: Linking remote sensing and social science. National Academy Press, Washington D.C., pp 145–163

Seybold CA, Herrick JE, Brejda JJ (1999) Soil resilience: A fundamental component of soil quality. Soil Sci 164(4):224–234

Shah M (2002) Food in the 21st century: Global climate of disparities. In: Steffen W, Jäger J, Carson D, Bradshaw C (eds) Challenges of a changing Earth. Proceedings of the Global Change Open Science Conference, Amsterdam, The Netherlands, 10–13 July 2001. (The IGBP Series), Springer, Berlin Heidelberg, pp 31–38

Shane DR (1986) Hoofprints in the forest: Cattle ranching and the destruction of Latin America's tropical forests. Institute for the Study of Human Issues, Philadelphia

Sharkawy MA, Chen X, Pretorious F (1995) Spatial trends of urban development in China. J Real Estate Lit 3:47–59

Sharma VP (1996) Ecological changes and vector borne diseases. Trop Ecol 37(1):57–65

Sharma VP, Srivastava A, Nagpal BN (1994) A study of relationship of rice cultivation and annual parasite incidence of malaria in India. Soc Sci Med 38:165–178

Sheehy DP (1992) A perspective on desertification of grazingland ecosystems in North China. Ambio 21(4):303–307

Shidong Z, Jiehua L, Hongqi Z, Yi Z, Wenhu Q, Zhiwu L, Taolin Z, Guiping L, Mingzhou Q, Leiwen J (2001a) Population, consumption, and land use in the Jitai Basin Region, Jiangxi Province. In: Indian National Science Academy, Chinese Academy of Sciences, and U.S. National Academy of Sciences (eds) Growing populations, changing landscapes: Studies from India, China, and the United States. National Academy Press, Washington D.C., pp 179–205

Shidong Z, Yi Z, Wanqi B, Jiehua L, Wenhu Q, Taolin Z, Guiping L, Mingzhou Q, Leiwen J (2001b) Population, consumption, and land use in the Pearl River Delta, Guangdong Province. In: Indian National Science Academy, Chinese Academy of Sciences, and U.S. National Academy of Sciences (eds) Growing populations, changing landscapes: Studies from India, China, and the United States. National Academy Press, Washington D.C., pp 207–230

Shively GE (2001) Agricultural change, rural labor markets, and forest clearing: An illustrative case from the Philippines. Land Econ 77(2):268–284

Shively GE, Martinez E (2001) Deforestation, irrigation, employment and cautious optimism in Southern Palawan, the Philippines. In: Angelsen A, Kaimowitz D (eds) Agricultural technologies and tropical deforestation. CAB International, Wallingford New York, pp 335–346

Shukla J, Nobre C, Sellers P (1990) Amazon deforestation and climate change. Science 247:1322–1325

Siegert TF, Ruecker G, Hinrichs A, Hoffmann AA (2001) Increased damage from fires in logged forests during droughts caused by El Niño. Nature 414:437–440

Sierra R (2000) Dynamics and patterns of deforestation in the western Amazon: The Nape deforestation front 1986–1996. Appl Geogr 20:1–16

Sierra R, Stallings J (1998) The dynamics and social organization of tropical deforestation in Northwest Ecuador 1983–1995. Hum Ecol 26(1):135–161

Silva JMC da, Bates JM (2002) Biogeographic patterns and conservation in the South American cerrado: A tropical savanna hotspot. BioScience 52:225–234

Silva EA, Clarke KC (2002) Calibration of the SLEUTH urban growth model for Lisbon and Porto, Portugal. Comput Environ Urban 26:525–552

Silver W, Ostertag A, Lugo E (2000) The potential for carbon sequestration through reforestation of abandoned tropical agricultural and pasture lands. Restor Ecol 8:394–407

Simon M, Plummer S, Fierens F, Hoeltzemann JJ, Arino O (2004) Burnt area detection at global scale using ATSR-2: The GlobScar products and their qualification. J Geophys Res 109, D14S02, DOI:10.1029/2003JD003622

Singh G, Geissler EA (1985) Late Cainozoic history of vegetation, fire, lake levels and climate, at Lake George, New South Wales, Australia. Philos Trans R Soc Lond B Biol Sci 311:379–447

Sisk TD (ed) (1998) Perspectives on the land use history of North America: A context for understanding our changing environment. U.S. Geological Survey, Biological Resources Division, Biological Sciences Report USGS/BRD/BSR – 1998-0003 (Revised 1999), 104 pp

Skole D, Tucker C (1993) Tropical deforestation and habitat fragmentation in the Amazon: Satellite data from 1978 to 1988. Science 260:1905–1910

Smalling EMA, Nandwa SM, Janssen BH (1997) Soil fertility in Africa is at stake. In: Buresh RJ, Sánchez PA, Calhoun F (eds) Replenishing soil fertility in Africa. Special Publication 51, American Society of Agronomy, Soil Science Society of America, Madison, pp 46–61

Smart A, Smart J (2001) Local citizenship: Welfare reform urban/rural status, and exclusion in China. Environ Plann A 33:1853–1869

Smil V (1995) Who will feed China? China Quart 143:801–813

Smith SE (1986) Drought and water management: The Egyptian response. J Soil Water Conserv 41:297–300

Smith SV, Buddemeier RW, Wulff F, Swaney DP (2005) C, N, P fluxes in the coastal zone. In: Crossland CJ, Kremer HH, Lindeboom HJ, Marshall Crossland JI, Le Tissier MDA (eds) Coastal fluxes in the Anthropocene: The Land-Ocean Interactions in the Coastal Zone Project of the International Geosphere-Biosphere Programme. (The IGBP Series), Springer, Berlin Heidelberg, pp 95–144

Sneath D (1998) State policy and pasture degradation in Inner Asia. Science 281:1147–1148

Snyder PK, Hitchman MH, Foley JA (2006) Analyzing the influence of tropical deforestation on the Northern Hemisphere climate through atmospheric teleconnections. J Climate, submitted

Soares-Filho B, Alencar A, Nepstad D, Cerqueira G, Diaz M, Rivero S, Solorzano L, Voll E (2004) Simulating the response of land-cover changes to road paving and governance along a major Amazon highway: The Santarém-Cuiabá corridor. Glob Change Biol 10:745–764

Soberon JM (2004) Translating life's diversity: Can scientists and policymakers learn to communicate better? Environment 46:10–21

Soderstrom B, Part T, Linnarsson E (2001) Grazing effects on between-year variation of farmland bird communities. Ecol Appl 11:1141–1150

Soderstrom B, Kiema S, Reid RS (2003) Intensified agricultural land-use and bird conservation in Burkina Faso. Agric Ecosyst Environ 99:113–124

Sohn YS, Moran E, Gurri F (1999) Deforestation in north-central Yucatán (1985–1995): Mapping secondary succession of forest and agricultural land use in Sotuta using the cosine of the angle concept. Photogramm Eng Remote Sens 65(8):947–958

Sojka RE, Upchurch DR (1999) Reservations regarding the soil quality concept. Soil Sci Soc Am J 63:1039–1054

Solbrig OT (1993) Ecological constraints to savanna land use. In: Young MD, Solbrig OT (eds) The world's savannas: Economic driving forces, ecological constraints and policy options for sustainable land use. Parthenon Press, Paris, pp 21–48

Solecki WD, Oliveri C (2004) Downscaling climate change scenarios in an urban land use change model. J Environ Manage 72:105–115

Solecki WD, Walker R (2001) Transformation of the South Florida landscape. In: Indian National Science Academy, Chinese Academy of Sciences, U.S. National Academy of Sciences (eds) Growing populations, changing landscapes: Studies from India, China, and the United States. National Academy Press, Washington D.C., pp 237–273

Sorrensen C (2004) Contributions of fire use study to land use/cover change frameworks: Understanding landscape change in agricultural frontiers. Hum Ecol 32(4):395–420

Southworth J (2004) An assessment of Landsat TM band 6 thermal data for analysing land cover in tropical dry forest regions. Int J Remote Sens 25:689–706

Southworth J, Tucker CM (2001) Forest cover change in western Honduras: The role of socio-economic and biophysical factors, local institutions, and land tenure. Mt Res Dev 21(3):276–283

Southworth J, Nagendra H, Tucker C (2002) Fragmentation of a landscape: Incorporating landscape metrics into satellite analyses of land cover change. Landscape Res 27(3):253–269

Sparling GP, Schipper LA, Hewitt AE, Degens BP (2000) Resistance to cropping pressure of two New Zealand soils with contrasting mineralogy. Aust J Soil Res 38:85–100

Sparling G, Parfitt RL, Hewitt AE, Schipper LA (2003) Three approaches to define desired soil organic matter contents. J Environ Qual 32:760–766

Spencer J (1966) Shifting cultivation in Southeast Asia. U.C. Publications in Geography 19, U.C. Press, Berkeley

Stafford-Smith DM, Reynolds JF (2002) Desertification: A new paradigm for an old problem. In: Reynolds JF, Stafford-Smith DM (eds) Global desertificaton: Do humans cause deserts? Dahlem Workshop Report No. 88, Dahlem University Press, Berlin, pp 403–424

Steffen W, Sanderson A, Tyson PD, Jäger J, Matson PA, Moore B III, Oldfield F, Richardson RJ, Schellnhuber HJ, Turner II BL, Wasson RJ (2004) Global change and the Earth system: A planet under pressure. (The IGBP Series), Springer, Berlin Heidelberg, 336 pp

Steininger MK, Tucker CJ, Ersts P, Killeen TJ, Villegas Z, Hecht SB (2001) Clearance and fragmentation of tropical deciduous forest in the Tierras Bajas, Santa Cruz, Bolivia. Conserv Biol 15:856–866

Stéphenne N, Lambin EF (2001) A dynamic simulation model of land-use changes in Sudano-sahelian countries of Africa (SALU). Agric Ecosyst Environ 85:145–161

Stewart X (1956) Fire as the first great force employed by man. In: Thomas WL Jr (ed) Man's role in changing the face of the Earth. University of Chicago Press, Chicago, pp 115–133

Stohlgren TJ, Chase TN, Pielke RA, Kittel TGF, Baron JS (1998) Evidence that local land use practices influence regional climate, vegetation, and stream flow patterns in adjacent natural areas. Glob Change Biol 4:495–504

Stolle F, Lambin EF (2003) Inter-provincial and inter-annual differences in the causes of forest fires in Sumatra, Indonesia. Environ Conserv 30(4):357–387

Stolle F, Chomitz K, Lambin EF, Tomich T (2003) Land use and vegetation fires in Jambi Province, Sumatra, Indonesia. Forest Ecol Manag 179(1–3):277–292

Stone GD (1996) Settlement ecology: The social and spatial organization of Kofyar agriculture. University of Arizona Press, Tucson

Stoorvogel JJ, Antle JM, Crissman CC (2004) Trade-off analysis in the Northern Andes to study the dynamics in agricultural land use. J Environ Manage 72(1/2):23–33

Straatman B, White R, Engelen G (2004) Towards an automatic calibration procedure for constrained cellular automata. Comput Environ Urban 28:149–170

Strahler A, Boschetti L, Foody GM, Fiedl MA, Hansen MC, Herold M, Mayaux P, Morisette JT, Stehman SV, Woodcock C (2006) Global land cover validation: Recommendations for evaluation and accuracy assessment of global land cover maps. Report of Committee of Earth Observation Satellites (CEOS), Working Group on Calibration and Validation (WGCV), forthcoming

Sturgeon JC, Sikor T (2004) Introduction: Post-socialist property in Asia and Europe. Conserv Soc 2:1–18

Sud YC, Walker GK, Kim JH, Liston GE, Sellers PJ, Lau WKM (1996) Biogeophysical consequences of a tropical deforestation scenario: A GCM simulation. J Climate 9:3225–3247

Sui DZ, Zeng H (2001) Modeling the dynamics of landscape structure in Asia's emerging desakota regions: A case study in Shenzhen. Landscape Urban Plan 53:37–52

Sukhinin AI, French NHF, Kasischke ES, Hewson JH, Soja AJ, Csiszar IA, Hyer EJ, Loboda T, C SG, Romasko VI, Pavlichenko EA, Miskiv SI, Slinkina OA (2004) AVHRR satellite-based mapping of fires in Russia: New products for fire management and carbon cycle studies. Remote Sens Environ 93:546–564

Suliman MM (1988) Dynamics of range plants and desertification monitoring in the Sudan. Desertic Control Bull 16:27–31

Sundberg J (2003) Strategies for authenticity and space in the Maya Biosphere Reserve, Petén, Guatemala. In: Zimmerer KS, Bassett TJ (eds) Political ecology: An integrative approach to geography and environment-development studies. The Guilford Press, New York London, pp 50–69

Sunderlin WD, Angelsen A, Resosudarmo DP, Dermawan A, Rianto E (2001) Economic crisis, small farmer well-being, and forest cover change in Indonesia. World Dev 29(5):767–782

Sutton R (1999) The policy process: An overview. Overseas Development Institute, London

Sweeney BW, Bott TL, Jackson JK, Kaplan LA, Newbold JD, Standley LJ, Cully Hession WC, Horwitz RJ (2004) Riparian deforestation, stream narrowing, and loss of stream ecosystem services. Proc Natl Acad Sci USA 101(39):14132–14137

Swift MJ, Vandermeer J, Ramakrishnan PS, Anderson JM, Ong CK, Hawkins BA (1996) Biodiversity and agroecosystem function. In: Mooney HA, Cushman JH, Medina E, Sala OE, Schulze ED (eds) Functional roles of biodiversity: A global perspective. John Wiley & Sons, Ltd., Chichester New York, pp 261–298

Swinton SM, German E, Reardon T (2003) Poverty and environment in Latin America: Concepts, evidence and policy implications. World Dev 31(11):1865–1872

Tainter J (1990) The collapse of complex societies: New studies in archaeology. Cambridge University Press, Cambridge, 260 pp

Tan ZX, Lal R, Smeck NE, Calhoun FG, Gehring RM, Parkinson B (2003) Identifying associations among soil and site variables using canonical correlation analysis. Soil Sci 168(5):376–382

Tan M, Li X, Xie H, Lu C (2006) Urban land expansion and arable land loss in China – A case study of Beijing-Tianjin-Hebei region. Land Use Policy 22(3):187–196

Tansey K, Grégoire J-M, Stroppiana D, Sousa A, Silva JMN, Pereira JMC, Boschetti L, Maggi M, Brivio PA, Fraser R, Flasse S, Ershov D, Binaghi E, Graetz D, Peduzzi P (2004) Vegetation burning in the year 2000: Global burned area estimates from SPOT VEGETATION data. J Geophys Res-Atmos 109: D14S03, doi:10.1029/2003JD003598

Tardin AT, Cunha RP (1989) Avaliação da alteração da cobertura florestal na Amazônia Legal utilizando Sensoriamento Remoto Orbital. Technical Report INPE-5010-RPE/607, Instituto Nacional de Pesquisas Espaciais, São José dos Campos

Tardin AT, Lee DCL, Santos RJR, de Assis OR, dos Santos de Barbosa MP, de Lourdes Moreira M, Pereira MT, Silva D, dos Santos Filho CP (1980) Subprojeto desmatamento: Convênio IBDF/CNPq – INPE. Technical report INPE-1649-RPE/103, Instituto de Pesquisas Espaciais, São José dos Campos

Taussig M (1978) Peasant economics and the development of capitalist agriculture in the Cauca Valley, Colombia. Lat Am Perspect 5(3):62–91

Taylor PJ, Watts MJ, Johnston RJ (2002a) Geography/globalization. In: Johnston RJ, Taylor PJ, Watts MJ (eds) Geographies of global change: Remapping the world, 2nd ed. Blackwell, Cambridge, pp 1–17

Taylor CM, Lambin EF, Stephenne N, Harding RJ, Essery RLH (2002b) The influence of land use change on climate in the Sahel. J Climate 15(24):3615–3629

Thakker PS (2001) Location potential archaeological sites in Gujarat using satellite data. In: Tripathi A (ed) Remote sensing and archaelogy. Sundeep Prakashan, New Delhi, pp 75–84

Theobald DM (2004) Placing exurban land use change in a human modification framework. Front Ecol Environ 3:139–144

Thomas WL Jr (ed) (1956) Man's role in changing the face of the Earth. Chicago University Press, Chicago, 1193 pp

Thompson M, Homewood K (2002) Entrepreneurs, elites and exclusion in Maasailand. Hum Ecol 30(1):107–138

Thornton PK, Galvin KA, Boone RB (2003) An agro-pastoral household model for the rangelands of East Africa. Agricult Sys 76:601–622

Tian H, Melillo J, Kicklighter D, McGuire AD, Helfrich J, Moore B III, Vörösmarty C (1998) Effect of interannual climate variability on carbon storage in Amazonian ecosystems. Nature 396: 664–667

Tiffen M (2003) Transition in Sub-Saharan Africa: Agriculture, urbanization and income growth. World Dev 31:1343–1366

Tiffen M, Mortimore M (1994) Malthus controverted: The role of capital and technology in growth and environment recovery in Kenya. World Dev 22(7):997–1010

Tiffen M, Mortimore M, Gichuki F (1994a) More people, less erosion: Environmental recovery in Kenya. John Wiley & Sons, Chichester New York, 311 pp

Tiffen M, Mortimore M, Gichuki F (1994b) Population growth and environmental degradation: Revising the theoretical framework. In: Tiffen M, Mortimore M, Gichuki F (eds) More people, less erosion. John Wiley & Sons, Chichester New York, pp 261–274

Tilman D (1999) Global environmental impacts of agricultural expansion: The need for sustainable and efficient practices. Proc Natl Acad Sci USA 96(11):5995–6000

Tole L (2004) A quantitative investigation of the population-land inequality-land clearance nexus. Popul Environ 26(2): 75–106

Tomich TP (1999) ICRAF's role in policy research, policy development, and advocacy: Opportunities and pitfalls. Unpublished msc. World Agroforestry Centre, SE Asia Regional Research Programme, Bogor, Indonesia

Tomich TP, Lewis J (2001) Putting community-based forest management on the map. Alternatives to Slash-and-Burn Programme, Nairobi

Tomich TP, Noordwijk M van, Vosti SA, Witcover J (1998) Agricultural development with rainforest conservation: Methods for seeking best bet alternatives to slash-and-burn, with applications to Brazil and Indonesia. Agricult Econ 19:159–174

Tomich TP, Chomitz K, Francisco H, Izac A-MN, Murdiyarso D, Ratner BD, Thomas DE, van Noordwijk M (2004a) Policy analysis and environmental problems at different scales: Asking the right questions. Agric Ecosyst Environ 104(1):5–18

Tomich TP, van Noordwijk MV, Thomas DE (2004b) On bridging gaps. Agric Ecosyst Environ 104:1–3

Tomich TP, Thomas DE, van Noordwijk M (2004c) Environmental services and land use change in Southeast Asia: From recognition to regulation or reward? Agric Ecosyst Environ 104(1): 229–244

Tomich TP, Cattaneo A, Chater S, Geist HJ, Gockowski J, Lambin EF, Lewis J, Palm C, Stolle F, Valentim J, van Noordwijk M, Vosti SA (2005) Balancing agricultural development and environmental objectives: Assessing trade-offs in the humid tropics. In: Palm CA, Sanchez SA, Vosti PJ, Ericksen PA, Juo ASR (eds) Slash-and-burn agriculture: The search for alternatives. University of Columbia Press, Columbia New York

Topp CFE, Mitchell M (2003) Forecasting the environmental and socioeconomic consequences of changes in the Common Agricultural Policy. Agricult Sys 76:227–252

Torrens PM, O'Sullivan D (2001) Editorial. Cellular automata and urban simulation: Where do we go from here? Environ Plann B 28:163–168

Toth FL (1988) Policy exercises. J Sim Game 19(3):235–276

Toth FL (1995) Simulation/gaming for long-term policy problems. In: Crookall D, Arai K (eds) Simulation/gaming across disciplines and cultures: ISAGA at a watershed. Sage Publications, Thousand Oaks, pp 134–142

Tri-Academy Panel (2001) Population and land use in India, China, and the United States: Context, observations, and findings. In: Indian National Science Academy, Chinese Academy of Sciences, and U.S. National Academy of Sciences (eds) Growing populations, changing landscapes: Studies from India, China, and the United States. National Academy Press: Washington D.C., pp 9–72

Trimble SW, Crosson P (2000) Land use: U.S. soil erosion rates. Myth and reality. Science 289:248–250

Tripathi JK, Bock B, Rajamani V, Eisenhauer A (2004) Is river Ghaggar, Saraswati? Geochemical constraints. Curr Sci 87:1141–1145

Tucker CM (1999) Private vs. communal forests: Forest conditions and tenure in a Honduran community. Hum Ecol 27:201–230

Tucker CJ, Nicholson SE (1999) Variations in the size of the Sahara desert from 1980 to 1997. Ambio 28:587–591

Tucker CJ, Townshend JRG (2000) Strategies for monitoring tropical deforestation using satellite data. Int J Remote Sens 21: 1461–147

Tucker CJ, Dregne HE, Newcomb WW (1991) Expansion and contraction of the Sahara desert from 1980 to 1990. Science 253: 299–301

Turkenburg WC (2000) Renewable energy technology. In: Goldemberg J (ed) World energy assessment: Energy and the challenge of sustainability. United Nations Development Programme (UNDP), United Nations Department of Economic and Social Affairs, World Energy Council, New York, pp 219–272

Turner FJ (1920) The frontier in American history. Henry Holt, New York

Turner BL II (1989) The human causes of global environmental change. In: DeFries RS, Malone TF (eds) Global change and our common future: Papers from a forum (Committee on Global Change, National Research Council). National Academy of Sciences, Washington D.C., pp 90–99

Turner BL II (1997) Spirals, bridges and tunnels: Engaging human-environment perspectives in geography. Ecumene 4:196–217

Turner MD (1999) Labor process and the environment: The effects of labor availability and compensation on the quality of herding in the Sahel. Hum Ecol 27:267–296

Turner BL II (2002) Toward integrated land-change science: Advances in 1.5 decades of sustained international research on land-use and land-cover change. In: Steffen W, Jäger J, Carson D, Bradshaw C (eds) Challenges of a changing Earth: Proceedings of the Global Change Open Science Conference, Amsterdam, The Netherlands, 10–13 July 2001. (The IGBP Series), Springer, Berlin Heidelberg, pp 21–26

Turner MD (2003) Environmental science and social causation in the analysis of Sahelian pastoralism. In: Zimmerer KS, Bassett TJ (eds) Political ecology: An integrative approach to geography and environment-development studies. The Guilford Press, New York London, pp 159–178

Turner BL II (2006) Land change as a forcing function in global environmental change. In: Geist HJ (ed) Our Earth's changing land: An encyclopedia of land-use and land-cover change, vol. 1 (A–K). Greenwood Press, Westport, London, pp xxv–xxxii

Turner BL II, Ali AMS (1995) Induced intensification: Agricultural change in Bangladesh with implications for Malthus and Boserup. Proc Natl Acad Sci USA 93:14984–14991

Turner BL II, McCandless S (2004) How humankind came to rival nature: A brief history of the human-environment condition and the lessons learned. In: Clark WC, Crutzen P, Schellnhuber H-J (eds) Earth system analysis for sustainability. Dahlem Workshop report No. 91, MIT Press, Cambridge, pp 227–243

Turner BL II, Meyer WB (1994) Global land-use and land-cover change: An overview. In: Meyer WB, Turner BL II (eds) Changes in land use and land cover: A global perspective. Press Syndicate of the University of Cambridge, Cambridge New York Melbourne, pp 3–10

Turner MD, Williams TO (2002) Livestock market dynamics and local vulnerabilities in the Sahel. World Dev 30:683–705

Turner BL II, Hanham RQ, Portararo AV (1977) Population pressure and agricultural intensity. Ann Assoc Am Geogr 67(3):384–396

Turner BL II, Clark WC, Kates RW, Richards JF, Mathews JT, Meyer WB (eds) (1990) The Earth as transformed by human action: Global and regional changes in the biosphere over the past 300 years. Cambridge University Press, Cambridge, 713 pp

Turner BL II, Hyden G, Kates RW (eds) (1993a) Population growth and agricultural change in Africa. University Press of Florida, Gainesville

Turner BL II, Moss RH, Skole DL (eds) (1993b) Relating land use and global land-cover change: A proposal for an IGBP-HDP core project. IGBP Report 24/HDP Report 5, International Geosphere-Biosphere Programme, Stockholm

Turner BL II, Meyer WB, Skole DL (1994) Global land-use/land-cover change: Towards an integrated study. Ambio 23(1):91–95

Turner BL II, Skole D, Sanderson S, Fischer G, Fresco L, Leemans R (1995) Land-use and land-cover change science/research plan. IGBP Glob. Change Rep. 35/HDP Rep. 7). Int. Geosph.-Biosph. Program, Hum. Dimens. Glob. Environ. Change Program, Stockholm/Geneva

Turner BL II, Kasperson RE, Matson PA, McCarthy JJ, Corell RW, Christensen L, Eckley N, Kasperson JX, Luers A, Martello ML, Polsky C, Pulsipher A, Schiller A (2003a) A framework for vulnerability analysis in sustanability science. Proc Natl Acad Sci USA 100(14):8074–8079

Turner BL II, Klepeis P, Schneider L (2003b) Three millennia in the southern Yucatán Peninsular region: Implications for occupancy, use, and carrying capacity. In: Gómez-Pompa A, Allen M, Fedick S, Jimenez-Osornio J (eds) The lowland Maya area: Three millennia at the human-wildland interface. Halworth Press, New York, pp 361–387

Turner BL II, Matson PA, McCarthy JJ, Corell RW, Christensen L, Eckley N, Hovelsrud-Broda G, Kasperson JX, Kasperson RE, Luers A, Martello ML, Mathiesen S, Naylor R, Polsky C, Pulsipher A, Schiller A, Selin H, Tyler N (2003c) Illustrating the coupled human-environment system for vulnerability: Three case studies. Proc Natl Acad Sci USA 100(14):8080–8085

Turner BL II, Geoghegan JG, Foster DR (2004) Integrated land-change science and tropical deforestation in the southern Yucatán: Final frontiers. Clarendon Press of Oxford University Press, Cambridge, 348 pp

U.S. CCSP/SGCR (U.S. Climate Change Science Program and the Subcommittee on Global Change Research) (2003) Strategic plan for the U.S. Climate Change Science Program. A Report by the Climate Change Science Program and the Subcommittee on Global Change Research. USCCSP, Washington D.C.

U.S. Department of Agriculture (1998) A history of American agriculture 1776–1990. Economic Research Service, United States Department of Agriculture, http://www.usda.gov/history2/front.htm

U.S. EPA (Environmental Protection Agency) (1999) U.S. methane emissions 1990–2020: Inventories, projections, and opportunities for reductions. EPA 430-R-99-013, EPA Office of Air and Radiation, Washington D.C.

U.S. EPA (Environmental Protection Agency) (2000) Projecting land-use change: A summary of models for assessing the effects of community growth and change on land-use patterns. U.S. Environmental Protection Agency, Office of Research and Development, Cincinnati

U.S. Geological Survey (1987) 1987-04-14. Landsat TM Scene, WRS-2 Path 200, Row 034, level orthorectified. USGS, Sioux Falls

U.S. Geological Survey (2000) 1987-04-25. Landsat ETM+ Scene, WRS-2 Path 200, Row 034, level orthorectified. USGS, Sioux Falls

U.S. National Research Council, Committee on Global Change Research, Board on Sustainable Development, Policy Division (1999) Global environmental change: Research pathways for the next decade. National Academy Press, Washington D.C., 595 pp

UNEP (United Nations Environment Programme) (1992) The world environment 1972–1992. United Nations Development Programme, Nairobi

UNEP (United Nations Environment Programme) (1994) United Nations Convention to Combat Desertification. United Nations Development Programme, Nairobi

UNEP (United Nations Environment Programme) (2002) Global environment outlook 3. Past, present and future perspectives. Earthscan, London

UNEP (United Nations Environment Programme) (2004a) Global environment outlook scenario framework. Background paper for UNEP's third global environment outlook report (GEO-3), UNEP, Nairobi

UNEP (United Nations Environment Programme) (2004b) The Convention on Biological Diversity. www.biodiv.org

United Nations Population Division (2002) World urbanization prospects: The 2001 revision. ESA/P/WP.173, Department of Economic and Social Affairs, UN, New York, 328 pp

Unruh J, Nagendra H, Green GM, McConnell WJ, Vogt N (2005) Cross-continental comparisons: Africa and Asia. In: Moran EF, Ostrom E (eds) Seeing the forest and the trees: Human-environment interactions in forest ecosystems. MIT Press, Cambridge London, pp 303–324

Uran O, Janssen R (2003) Why are spatial decision support systems not used? Some experiences from The Netherlands. Comput Environ Urban 27:511–526

Vallejos A, Pulido-Bosch A, Molina Sánchez L, Sánchez Martos F, Gisbert J (2003) Observaciones hidroquimicas en un área afectada por intrusion marina: Unidad de Aguadulce (Almería). In: IGME (ed) Tecnologia de la intrusión de agua de mar en acuiferos costeros: Países mediterráneos. IGME, Madrid

van Benthem BHB, Vanwambeke SO, Khantikul N, Burghoorn-Maas C, Panart K, Oskam L, Lambin EF, Somboon P (2005) Spatial patterns of and risk factors for seropositivity for dengue infection. Am J Trop Med Hyg 72:201–208

van der Leeuw SE (2004) Why model? Cybernet Syst 35:117–128

van der Veen A, Rotmans J (2001) Dutch perspectives on "Agents, Regions and Land Use Change". Environ Monit Assess 6:83–86

van Ittersum MK, Roetter RP, van Keulen H, de Ridder N, Hoanh CT, Laborte AG, Aggarwal PK, Ismail AB, Tawang A (2004) A systems network (SysNet) approach for interactively evaluating strategic land use options at sub-national scale in south and south-east Asia. Land Use Policy 21:101–113

Van Laake PE, Sánchez-Azofeifa GA (2004) Focus on deforestation: Zooming in on hot spots in highly fragmented ecosystems in Costa Rica. Agric Ecosyst Environ 102:3–15

van Lynden GWJ (2000) Soil degradation in central and eastern Europe. Food and Agriculture Organization of the United Nations, Rome, International Soil Reference and Information Centre, ISRIC, Wageningen, 39 pp

van Meijl H, van Rheenen T, Tabeau A, Eickhout B (2006) The impact of different policy environments on land use in Europe. Agric Ecosyst Environ 114:20–38

van Noordwijk M (2002) Scaling trade-offs between crop productivity, carbon stocks and biodiversity in shifting cultivation landscape mosaics: The FALLOW model. Ecol Model 149:113–126

van Noordwijk M, Tomich TP, Verbist B (2001) Negotiation support models for integrated natural resource management in tropical forest margins. Conserv Ecol 5(2), www.consecol.org/vol5/iss2/art21

van Noordwijk M, Poulsen JG, Ericksen PJ (2004) Quantifying off-site effects of land use change: Filters, flows and fallacies. Agric Ecosyst Environ 104(1):19–34

van Wey LK, Ostrom E, Meretsky V (2005) Theories underlying the study of human-environment interactions. In: Moran EF, Ostrom E (eds) Seeing the forest and the trees: Human-environment interactions in forest ecosystems. MIT Press, Cambridge Massachusetts London, pp 23–56

Vanacker V, Govers G, Barros S, Poesen J, Deckers J (2003) The effect of short-term socio-economic and demographic change on land use dynamics and its corresponding geomorphic response with relation to water erosion in a tropical mountainous catchment, Ecuador. Landscape Ecol 18:1–15

Vance C, Geoghegan J (2004) Modeling the determinants of semi-subsistent and commercial land uses in an agricultural frontier of southern Mexico: A swithcing regression approach. Int Regional Sci Rev 27:326–347

Vanwambeke SO (2005) Impacts of land-use change on mosquito-borne diseases in Northern Thailand. PhD thesis. Université catholique de Louvain, Department of Geography

Vashishta PS, Sharma RK, Malik RPS, Bathla S (2001) Population and land use in Haryana. In: Indian National Science Academy, Chinese Academy of Sciences, and U.S. National Academy of Sciences (eds) Growing populations, changing landscapes: Studies from India, China, and the United States. National Academy Press, Washington D.C., pp 107–144

Veiga MM, Meech JA, Oñate N (1994) Mercury pollution from deforestation. Nature 368:816–817

Veiga JB, Tourrand JF, Piketty MG, Poccard-Chapuis R, Alves AM, Thales MC (2004) Expansão e trajetórias da pecuária na Amazônia: Pará, Brasil. Editora Universidade de Brasilia, Brazília

Veldkamp A, Fresco LO (1996) CLUE-CR: An integrated multi-scale model to simulate land use change scenarios in Costa Rica. Ecol Model 91(1):231–248

Veldkamp A, Fresco LO (1997) Reconstructing land use drivers and their spatial scale dependence for Costa Rica. Agricult Sys 55:19–43

Veldkamp A, Lambin EF (2001) Predicting land-use change: Editorial. Agric Ecosyst Environ 85(1–3):1–6

Veldkamp A, Verburg PH (2004) Modeling land use change and environmental impact. J Environ Manage 72(1/2):1–4

Venetoulis J, Chazan D, Gaudet C (2004) Ecological footprint of nations. Redefining Progress, Oakland, 22 pp

Verburg PH, Chen YQ (2000) Multi-scale characterization of land-use patterns in China. Ecosystems 3:369–385

Verburg PH, Veldkamp A (2004) Projecting land use transitions at forest fringes in the Philippines at two spatial scales. Landscape Ecol 19:77–98

Verburg PH, Veldkamp A (2005) Editorial: Spatial modeling to explore land use dynamics. Int J Geogr Inf Sci 19:99–102

Verburg PH, Veldkamp A, de Koning GHJ, Kok K, Bouma J (1999) A spatial explicit allocation procedure for modelling the pattern of land use change based upon actual land use. Ecol Model 116:45–61

Verburg PH, Soepboer W, Limpiada R, Espaldon MVO, Sharifa M, Veldkamp A (2002) Land use change modelling at the regional scale: The CLUE-S model. Environ Manage 30:391–405

Verburg PH, de Groot WT, Veldkamp A (2003) Methodology for multi-scale land-use change modelling: Concepts and challenges. In: Dolman AJ, Verhagen A, Rovers CA (eds) Global environmental change and land use. Kluwer Academic, Dordrecht Boston London, pp 17–51

Verburg PH, de Nijs TCM, Ritsema van Eck J, Visser H, de Jong K (2004a) A method to analyse neighbourhood characteristics of land use patterns. Comput Environ Urban 28:667–690

Verburg PH, Schot P, Dijst M, Veldkamp A (2004b) Land use change modelling: Current practice and research priorities. GeoJournal 61:309–324

Verburg PH, Veldkamp A, Willemen L, Overmars KP, Castella JC (2004c) Landscape level analysis of the spatial and temporal complexity of land-use change. In: DeFries RS, Asner GP, Houghton RA (eds) Ecosystems and land use. Geophysical Monograph Series 153, American Geophysical Union, Washington D.C., pp 217–230

Verburg PH, Overmars KP, Witte N (2004d) Accessibility and land use patterns at the forest fringe in the Northeastern part of the Philippines. Geogr J/RGS 170:238–255

Verburg PH, Schulp CJE, Witte N, Veldkamp A (2006a) Downscaling of land use scenarios to assess the dynamics of European landscapes. Agric Ecosyst Environ 114:39–56

Verburg PH, Rounsevell MDA, Veldkamp A (2006b) Scenario-based studies of future land use in Europe. Agric Ecosyst Environ 114:1–6

Vernberg FJ, Vernberg WB (2001) The coastal zone: Past, present, and future. University of South Carolina Press, Columbia

Visser H (ed) (2004) The map comparison kit: Methods, software and applications. RIVM report 550002005/2004, Bilthoven

Vitousek PM, Ehrlich PR, Ehrlich AH, Matson PA (1986) Human appropriation of the products of photosynthesis. BioScience 36:368–373

Vitousek PM, Mooney HA, Lubchenco J, Melillo JM (1997) Human domination of Earth's ecosystems. Science 277(5335):494–499

Vlek PLG (2005) Nothing begets nothing: The creeping disaster of land degradation. Interdisciplinary Security Connections Publication Series of United Nations University Institute for Environment and Human Security, Bonn, 28 pp

Vlek PLG, Kühne RF, Denich M (1997) Nutrient resources for crop production in the tropics. Philos Trans R Soc Lond B Biol Sci 352:975–985

Vogel CH (2006) People at risk. In: Geist HJ (ed) Our Earth's changing land: An encyclopedia of land-use and land-cover change, vol. 2 (L–Z). Greenwood Press, Westport, London, pp 466–471

von Neumann J (1966) Theory of self-reproducing automata. University of Ilinois Press, Illinois

von Thünen JH (1966) Der Isolierte Staat in Beziehung auf Landwirtschaft und Nationalökonomie. In: Hall P (ed) Von Thünen's isolated state. Pergamon, Oxford

Vörösmarty CJ, Meybeck M (2004) Responses of continental aquatic systems at the global scale: New paradigms, new methods. In: Kabat Pl, Claussen M, Dirmeyer, Paul A, Gash, John HC, Bravo de Guenni L, Meybeck M, Pielke RA Sr, Vörösmarty CJ, Hutjes RWA, Lütkemeier S (eds) Vegetation, water, humans and the climate: A new perspective on an interactive system. (The IGBP Series), Springer, Berlin Heidelberg, pp 375–414

Vörösmarty CJ, Sahagian D (2000) Anthropogenic disturbance of the terrestrial water cycle. BioScience 50:573–765

Vosti SA, Witcover J, Carpentier C (2003) Agricultural intensification by smallholders in the western Brazilian Amazon: From deforestation to sustainable use. IFPRI Research Report 130, International Food Policy Research Institute, Washington D.C.

Wackernagel M, Onisto L, Linares AC, Falfan ISL, Garcia JM, Guerrero AIS, Guerrero MGS (1997) Ecological footprints of nations: How much nature do they use? How much nature do they have? International Council for Local Environmental Initiatives, Toronto

Waggoner PE, Ausuber JH (2001) How much will feeding more and wealthier people encroach on forests? Popul Dev Rev 27(2):239–257

Wagner W, Luckman A, Vietmeier J, Tansey K, Balzter H, Schmullius C, Davidson M, Gaveau D, Gluck M, Toan T, Quegan S, Shvidenko A, Wiesmann A, Yu J (2003) Large-scale mapping of boreal forest in SIBERIA using ERS tandem coherence and JERS backscatter data. Remote Sens Environ 85:125–144

Waisanen PJ, Bliss NB (2002) Changes in population and agricultural land in conterminous United States counties 1790 to 1997. Global Biogeochem Cy 16(4):doi:10.1029/2001GB001843

Walker HJ (1990) The coastal zone. In: Turner BL II, Clark WC, Kates RW, Richards JF, Mathews JT, Meyer WB (eds) The Earth as transformed by human action: Global and regional changes in the biosphere over the past 300 years. Cambridge University Press, Cambridge, pp 271–294

Walker BH (1993) Rangeland ecology: Understanding and managing change. Ambio 22:80–87

Walker R (2003) Evaluating the performance of spatially explicit models. Photogramm Eng Remote Sens 69:1271–12788

Walker R (2004) Theorizing land-cover and land-use change: The case of tropical deforestation. Int Regional Sci Rev 27(3):247–270

Walker R, Solecki WD (2004) Theorizing land-cover and land-use change: The case of the Florida Everglades and its degradation. Ann Assoc Am Geogr 94:311–238

Walker R, Homma A, Oyama K (1996) Land use and land cover dynamics in the Brazilian Amazon: An overview. Ecol Econ 18(1):67–80

Walker B, Steffen W, Canadell J, Ingram JS (eds) (1999) The terrestrial biosphere and global change: Implications for natural and managed ecosystems. The IGBP Book Series, Cambridge University Press, Cambridge

Walker R, Moran E, Anselin L (2000) Deforestation and cattle ranching in the Brazilian Amazon: External capital and household processes. World Dev 28(4):683–699

Walker R, Perz S, Caldas M, Silva LGT (2002) Land use and land cover change in forest frontiers: The role of household life cycles. Int Regional Sci Rev 25(2):169–199

Wallace JS (2000) Increasing agricultural water use efficiency to meet future food production. Agric Ecosyst Environ 82:105–199

Walsh SJ, Crews-Meyer KA (eds) (2002) Linking people, place, and policy: A GIScience approach. Kluwer Academic, Dordrecht Boston London, pp 187–214

Walsh SJ, Evans TP, Welsh WF, Entwisle B, Rindfuss RR (1999) Scale-dependent relationships between population and environment in northeastern Thailand. Photogramm Eng Remote Sens 65(1):97–105

Walsh SJ, Crawford TW, Crews-Meyer KA, Welsh WF (2001) A multi scale analysis of land use land cover change and NDVI variation in Nang Rong District, northeast Thailand. Agric Ecosyst Environ 85:47–64

Walsh SJ, Messina JP, Crews-Meyer KA, Bilsborrow RE, Pan WK (2002) Characterizing and modeling patterns of deforestation and agricultural extensification in the Ecuadorian Amazon. In: Walsh S, Crews-Meyer K (eds) Linking people, place, and policy: A GIScience approach. Kluwer Academic, Dordrecht Boston London, pp 187–214

Walsh SJ, Bilsborrow RE, McGregor SJ, Frizelle BG, Messina JP, Pan WKT, Crews-Meyer KA, Taff GM, Baquero F (2003) Integration of longitudinal surveys, remote sensing time series, and spatial analyses: Approaches for linking people and place. In: Fox J, Rindfuss RR, Walsh SJ, Mishra V (eds) People and the environment: Approaches for linking household and community surveys to remote sensing and GIS. Kluwer Academic, Dordrecht Boston London, pp 91–130

Wang G, Eltahir EAB (2000) Ecosystem dynamics and the Sahel drought. Geophys Res Lett 27:795–798

Ward DP, Murray AT, Phinn SR (2000) A stochastically constrained cellular model of urban growth. Comput Environ Urban 24: 539–558

Warren A (2002) Land degradation is contextual. Land Degrad Dev 13:449–459

Warren-Rhodes K, Koenig A (2001) Escalating trends in the urban metabolism of Hong Kong: 1971–1997. Ambio 30:429–438

Wasserman JC, Hacon S, Wasserman MA (2003) Biogeochemistry of mercury in the Amazonian environment. Ambio 32(5): 336–342

Watson MK (1978) The scale problem in human geography. Geogr Ann Hum Geogr 60B:36–47

Watson RT, Zinyowera MC, Moss RH (eds) (1996) Climate change 1995: Impacts, adaptations and mitigation of climate change. Scientific-technical analyses. Contribution to the Intergovernmental Panel on Climate Change (IPCC), Cambridge University Press, Cambridge, 878 pp

Watts MJ (1989) The agrarian question in Africa: Debating the crisis. Prog Hum Geog 13(1):1–41

Watts MJ (1994) Development II: The privatization of everything. Prog Hum Geog 18(3):371–384

Watts MJ (1996) Development III: The global agrofood system and late twentieth-century development (or Kautsky redux). Prog Hum Geog 20(2):230–245

Watts MJ, Bohle H-G (1993a) Hunger, famine and the space of vulnerability. GeoJournal 30(2):117–125

Watts MJ, Bohle H-G (1993b) The space of vulnerability: The causal structure of hunger and famine. Prog Hum Geog 17(1):43–67

Welch R, Pannell CW (1982) Mapping recent agricultural developments in China from satellite data. Adv Space Res 2:111–125

Wells KF, Wood NH, Laut P (1984) Loss of forests and woodlands in Australia: A summary by state, based on rural local government areas, CSIRO Technical Memorandum 84/4. CSIRO (Commonwealth Scientific and Industrial Research Organization), Institute of Resources, Division of Water and Land Resources, Canberra

Werth D, Avissar R (2002) The local and global effects of Amazon deforestation. J Geophys Res-Atmos 107: doi:1029/2001JD000717

Werth D, Avissar R (2004) The regional evapotranspiration of the Amazon. J Hydrometeorol 5(1):100–109

West TO, Post WM (2002) Soil organic carbon sequestration rates by tillage and crop rotation: A global data analysis. Soil Sci Soc Am J 66:1930–1946

White R, Engelen G (2000) High-resolution integrated modelling of the spatial dynamics of urban and regional systems. Comput Environ Urban 24:383–400

White MA, Mladenoff DJ (1994) Old-growth forest landscape transitions from pre-European settlement to present. Landscape Ecol 9:191–205

White R, Engelen G, Uijee I (1997) The use of constrained cellular automata for high-resolution modelling of urban land-use dynamics. Environ Plann B 24:323–343

White R, Straatman B, Engelen G (2004) Planning scenario visualization and assessment: A cellular automata based integrated spatial decision support system. In: Goodchild MF, Janelle D (eds) Spatially integrated social science. Oxford University Press, New York, pp 420–442

Wiegers ES, Hijmans RJ, Herve D, Fresco LO (1999) Land use intensification and disintensification in the Upper Canete Valley, Peru. Hum Ecol 27(2):319–339

Wiggins S (2000) Interpreting changes from the 1970s to the 1990s in African agriculture through village studies. World Dev 28(4):631–662

Williams M (1990) Forests. In: Turner BL II, Clark WC, Kates RW, Richards JF, Mathews JT, Meyer WB (eds) The Earth as transformed by human action. Cambridge University Press, New York, pp 179–201

Williams M (1994) Forests and tree cover. In: Meyer WB, Turner BL II (eds) Changes in land use and land cover: A global perspective. Press Syndicate of the University of Cambridge, New York Melbourne, pp 97–124

Williams DM (1996) Grassland enclosures: Catalyst of land degradation in Inner Mongolia. Hum Organ 55:307–313

Williams M (2000) Dark ages and dark areas: Global deforestation in the deep past. J Hist Geogr 26:28–46

Williams M (2003) Deforesting the Earth: From prehistory to global crisis. University of Chicago Press, Chicago, 715 pp

Wilson EO (1992) The diversity of life. Belknap Press of Harvard University Press, Cambridge

Wilson CJ, Reid RS, Stanton NL, Perry BD (1997) Ecological consequences of controlling the tsetse fly in southwestern Ethiopia: Effects of land-use on bird species diversity. Conserv Biol 11:435–447

Wissmar RC, Timm RK, Logsdon MG (2004) Effects of changing forest and impervious land covers on discharge characteristics of watersheds. Environ Manage 34(1):91–98

Wood S, Sebastian K, Scherr SJ (2000) Pilot analysis of global ecosystems: Agroecosystems. International Food Policy Research Institute, World Resources Institute, Washington D.C.

Woodcock CE, Ozdogan M (2004) Trends in land cover mapping and monitoring. In: Gutman G, Janetos AC, Justice CO, Moran EF, Mustard JF, Rindfuss RR, Skole D, Turner BL II, Cochrane MA (eds) Land change science: Observing, monitoring and understanding trajectories of change on the Earth's surface. Remote Sensing and Digital Image Processing Series 6, Kluwer Academic, Dordrecht Boston London, pp 367–377

Woodwell GM, Hobbie JE, Houghton RA, Melillo JM, Moore B, Peterson BJ, Shaver GR (1983) Global deforestation: Contribution to atmospheric carbon dioxide. Science 222:1081–1086

World Bank (1986) Poverty and hunger: Issues and options for food security in developing countries. World Bank, Washington D.C.

World Health Organization (WHO) (1999) The world health report. Making a difference. World Health Organization, Geneva

World Health Organization (WHO) (2004) Roll back malaria. www.rbm.who.int

Worster D (1979) Dust bowl: The Southern Great Plains in the 1930s. Oxford University Press, Oxford

WRR (Wetenschappelijke Raad voor het Regeringsbeleid) (1992) Ground for choices: Four perspectives for the rural areas in the European Community. Sdu Uitgevers, The Hague

Wu F (1999) GIS-based simulation as an exploratory analysis for space-time processes. J Geogr Sys 1:199–218

Wu R, Tiessen H (2002) Effect of land use on soil degradation in alpine grassland soil, China. Soil Sci Soc Am J 66:1648–1655

Xu JC, Mikesell S (eds) (2003) Landscapes of diversity: Indigenous knowledge, sustainable livelihoods and resource governance in montane mainland Southeast Asia. Yunnan Science and Technology Press, Kunming Yunnan

Xu JC, Rana G (2005) Living in the mountains. In: Jeggle T (ed) Know risk. United Nations Interagency Secretariat of the International Strategy for Disaster Reduction, pp 195–199

Xu JC, Salas M (2003) Moving the periphery to the centre: Indigenous people, culture and knowledge in a changing Yunnan. In: Kaosaard M, Dore J (eds) Social challenges for the Mekong Region. White Lotus, Bangkok, pp 123–145

Xu JC, Wilkes A (2004) Biodiversity impact analysis in northwest Yunnan, southwest China. Biodivers Conserv 13:959–983

Xu JC, Fox J, Xing L, Podger N, Leisz S, Xihui A (1999) Effects of swidden cultivation, population growth, and state policies on land cover in Yunnan, China. Mt Res Dev 19(2):123–132

Xu JC, Ai XH, Deng XQ (2005a) Exploring the spatial and temporal dynamics of land use in Xizhuang watershed of Yunnan, southwest China. Int J Appl Earth Obs Geoinformation 7(2005):299–309

Xu JC, Fox J, Vogler JB, Zhang PF, Fu YS, Yang LX, Qian J, Leisz S (2005b) Land-use and land-cover change and farmer vulnerability in Xishuangbanna Prefecture. SW China Env Manage 36(3):404–413

Xu JC, Ma ET, Tashi D, Fu Y, Lu Z, Melick D (2005c) Integrating sacred knowledge for conservation: Cultures and landscapes in southwest China. Ecol Soc 10(2):7, http://www.ecologyandsociety.org/vol10/iss2/art7/

Xue Y, Hutjes RWA, Harding RJ, Claussen M, Prince SD, Lambin EF, Allen SJ, Dirmeyer PA, Oki T (2003) The Sahelian climate. In: Kabat P, Claussen M, Dirmeyer PA, Gash JHC, Bravo de Guenni L, Meybeck M, Vörösmarty CJ, Hutjes RWA, Lütkemeier S (eds) Vegetation, water, humans and the climate. (The IGBP Series), Springer, Berlin Heidelberg, pp 59–78

Yamano T, Jayne TS (2004) Measuring the impacts of working-age adult mortality on small-scale farm households in Kenya. World Dev 32(1):91–119

Yang X (2001) The oases along the Keriya River in the Taklamakan Desert, China, and their evolution since the end of the last glaciation. Environ Geol 41:314–320

Yang Y, Billings SA (2000a) Extracting Boolean rules from CA patterns. IEEE T Syst Man Cy A 30:573–581

Yang Y, Billings SA (2000b) Neighborhood detection and rule selection from cellular automata pattern. IEEE T Syst Man Cy A 30:840–847

Yang H, Li X (2000) Cultivated land and food supply in China. Land Use Policy 17:73–88

Yasunari T (2002) The role of large-scale vegetation and land use in the water cycle and climate in monsoon Asia. In: Steffen W, Jäger J, Carson D, Bradshaw C (eds) Challenges of a changing Earth: Proceedings of the Global Change Open Science Conference, Amsterdam, The Netherlands, 10–13 July 2001. (The IGBP Series), Springer, Berlin Heidelberg, pp 129–132

Yates PL (1981) Mexico's agricultural dilemma. The University of Arizona Press, Tucson, 291 pp

Yeh ET (2000) Forest claims, conflicts, and commodification: The political ecology of Tibetan mushroom-harvesting villages in Yunnan Province, China. China Quart 161:212–226

Yeh AG-O, Li X (2006) Errors and uncertainties in urban cellular automata. Comput Environ Urban 30:10–28

York R, Rosa EA, Dietz T (2003) STIRPAT, IPAT and ImPACT: Analytic tools for unpacking the driving forces of environmental impacts. Ecol Econ 46:351–365

Young A (1994) Land degradation in South Asia: Its severity, causes, and effects upon the people. World Soil Resources Report 78, United Nations Development Programme, United Nations Environment Programme, Food and Agriculture Organization of the United Nations, 100 pp

Young A (1999) Is there really spare land? A critique of estimates of available cultivable land in developing countries. Environ Dev Sustainability 1:3–18

Young OR (2002a) Are institutions intervening variables or basic causal forces? Causal clusters *versus* causal chains in international society. In: M Brecher, F Harvey (eds) Millennial reflections on international studies. Unversity of Michigan Press, Ann Arbor, pp 176–191

Young OR (2002b) The institutional dimensions of environmental change. MIT Press, Cambridge

Young OR (2003) Environmental governance: The role of institutions in causing and confronting environmental problems. International Environmental Agreements: Polit Law Econ 3:377–393

Zak MR, Cabido M (2002) Spatial patterns of the Chaco vegetation of central Argentina: Integration of remote sensing and phytosociology. Appl Veg Sci 5:213–226

Zanne AE, Chapman CA (2001) Expediting reforestation in tropical grasslands: Distance and isolation from seed sources in plantations. Ecol Appl 11:1610–1621

Zeng N, Neelin JD, Lau KM, Tucker CJ (1999) Enhancement of interdecadal climate variability in the Sahel by vegetation interaction. Science 286:1537–1540

Zhang H, Henderson-Sellers A, McGuffie K (1995) Impacts of tropical deforestation. Part I: Process analysis of local climatic change. J Climate 9:1497–1517

Zhao M, Pitman AJ, Chase TN (2001) The impact of land cover change on the atmospheric circulation. Clim Dynam 17:467–477

Zhou LM, Dickinson RE, Tian YH, Fang JY, Li QX, Kaufmann RK, Tucker CJ, Myneni RB (2004) Evidence for a significant urbanization effect on climate in China. Proc Natl Acad Sci USA 101:9540–9544

Index

G